AF356607

Impact of Climate Warming and Disturbances on Forest Ecosystems

Impact of Climate Warming and Disturbances on Forest Ecosystems

Editors

Any Mary Petritan
Mirela Beloiu

Basel • Beijing • Wuhan • Barcelona • Belgrade • Novi Sad • Cluj • Manchester

Editors
Any Mary Petritan
Forest Ecology
National Institute for
Research-Development in
Forestry "Marin Drăcea"
Brasov
Romania

Mirela Beloiu
Department of Environmental
Systems Science
ETH Zurich
Zurich
Switzerland

Editorial Office
MDPI
St. Alban-Anlage 66
4052 Basel, Switzerland

This is a reprint of articles from the Special Issue published online in the open access journal *Forests* (ISSN 1999-4907) (available at: www.mdpi.com/journal/forests/special_issues/climate_biomass).

For citation purposes, cite each article independently as indicated on the article page online and as indicated below:

Lastname, A.A.; Lastname, B.B. Article Title. *Journal Name* **Year**, *Volume Number*, Page Range.

ISBN 978-3-7258-0162-6 (Hbk)
ISBN 978-3-7258-0161-9 (PDF)
doi.org/10.3390/books978-3-7258-0161-9

Contents

About the Editors

Any Mary Petritan

Dr. Any Mary Petritan's research topics are related to forest ecology, biodiversity, and the impact of climate change on forest ecosystems. Her studies in virgin and old-growth forests contribute to a better understanding of the stand structure, biodiversity, carbon storage capacity, and resilience of these valuable forests, which are considered a reference for managed forests within the framework of a sustainable forest management concept.

Mirela Beloiu

Dr. Mirela Beloiu Schwenke's research is centred around the intersection of forest ecology, remote sensing, and climate change. In particular, she seeks to understand the future of Earth's forests in a changing climate. She investigates how human-caused climate change and drought affect forest ecosystems, including biodiversity, species distribution, tree resilience, recovery, and mortality. This, in turn, can help us better monitor and manage forests. In 2023, she received the Young Scientist Award for the best oral presentation (1st prize) at the 42nd EARSel Symposium.

Preface

The SI titled "Impact of climate warming and disturbances on forest ecosystems" underscores the critical importance of understanding how forests respond to environmental challenges and past management. The rapid pace of climate change is altering disturbance patterns and the adaptability of forests. Some key outputs of this SI include evidence on how climate change is already impacting forest ecosystems. The climatic envelope of many forest species has shifted due to global warming, making species more vulnerable, especially those in lower elevations and those at the edges of their distribution. Urgent adaptive measures in forest management are necessary in addressing this challenge. Climate change also affects vegetation phenology, tree growth, stand productivity, reproduction rates, and stand regeneration. Remote sensing data and ecological modelling techniques play a crucial role in monitoring and understanding these changes, especially in remote regions where field measurements are limited. The rising frequency and intensity of extreme events require enhanced prediction and automatic monitoring. Leveraging machine learning tools and remote sensing data is imperative. This SI provides insights into the intricate relationships among forests, climate change, and human interventions. We provide further research recommendations for the quantification and automated monitoring of forest fires and the management of forests to better withstand storms and increase their resilience to climate change.

Any Mary Petritan and Mirela Beloiu
Editors

Editorial

Forest Functioning under Climate Warming and Future Perspectives on Forest Disturbances

Any Mary Petritan [1,*] and **Mirela Beloiu Schwenke [2,*]**

[1] Department of Ecology, National Institute of Research and Development in Forestry 'Marin Dracea', Closca 13, 500040 Brasov, Romania
[2] Department of Environmental Systems Science, Institute of Terrestrial Ecosystems, ETH Zurich, 8096 Zurich, Switzerland
[*] Correspondence: apetritan@gmail.com (A.M.P.); mirela.beloiu@usys.ethz.ch (M.B.S.)

Abstract: The Special Issue "Impact of climate warming and disturbances on forest ecosystems" underscores the critical importance of understanding how forests respond to these environmental challenges and the legacy of past management practices. Forest ecosystems are facing significant challenges due to ongoing climate change, characterized by rising temperatures and increased frequency of extreme events. The rapid pace of climate change is altering disturbance patterns and the adaptability of forests, which have a direct impact on ecosystem services that contribute to human well-being. This Special Issue features 11 research papers from nine countries. Some key outputs from these research papers include evidence on how climate change is already impacting forest ecosystems. For instance, the climatic envelope of many forest species has shifted due to global warming, making species more vulnerable, especially in lower elevations and at the edges of their distribution. Urgent adaptive measures in forest management are necessary to address this challenge. Climate change also affects vegetation phenology, tree growth, stand productivity, reproduction rates, and stand regeneration. Remote sensing data and ecological modeling techniques play a crucial role in monitoring and understanding these changes, especially in remote regions where field measurements are limited. The rising frequency and intensity of extreme events like droughts, windstorms, and forest fires require enhanced prediction and automatic monitoring. Leveraging machine learning tools and remote sensing data is imperative. This Special Issue provides insights into the intricate relationships among forests, climate change, and human interventions. We provide further research recommendations for the quantification and automated monitoring of forest fires and the management of forests to better withstand storms and increase their resilience to climate change.

Keywords: climate change; forest management; forest resilience; drought; old-growth forests; forest fires

Citation: Mary Petritan, A.; Beloiu Schwenke, M. Forest Functioning under Climate Warming and Future Perspectives on Forest Disturbances. *Forests* **2023**, *14*, 2302. https://doi.org/10.3390/f14122302

Received: 10 November 2023
Accepted: 16 November 2023
Published: 24 November 2023

1. Introduction

Forests are invaluable ecosystems that have long been recognized for their multifaceted contributions to biodiversity and society. Forest dynamics are generally determined by changes in forest structure and composition over time, including their response to anthropogenic and natural disturbances. Yet, global climate change is affecting disturbance regimes and the adaptive capacity of forests at an accelerating rate [1,2]. (Disturbances such as droughts, storms, and fires have significant impacts on forests and the ecosystem services that they offer. Although there are numerous frameworks proposed for the automatic monitoring of forests [3–6], there is still a pressing need to enhance our capacity for actively detecting ongoing disturbances and predicting the future impact on forests and tree species. In this context, ground truth data and new technologies such as remote sensing, machine learning, and ecological modeling offer new opportunities for monitoring, quantifying, and predicting the impacts of climate change and disturbances on forest ecosystems [7,8]. Further, rapid climate-induced changes in disturbance regimes, combined with the ongoing legacy of past forest management (e.g., plantation of tree species outside of their natural distribution and different management types), have resulted in a complex and poorly understood interplay of factors that affect tree growth [9].

2. Overview of Contributions

The Special Issue "Impact of climate warming and disturbances on forest ecosystems" aimed to collect scientific research covering the following topics:

- Forest disturbance regimes (e.g., drought, forest fire, storms, and pathogens);
- Climate change impacts;
- Forest resilience and dynamics;
- Forest management practices;
- Old-growth forests.

This Special Issue, entitled *"Impact of Climate Warming and Disturbances on Forest Ecosystems"*, features 11 original research papers by 67 authors from nine countries on three continents: Asia (China, India, Syria, Saudi Arabia), Europe (Belgium, Poland, Romania, Greece), and North America (USA). They provide examples of different regions of the globe on how climate changes and anthropogenic pressure could affect the structure, growth dynamic, and functionality of the forests and thus their ability to continue to provide the ecosystem goods and services and contribute to the well-being of society.

3. Summary of Main Outputs and Findings

The already changed climate has been highlighted by Mihai et al. [10] for different provenances and regions in Romania, by different forest types. The authors showed that the climatic envelope of many forest species has already changed and become less suitable due to climate warming, which increases the vulnerability of species, especially at lower elevations or at the edges of their distribution. For instance, the annual mean temperature has risen by 1.33 °C nationwide, with a clear warming pattern in the plains (Figure 1). The authors warned that climate change appears to be occurring faster than species can adapt or migrate and argue that adaptative measures in forest management are urgently needed. Chandra et al. [11] analyzed changes in the habitat suitability of vulnerable species such as orchids under different climatic scenarios using ecological niche modeling. Their results indicated a reduction in highly suitable areas under future climate conditions and acknowledged the importance of identifying viable habitats to conserve vulnerable species and prevent local extinctions.

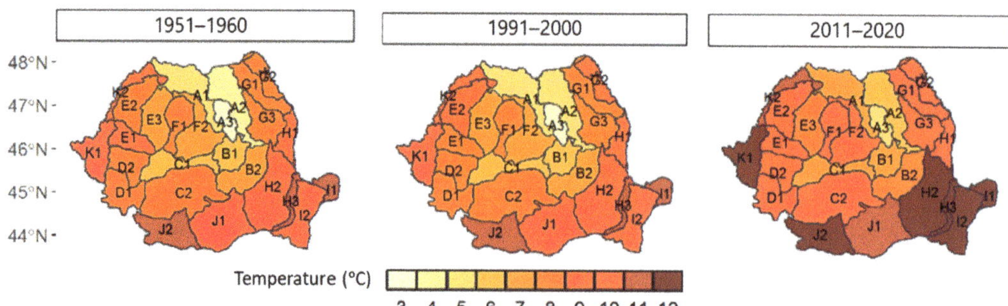

Figure 1. Changes in the mean annual temperature in Romania across different regions. Adapted from Mihai et al. [10].

Other papers published in this Special Issue have shown that climate changes affect vegetation phenology [12], tree growth [13] and stand productivity [14] but also the reproductivity rate [15] or stand regeneration [16]. Using remote sensing data, such as MODIS together with vegetation indices and vegetation gross primary productivity [12,14], allows the identification of changes in vegetation changes and the link to climate change. This is of great relevance, particularly in remote regions where field measurements are limited. To this end, using the MODIS gross primary productivity model, moving average trend analysis, linear regression analysis, and the climate tendency rate method, Aili et al. [14], analyzed the variation in the vegetation gross productivity of forests in Altay Mountains, China, in response to the daily temperature and precipitations. Dugesar et al. [12] investigated the change in vegetation phenology along an elevation gradient in the main forest types of the western Himalayas and observed a delayed trend for all vegetation types. A Normalized Difference Vegetation Index time series was used for this purpose. Because phenology responded differently in heterogeneous mountain landscapes to climate change, the authors concluded the importance of comprehensive observations at the local level to our improve understanding of forest response to climate change.

Using dendrochronological methods, Qi et al. [13] attempted to determine the main climate factors that influence the radial growth of two coniferous tree species. The study showed the growth response is species-specific and the drought during the growing season is a decisive negative factor. In addition, the authors stressed the importance of studying the effect of climatic factors not only over long periods but also over short periods of time, such as the decadal scale. The results of an experimental study in a common garden [15] indicated that water limitation led to the reallocation of resources toward reproduction rather than growth, an effect that is still present two years later. Respectively, treated shrubs showed a higher germination rate, an adaptive trait that should ensure survival under stress conditions. The authors concluded that the delayed effects of drought on reproductive traits in woody species may add more complexity to the consequences of climate change on species distributions and forest survival.

The rising global temperature together with the increase in the frequency, severity, and duration of extreme events such as droughts, windstorms or fires, followed by various associated disturbances, have already negatively affected forest ecosystems worldwide [17–19]. (An example of such events is the extreme and long-lasting hotter drought and heatwaves from 2018 and 2019 that affected forests in Central Europe [20,21] or windstorm Klaus, one of the most devasting windstorms in Europe [22]. As such events left a significant footprint in forest stands, followed by large economic and ecological losses, modern tools such as machine learning [22] or remote sensing and geographic information systems [23] are very useful to build better models to predict forest damage caused by windstorms [22] or to accurately monitor fire intensity and damages [23]. As such extreme events are expected to occur more frequently in the future, improving their prediction and monitoring will be essential for better planning.

In addition to climate change, human disturbance can negatively affect stand characteristics. As reported in another paper published in this Special Issue, some structural stand traits and soil microclimates were altered by the first silvicultural interventions carried out in a former virgin forest [24]. The changed microclimate conditions in the former virgin forest led to greater CO_2 soil emissions (Figure 2). Moreover, stand dynamic trough canopy gaps are influenced by topography, with forest gaps occurring more on slopes between 10 and 20 degrees and at higher elevation [25]. Topography also plays an important role in the relationships between climate and different vegetation traits [12].

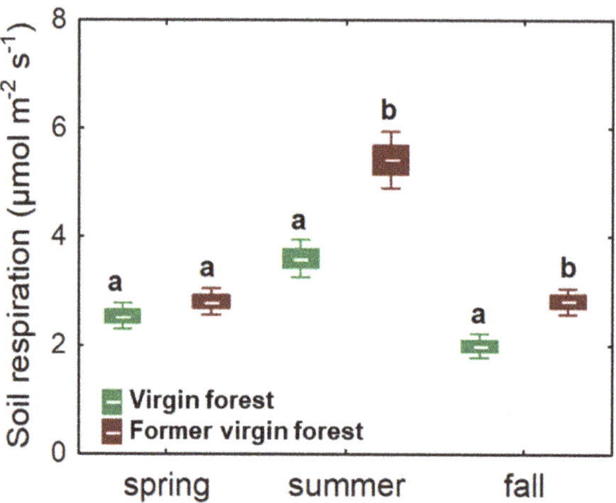

Figure 2. Differences in soil respiration between virgin and former virgin forests. Adapted from Braga et al. [24].

4. Challenges and Future Work: Forest Fires and Storm Disturbances

4.1. Forest Fires

In recent years, forest fires have captured global attention due to their increasing frequency, severity, and their far-reaching ecological, economic, and social impacts. Forest fires have left their mark on diverse ecosystems across the world. Millions of hectares burned in the last few years, ranging from the Amazon rainforest (e.g., in 2019–2020, 0.9 mil. ha) in South America to the boreal forests of Siberia (e.g., in 2019, 3 mil. ha), and the wildfire-prone regions of British Columbia (e.g., in 2017 and 2018, a total of 2.6 mil. ha were burnt), California (e.g., in 2018, 0.7 mil. ha), and Australia (e.g., in 2019–2020, 18.6 mil. ha) [26,27]. These mega-fires are often exacerbated by climate change, prolonged droughts, heatwaves, increasing aridity [28–30], forest management, deforestation [31], and shifts in weather patterns, creating a formidable challenge for forest ecosystems and the communities that depend on them. The occurrence of forest fires varied over time, with 2021 being the worst year in terms of the area burnt. In 2020, forest fires occurred mainly in southern Europe, while in 2021, they also occurred in northern regions and had a higher intensity and burnt area compared to 2000 (Figure 3). Both field observations and remote sensing data are essential to map forest fires. Forest fires ($\geq$30 ha) are mapped globally using MODIS satellite sensors. Since 2018, Sentinel-2 imagery was incorporated, which allowed mapping fires $\geq$ 5 ha in Europe [32].

To reduce the destructive effects of forest fires, they must be detected as early as possible. This requires early warning systems to predict the areas at high risk [33] and an active fire detection system to identify areas with forest fires [34,35]. This urges us to develop and refine fire behavior models to better predict fire spread, intensity, and behavior under varying climate conditions, including the influence of extreme events like heatwaves and prolonged droughts. Pezzatti et al. [36] proposed a novel metric to predict the risk of future forest fires. Early warning forest fire systems would help to identify regions at increased risk of forest fires and provide insights into adapting forest management strategies to mitigate these risks. Further, public awareness and cross-boundary collaboration are needed to mitigate risks and enhance community resilience.

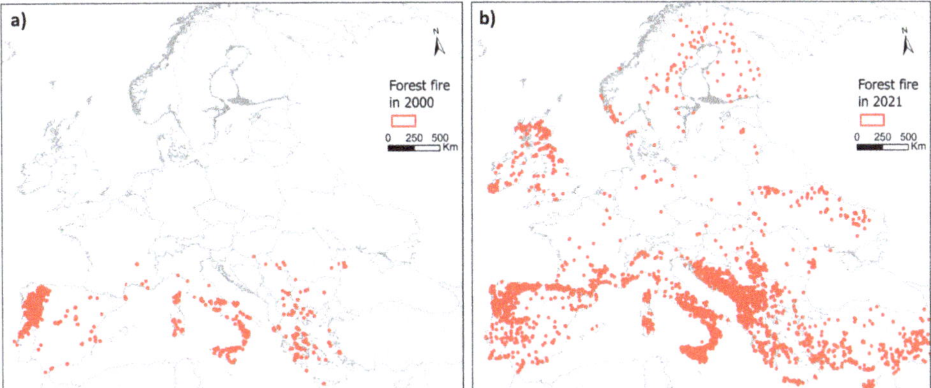

Figure 3. Burnt areas ($\geq$30 ha) produced by forest fires in (**a**) 2000 and (**b**) 2021 based on the European Forest Fire Information System (EFFIS) data.

4.2. Storms and Forest Resilience

Storms induce long-term ecological impacts, including changes in forest structure and composition [37,38]. To face the upcoming storms, we need to identify strategies for enhancing forest resilience to storms, including silvicultural practices and landscape-level planning. The measurements in this regard are at the stand or landscape level. For instance, it is recommended to avoid heterogeneous crowns, gaps, and edges between stands [39]. Yet, with changing climate patterns, understanding the influence of climate change on the frequency and intensity of storms is essential for proactive management.

Forests and how they are managed are very diverse, depending on many factors such as tree species, ecological and environmental conditions, as well as societal demands. As also shown in this Special Issue, climate change affects forest ecosystems and its impact is modulated by forest management [9,40,41]. Therefore, in addition to increasing research that considers both past climate and forest management in modeling forest growth, productivity, and response, there is a strong need to identify such win–win forest management practices (including non-intervention, climate-smart forestry, adaptative management, assisted migration) to better cope with climate change.

Author Contributions: The entire paper development process (conceptualization, visualization, and writing) was carried out by both authors, A.M.P. and M.B.S. All authors have read and agreed to the published version of the manuscript.

Funding: This research received no external funding.

Conflicts of Interest: The authors declare no conflict of interest.

References

1. Anderson-Teixeira, K.J.; Miller, A.D.; Mohan, J.E.; Hudiburg, T.W.; Duval, B.D.; DeLucia, E.H. Altered dynamics of forest recovery under a changing climate. *Glob. Chang. Biol.* **2013**, *19*, 2001–2021. [CrossRef] [PubMed]
2. Beloiu, M.; Stahlmann, R.; Beierkuhnlein, C. Drought impacts in forest canopy and deciduous tree saplings in Central European forests. *For. Ecol. Manag.* **2022**, *509*, 120075. [CrossRef]
3. Beloiu, M.; Heinzmann, L.; Rehush, N.; Gessler, A.; Griess, V.C. Individual tree-crown detection and species identification in heterogeneous forests using aerial RGB imagery and Deep Learning. *Remote Sens.* **2023**, *15*, 1463. [CrossRef]
4. Buras, A.; Rammig, A.; Zang, C.S. The European forest condition monitor: Using remotely sensed forest greenness to identify hot spots of forest decline. *Front. Plant Sci.* **2021**, *12*, 2355. [CrossRef] [PubMed]
5. Coops, N.C.; Tompalski, P.; Goodbody, T.R.H.; Achim, A.; Mulverhill, C. Framework for near real-time forest inventory using multi source remote sensing data. *For. Int. J. For. Res.* **2023**, *96*, 1–19. [CrossRef]
6. Ecke, S.; Dempewolf, J.; Frey, J.; Schwaller, A.; Endres, E.; Klemmt, H.-J.; Tiede, D.; Seifert, T. UAV-based forest health monitoring: A systematic review. *Remote Sens.* **2022**, *14*, 3205. [CrossRef]

7. Achim, A.; Moreau, G.; Coops, N.C.; Axelson, J.N.; Barrette, J.; Bédard, S.; Byrne, K.E.; Caspersen, J.; Dick, A.R.; D'Orangeville, L.; et al. The changing culture of silviculture. *For. Int. J. For. Res.* **2022**, *95*, 143–152. [CrossRef]
8. Zweifel, R.; Pappas, C.; Peters, R.L.; Babst, F.; Balanzategui, D.; Basler, D.; Bastos, A.; Beloiu, M.; Buchmann, N.; Bose, A.K.; et al. Networking the forest infrastructure towards near real-time monitoring—A white paper. *Sci. Total Environ.* **2023**, *872*, 162167. [CrossRef]
9. Marqués, L.; Peltier, D.M.P.; Camarero, J.J.; Zavala, M.A.; Madrigal-González, J.; Sangüesa-Barreda, G.; Ogle, K. Disentangling the legacies of climate and management on tree growth. *Ecosystems* **2022**, *25*, 215–235. [CrossRef]
10. Mihai, G.; Alexandru, A.-M.; Nita, I.-A.; Birsan, M.-V. Climate change in the provenance regions of Romania over the last 70 years: Implications for forest management. *Forests* **2022**, *13*, 1203. [CrossRef]
11. Chandra, N.; Singh, G.; Rai, I.D.; Mishra, A.P.; Kazmi, M.Y.; Pandey, A.; Jalal, J.S.; Costache, R.; Almohamad, H.; Al-Mutiry, M.; et al. Predicting Distribution and Range Dynamics of Three Threatened Cypripedium Species under Climate Change Scenario in Western Himalaya. *Forests* **2023**, *14*, 633. [CrossRef]
12. Dugesar, V.; Satish, K.V.; Pandey, M.K.; Srivastava, P.K.; Petropoulos, G.P.; Anand, A.; Behera, M.D. Impact of Environmental Gradients on Phenometrics of Major Forest Types of Kumaon Region of the Western Himalaya. *Forests* **2022**, *13*, 1973. [CrossRef]
13. Qi, C.; Jiao, L.; Xue, R.; Wu, X.; Du, D. Timescale Effects of Radial Growth Responses of Two Dominant Coniferous Trees on Climate Change in the Eastern Qilian Mountains. *Forests* **2022**, *13*, 72. [CrossRef]
14. Aili, A.; Xu, H.; Zhao, X.; Zhang, P.; Yang, R. Dynamics of Vegetation Productivity in Relation to Surface Meteorological Factors in the Altay Mountains in Northwest China. *Forests* **2022**, *13*, 1907. [CrossRef]
15. Vander Mijnsbrugge, K.; Schouppe, M.; Moreels, S.; Aguas Guerreiro, Y.; Decorte, L.; Stessens, M. Severe Drought Still Affects Reproductive Traits Two Years Later in a Common Garden Experiment of *Frangula alnus*. *Forests* **2023**, *14*, 857. [CrossRef]
16. Liu, K.; Sun, H.; He, H.S.; Guan, X. Seed Harvesting and Climate Change Interact to Affect the Natural Regeneration of Pinus koraiensis. *Forests* **2023**, *14*, 829. [CrossRef]
17. Allen, C.D.; Macalady, A.K.; Chenchouni, H.; Bachelet, D.; McDowell, N.; Vennetier, M.; Kitzberger, T.; Rigling, A.; Breshears, D.D.; Hogg, E.H.; et al. A global overview of drought and heat-induced tree mortality reveals emerging climate change risks for forests. *For. Ecol. Manag.* **2010**, *259*, 660–684. [CrossRef]
18. Beloiu, M. Forest Response to Climate Warming and Drought in Europe, Bayreuth. Doctoral Thesis, University of Bayreuth, Faculty of Biology, Chemistry and Earth Sciences, Bayreuth, Germany, 2022. [CrossRef]
19. Hammond, W.M.; Williams, A.P.; Abatzoglou, J.T.; Adams, H.D.; Klein, T.; López, R.; Sáenz-Romero, C.; Hartmann, H.; Breshears, D.D.; Allen, C.D. Global field observations of tree die-off reveal hotter-drought fingerprint for Earth's forests. *Nat. Commun.* **2022**, *13*, 1761. [CrossRef]
20. Beloiu Schwenke, M.; Schönlau, V.; Beierkuhnlein, C. Tree sapling vitality and recovery following the unprecedented 2018 drought in central Europe. *For. Ecosyst.* **2023**, *10*, 100140. [CrossRef]
21. Buras, A.; Rammig, A.; Zang, C.S. Quantifying impacts of the 2018 drought on European ecosystems in comparison to 2003. *Biogeosciences* **2020**, *17*, 1655–1672. [CrossRef]
22. Pawlik, Ł.; Godziek, J.; Zawolik, Ł. Forest Damage by Extra-Tropical Cyclone Klaus-Modeling and Prediction. *Forests* **2022**, *13*, 1991. [CrossRef]
23. Singh, S.; Singh, H.; Sharma, V.; Shrivastava, V.; Kumar, P.; Kanga, S.; Sahu, N.; Meraj, G.; Farooq, M.; Singh, S.K. Impact of Forest Fires on Air Quality in Wolgan Valley, New South Wales, Australia—A Mapping and Monitoring Study Using Google Earth Engine. *Forests* **2022**, *13*, 4. [CrossRef]
24. Braga, C.I.; Crisan, V.E.; Petritan, I.C.; Scarlatescu, V.; Vasile, D.; Lazar, G.; Petritan, A.M. Short-Term Effects of Anthropogenic Disturbances on Stand Structure, Soil Properties, and Vegetation Diversity in a Former Virgin Mixed Forest. *Forests* **2023**, *14*, 742. [CrossRef]
25. Tudose, N.C.; Petritan, I.C.; Toiu, F.L.; Petritan, A.-M.; Marin, M. Relation between Topography and Gap Characteristics in a Mixed Sessile Oak–Beech Old-Growth Forest. *Forests* **2023**, *14*, 188. [CrossRef]
26. Collins, L.; Bradstock, R.A.; Clarke, H.; Clarke, M.F.; Nolan, R.H.; Penman, T.D. The 2019/2020 mega-fires exposed Australian ecosystems to an unprecedented extent of high-severity fire. *Environ. Res. Lett.* **2021**, *16*, 044029. [CrossRef]
27. Haque, M.K.; Azad, M.A.K.; Hossain, M.Y.; Ahmed, T.; Uddin, M.; Hossain, M.M. Wildfire in Australia during 2019-2020, Its Impact on Health, Biodiversity and Environment with Some Proposals for Risk Management: A Review. *J. Environ. Prot.* **2021**, *12*, 391–414. [CrossRef]
28. Abatzoglou, J.T.; Battisti, D.S.; Williams, A.P.; Hansen, W.D.; Harvey, B.J.; Kolden, C.A. Projected increases in western US forest fire despite growing fuel constraints. *Commun. Earth Environ.* **2021**, *2*, 227. [CrossRef]
29. Grünig, M.; Seidl, R.; Senf, C. Increasing aridity causes larger and more severe forest fires across Europe. *Glob. Chang. Biol.* **2023**, *29*, 1648–1659. [CrossRef] [PubMed]
30. Juang, C.S.; Williams, A.P.; Abatzoglou, J.T.; Balch, J.K.; Hurteau, M.D.; Moritz, M.A. Rapid Growth of Large Forest Fires Drives the Exponential Response of Annual Forest-Fire Area to Aridity in the Western United States. *Geophys. Res. Lett.* **2022**, *49*, e2021GL097131. [CrossRef]
31. Lindenmayer, D.B.; Kooyman, R.M.; Taylor, C.; Ward, M.; Watson, J.E.M. Recent Australian wildfires made worse by logging and associated forest management. *Nat. Ecol. Evol.* **2020**, *4*, 898–900. [CrossRef]

32. San-Miguel-Ayanz, J.; Durrant, T.; Boca, R.; Maianti, P.; Libertà, G.; Oom, D.; Branco, A.; De Rigo, D.; Ferrari, D.; Roglia, E.; et al. *Advance Report on Forest Fires in Europe, Middle East and North Africa 2022*; Publications Office of the European Union, EU: Luxembourg, 2023.
33. Zhang, Y.; Geng, P.; Sivaparthipan, C.B.; Muthu, B.A. Big data and artificial intelligence based early risk warning system of fire hazard for smart cities. *Sustain. Energy Technol. Assess.* **2021**, *45*, 100986. [CrossRef]
34. Barmpoutis, P.; Papaioannou, P.; Dimitropoulos, K.; Grammalidis, N. A Review on Early Forest Fire Detection Systems Using Optical Remote Sensing. *Sensors* **2020**, *20*, 6442. [CrossRef] [PubMed]
35. Hong, Z.; Tang, Z.; Pan, H.; Zhang, Y.; Zheng, Z.; Zhou, R.; Ma, Z.; Zhang, Y.; Han, Y.; Wang, J.; et al. Active Fire Detection Using a Novel Convolutional Neural Network Based on Himawari-8 Satellite Images. *Front. Environ. Sci.* **2022**, *10*, 794028. [CrossRef]
36. Pezzatti, G.B.; De Angelis, A.; Bekar, İ.; Ricotta, C.; Bajocco, S.; Conedera, M. Complementing daily fire-danger assessment using a novel metric based on burnt area ranking. *Agric. For. Meteorol.* **2020**, *295*, 108172. [CrossRef]
37. Holzwarth, S.; Thonfeld, F.; Abdullahi, S.; Asam, S.; Da Ponte Canova, E.; Gessner, U.; Huth, J.; Kraus, T.; Leutner, B.; Kuenzer, C. Earth observation based monitoring of forests in Germany: A review. *Remote Sens.* **2020**, *12*, 3570. [CrossRef]
38. Senf, C.; Seidl, R. Storm and fire disturbances in Europe: Distribution and trends. *Glob. Chang. Biol.* **2021**, *27*, 3605–3619. [CrossRef] [PubMed]
39. Thom, D.; Spathelf, P. Adaptive Waldbewirtschaftung zur Minderung von Störungen. *Schweiz. Z. Forstwes.* **2023**, *174*, 70–75. [CrossRef]
40. Petritan, A.M.; Petritan, I.C.; Hevia, A.; Walentowski, H.; Bouriaud, O.; Sánchez-Salguero, R. Climate warming predispose sessile oak forests to drought-induced tree mortality regardless of management legacies. *For. Ecol. Manag.* **2021**, *491*, 119097. [CrossRef]
41. Stojanović, M.; Sánchez-Salguero, R.; Levanič, T.; Szatniewska, J.; Pokorný, R.; Linares, J.C. Forecasting tree growth in coppiced and high forests in the Czech Republic. The legacy of management drives the coming Quercus petraea climate responses. *For. Ecol. Manag.* **2017**, *405*, 56–68. [CrossRef]

forests

MDPI

Article

Climate Change in the Provenance Regions of Romania over the Last 70 Years: Implications for Forest Management

Georgeta Mihai [1], Alin-Madalin Alexandru [1,2], Ion-Andrei Nita [3,*] and Marius-Victor Birsan [3,4]

[1] Department of Forest Genetics and Tree Breeding, "Marin Dracea" National Institute for Research and Development in Forestry, 077190 Bucharest, Romania; gmihai_2008@yahoo.com (G.M.); alexandru.alin06@yahoo.com (A.-M.A.)
[2] Faculty of Silviculture and Forest Engineering, Transilvania University of Brasov, 500036 Brasov, Romania
[3] VisualFlow, 020099 Bucharest, Romania; marius.birsan@gmail.com
[4] Department of Research and Meteo Infrastructure Projects, Meteo Romania (National Meteorological Administration), 013686 Bucharest, Romania
* Correspondence: nitaandru@gmail.com

Abstract: The recent climate change scenarios show significant increases in temperature and extreme drought events in Southern and Eastern Europe by the end of the 21st century, which will have a serious impact on forest growth and adaptation, and important consequences for forest management. The system of provenance regions, according to the OECD Scheme and EU Directive, was thought to encourage the use of the local seed sources, under the concept 'local is the best'. However, climate is changing faster than some species or populations can adapt or migrate, which raises some uncertainties with respect to the future performance of local populations. In Romania, as in other countries, the delimitation of provenance regions is based on geographical, ecological and vegetation criteria. The aim of this study is to evaluate: (1) the climate change that has occurred at the level of the provenance regions; (2) which regions will be most vulnerable to climate change; (3) which forest types will be the most vulnerable in a certain region; and (4) changes in the climatic envelope of forest species. Several climatic parameters and an ecoclimatic indices have been calculated and analyzed at the level of provenance regions, subregions and ecological sectors (forest types) in Romania, during the period 1951–2020. The results highlight a general shift towards warmer and drier conditions in the last 30 years, the mean annual temperature increasing with 0.3–1.1 °C across the provenance subregions. The De Martonne aridity index for the vegetation season shows that 86% of the ecological sectors fell into the arid and semiarid categories, which indicates a very high degree of vulnerability for forest species. On the Lang rainfall index, forest steppe climatic conditions occurred in all pure or mixed pedunculate oak forests, thermophile oak species, meadow forests, poplar and willow, Turkey oak and Hungarian oak forests. The Ellenberg coefficient highlights that the warming process is more evident along the altitude and the degree of vulnerability increase at lower altitude or at the edge of species distribution. The climate envelopes of many forest species have already shifted to another ecosystem's climate. This paper presents the importance of re-delineation the provenance regions for the production and deployment of forest reproductive materials according to the climate change occurred in the last decades, as a fundamental tool for an adaptive forest management.

Keywords: provenance regions; forest ecosystems; climate envelope; climate change; forest management; south-eastern Europe

Citation: Mihai, G.; Alexandru, A.-M.; Nita, I.-A.; Birsan, M.-V. Climate Change in the Provenance Regions of Romania over the Last 70 Years: Implications for Forest Management. *Forests* **2022**, *13*, 1203. https://doi.org/10.3390/f13081203

Academic Editor: Brian Buma

Received: 30 June 2022
Accepted: 28 July 2022
Published: 31 July 2022

1. Introduction

In the last few decades, the choice, use and transfer of forest reproductive material, in the context of climate change, have been hot topics discussed by several international organizations. The Ministerial Conference on the Protection of Forests in Europe (held in Strasbourg 1990, Helsinki 1993, Lisbon 1998, Vienna 2003, Warsaw 2007, etc.), in its resolutions, has called for the member states to initiate activities to monitor climate change, assess its impact on forests and establish strategies at the national and regional levels to reduce its effects and increase forest adaptation. The international regulations on forest reproductive material (EU directive 1999/105/EC and the Organization for Economic Co-operation and Development [OECD] Seed and Seedling Scheme) also require member states to develop a national certification system and transfer guidelines for forest reproductive materials.

In this context, the delimitation of provenance regions is the basic step towards the selection, usage and transfer of forest reproductive material and the conservation and improvement of forest genetic resources. According to the definition of the 2021 OECD Scheme and the 1999 EU Directive, the region of provenance for a tree species or subspecies is an area or group of areas with sufficiently uniform ecological conditions in which the stands or seed sources have similar phenotypic or genetic characteristics. The delimitation of provenance regions is mandatory for each EU member state, which must establish them for the main forest species.

The delimitation of regions of provenance is based on the existence of intraspecific genetic diversity and using local seed sources in reforestation, which are optimally adapted to those environmental conditions. Through the delimitation of provenance regions one seeks to maximize tree growth and minimize the risk of maladaptation [1,2], because within a provenance region the forest reproductive material can be moved with little risk of maladaptation due to low environmental variation. From this perspective, the delineation of provenance regions has great importance in reforestation and restoration programs. Furthermore, using forest reproductive material that performs well and is well-adapted represents a fundamental step in forest management.

Generally, the regions of provenance of forest species are delimited on the basis of environmental homogeneity and the criteria used are mainly geographical and ecological [3–6]. In recent years, new approaches regarding the delimitation of provenance regions have appeared that use climate variables instead of geographical ones [7–11]. This is because it is considered that climate is the main driving factor behind the adaptation of several processes in forest species [12], especially at altitudinal and latitudinal treelines [13].

In recent decades, various studies have highlighted global climate change, which is certain to affect forest ecosystems [14–16]. There have already been changes to local seasonal conditions experienced throughout the range of forest species, and although it is believed that not all species will be affected by climate change, rising temperatures, extended periods of drought and changes in winter precipitation patterns, together with the biotic stressors, will pose significant threats to forest species [17,18]. The effects of climate change are generally expected to reduce forest growth and survival and to change forest structure and composition at the landscape scale.

Forest species will survive under new climates through migration to suitable habitats, or persist in situ by changes in their genetic composition or adjusting to environmental changes using phenotypic plasticity [19,20]. Chen used a meta-analysis and estimated that the distributions of species have recently shifted to higher elevations at a median rate of 11.0 m per decade, and to higher latitudes at a median rate of 16.9 km per decade [21]. Genetic adaptation to climate change involves changes in allele frequencies that occur over several trees' generations [22]. Furthermore, many forests are so fragmented that natural migration will be limited [23]. Therefore, climate is changing faster than some species or populations can adapt or migrate, which will have negative consequences for forest management and conservation [24].

This mismatch between climate change and tree adaptation raises some uncertainties with respect to the future performance of local populations, and whether the axiom 'local population is the best' will still be valid under new environmental conditions. Adapting to climate change involves monitoring and anticipating change, undertaking actions to avoid negative consequences, and allowing adaptation to occur [25]. Therefore, an adaptive forest management involves an understanding of the effects of climate on forests, predictions of how these effects might change over time, and the incorporation of this knowledge into management decisions [26]. In this context, a reliable procedure to the delimitation of provenance regions for selection and use of forest reproductive material are fundamental for forests to be able to adapt and achieve resilience to new climatic conditions [27–29].

In Romania, the delimitation of provenance regions has been approved by Minister's order No. 1028/2010 [30] and is based on geographical, ecological and vegetation criteria. The climatic variables that characterize and underlie the current provenance regions were determined 70 years ago [31]. However, a number of studies have since highlighted that the climate has changed in Romania over the last few decades, and that Southern Europe will be considerably more vulnerable than Northern Europe, especially in terms of ecosystem services [32–35].

Recent climatic changes in Romania show increasing temperature extremes [36–38], declining snow pack [39,40] and wind speed [41,42] and an increase in rain shower frequency [43,44], due to changes in large-scale atmospheric circulation [45–47]. These changes were proven to affect drought [48,49], water resources [50], human health [51,52] and terrestrial ecosystems [53,54] in various ways.

The aim of this study was to evaluate: (1) the climate change that has occurred over the last 70 years at the level of the provenance region for the main tree species in Romania; (2) which regions will be most vulnerable to climate change; (3) which forest types will be the most vulnerable in a certain region; (4) changes in the geographic distribution of climatic optimum (climatic envelope) of forest species. The findings of this study will be of great importance to forest management, in terms of providing an evaluating of the effects of climate change on forest ecosystems and developing efficient adaptation and mitigation strategies.

2. Materials and Methods

Romania is located in the south-eastern Europe and is characterized by a diverse relief (Figure S1). Of the entire surface of the country, 31% is occupied by mountains (with altitudes between 800 and 2543 m), 36% by hills and plateaus, and 33% by plains (below 200 m altitude). The relief is focused on the arc of the Carpathian Mountains. In the center of the territory is the Transylvanian Plateau, surrounded by the mountain ranges of the Carpathians. The intermediate relief step between the mountains and the plains is the hills and plateaus, located inside and outside the Carpathian arc. Romania's climate is continental-temperate, with oceanic influences in the west, Mediterranean influences in the south-west and continental-excessive influences in the north-east.

The system of provenance regions in Romania divides the land into 11 broad geographical regions, which are further subdivided into 26 subregions and 155 ecological sectors (Figure 1). The ecological sectors are the basic units of the provenance regions because has the most pronounced climatic homogeneity and actually represent the main forest types. The system of delimitation of the provenance regions has been applied uniformly to all species and was based on the Romanian vegetation map [55]. This approach was justified by the huge environmental variability existing in Romania, the climatic variability being reflected in the forest composition and distribution [31]. Therefore, we conducted analyses on different spatial scales—regional, subregional and local (forest types within a subregion).

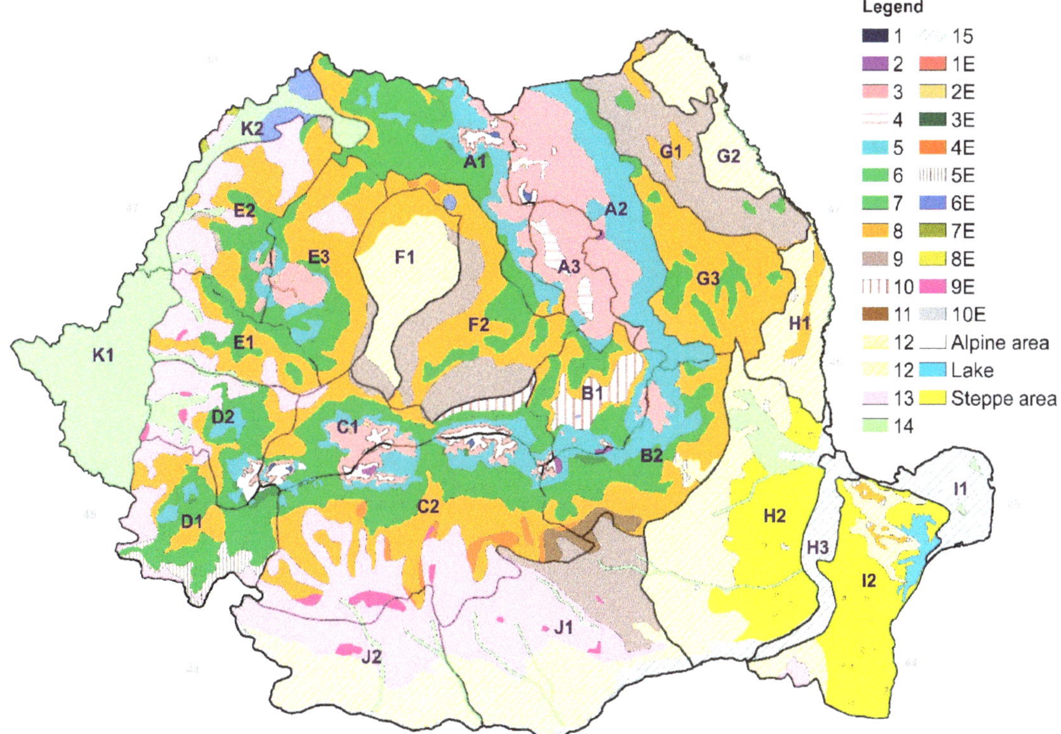

Figure 1. The map of provenance regions in Romania (as defined in [30]): A1: Eastern Carpathians, western cline; A2: Eastern Carpathians, eastern cline; A3: Giurgeu–Ciuc depression; B1: Brașov depression; B2: Curvature Carpathians, outer cline; C1: Southern Carpathians, northern cline; C2: Southern Carpathians, southern cline; D1: Mehedinți/Cerna/Semenic Mountains; D2: Țarcu/Poiana Ruscă Mountains; E1: Zarand/Metaliferi Mountains; E2: Western Apuseni Mountains; E3: Eastern Apuseni Mountains; F1: Transylvania Plain; F2: Transylvania Plateau; G1: Suceva/Siret/Iasi Hills; G2: Jijia Plain; G3: Bârlad Plateau; H1: Covur Plateau; H2: Siret and Bărăgan Plains; H3: Danube water holes; I1: Danube Delta; I2: Dobrogea Plateau; J1: Bucharest Plain; J2: Oltenia Plain; K1: Timiș and Arad Plain; K2: Cris/Carei/Someș Plain; 1–15: Forest types under normal ecological conditions; 1E–10E: Forest types under extreme ecological conditions.

For the analysis of the temporal and spatial variability of the climate in Romania over the last seven decades, we calculated four climatic parameters: (1) mean annual air temperature (MAT); (2) mean temperature during the vegetation season (April–September) (MTvs); (3) annual amount of precipitation (AAP); and (4) amount of precipitation during the vegetation season (April–September) (APVS). The climate variables were estimated using the ROCADA database [56], at a spatial resolution of $0.1 \times 0.1°$. ROCADA is a daily gridded observational dataset for precipitation, temperature, soil surface temperature, etc. in Romania, based on station information. The dataset used has a high spatial resolution (1×1 km) and was built using improved spatial interpolation techniques to better assess the spatial variability of climatic parameters. The period analyzed was 1951–2020, with the decade 1951–1960 considered as a reference because the current provenance regions are based on the climatic data from this decade. In addition, the climatological norm was calculated according to criteria recommended by the World Meteorological Organization (WMO), based on averages of the meteorological parameters, for the period 1991–2020.

In order to highlight the recent climatic changes and vulnerability of forest types, three ecoclimatic index were also calculated, including the De Martonne aridity index (DMAI), the Lang rainfall index (LRI) and the Ellenberg coefficient (EC).

The DMAI shows the degree of aridity in a certain area, and classifies the types of climates in relation to the availability of water. It is calculated based on air temperature and precipitation. Thus, the formula for the DMAI is: $DMAI = P/(T + 10)$, where P is the AAP (mm) and T is the average annual air temperature ($^{\circ}$C). The DMAI was calculated for the whole year and for the vegetation season (DMAIVS). The classification of the DMAI is given in Table 1 [57].

Table 1. The De Martonne aridity index classification (DMAI).

Values of DMAI	Type of Climate	Vegetation
5–10	Arid	Desert
11–25	Semiarid	Steppe
26–30	Moderate arid	Steppe
31–35	Semihumid	Forest steppe
36–40	Moderate humid	Oak forests
41–45	Humid	Beech forests
46–50		Coniferous forests
51–55	Very humid	Subalpine
56–60		Alpine
>60	Extremely humid	

The LRI—also known as the pluviothermal index—indicates the degree of atmospheric humidity and its variation (Table 2). This was calculated for the whole year: $LRI = P/T$, where P is the annual amount of precipitation (mm) and T is the average annual air temperature ($^{\circ}$C) [58].

Table 2. The Lang rainfall index classification.

Values of LRI	Type of Climate
>160	Wet
160–100	Temperate wet
100–60	Temperate warm
60–40	Semiarid
40–20	Arid
20–0	Desert

The EC [59] indicates the degree of compatibility of a forest species with a given climate. The formula is: $EC = Tmax/P \times 1000$, where Tmax is the temperature of the warmest month of the year and P is the annual amount of precipitation.

The ecoclimatic indices (DMAI, LRI and EC) were calculated for two periods, 2011–2020 and 1991–2020 annually, and for the vegetation season (DMAI). The monthly values of the climatic parameters from 1991 to 2020 were used as the basic data.

The variation in climatic parameters was analyzed at the level of (1) subregions, as the more homogeneous regional unit, in terms of climatic influence; and (2) ecological sectors (i.e., forest types), as the elementary units of provenance regions, which are homogeneous both ecologically and in terms of forest composition. In total, the climatic parameters were calculated for 26 subregions (Table 3) and 25 ecological sectors (forest types within subregions), as indicators of the differentiation and separation of provenance regions

(Table 4). Of the total number of ecological sectors, 10 represent forest types growing in extreme vegetation conditions (located in extreme site conditions such as climate and soil), corresponding to the basic material included in the category 'Identified source', and 15 forest types under normal vegetation conditions, corresponding to the category 'Selected source' [30].

Table 3. Provenance subregions.

Code	Subregion	Code	Subregion
A1	Eastern Carpathians, western cline	F2	Transylvania Plateau
A2	Eastern Carpathians, eastern cline	G1	Suceva/Siret/Iasi Hills
A3	Giurgeu–Ciuc depression	G2	Jijia Plain
B1	Brasov depression	G3	Barlad Plateau
B2	Curvature Carpathians, outer cline	H1	Covur Plateau
C1	Southern Carpathians, northern cline	H2	Siret and Baragan Plains
C2	Southern Carpathians, southern cline	H3	Danube water holes
D1	Mehedinti/Cerna/Semenic Mountains	I1	Danube Delta
D2	Tarcu/Poiana Rusca Mountains	I2	Dobrogea Plateau
E1	Zarand/Metaliferi Mountains	J1	Bucharest Plain
E2	Western Apuseni Mountains	J2	Oltenia Plain
E3	Eastern Apuseni Mountains	K1	Timis and Arad Plain
F1	Transylvania Plain	K2	Cris/Carei/Somes Plain

Table 4. The main forest types.

Type Code	Extreme Ecological Conditions	Type Code	Normal Ecological Conditions
1E	Norway spruce or mixed with other species at upper altitudinal limit	1	Swiss pine and mixed with Norway spruce
2E	Norway spruce on marshland	2	Larch and mixed larch and Norway spruce
3E	Beech at upper altitudinal limit	3	Norway spruce
4E	Sessile and pedunculate oak forests on pseudo-gleyic soils	4	Norway spruce in mountain depression
5E	Thermophile sessile oak and mixed with other broadleaf species	5	Mixed beech and coniferous species
6E	Pedunculate oak on marshland	6	Mountain beech
7E	Thermophile pedunculate oak on sand	7	Hilly pure beech and mixed beech with broadleaf species
8E	*Quercus pedunculiflora* on sand	8	Pure sessile oak and mixed with broadleaf species
9E	Turkey oak and Hungarian oak forests on pseudo-gleyic soils	9	Pure pedunculate oak and mixed with other oaks species
10E	Steppe island forests	10	Pedunculate oak in depression
		11	Pedunculate oak on piedmont
		12	Thermophile oak species
		13	Turkey oak and Hungarian oak forests
		14	Meadow forests (alder, ash, oak)
		15	Poplar and willow forest

The MAT and AAP for the period 1991–2020 (climatological norm) were used to develop the climate envelope for eight species: Norway spruce, silver fir, European beech, sessile oak, pedunculate oak, Hungarian oak, pubescent oak and *Quercus pedunculiflora*. They are the main tree species of mountain, hilly and plain forest ecosystems in Romania. The climate envelope was generated based on the ecological requirements of each species [60], considering that the climatic factors are among the most important factors in the context of climate change. Shifts in species climate envelope were used to evaluate the impact of climate change on species distribution. Vulnerability was discussed as the degree to which an ecosystem is susceptible to, or unable to adapt to, adverse effects of climate change [61].

3. Results

3.1. Climate Change at the Provenance Subregion Level

At the level of the provenance subregions, the MAT has increased over the last 70 years, from a multiannual average of 8.70 °C in 1951–1960 to 9.37 °C in 1991–2020. In the last decade (2011–2020), however, the temperature rise has been more pronounced, averaging 1.33 °C higher than in 1951–1960.

In addition, during the analyzed period, the increase in MAT has varied greatly at the region and subregion levels. The MATs on the decades for each provenance subregion are presented in Figure 2. In the last decade, the MAT has varied between 5.13 °C in the A3 subregion to 12.45 °C in the I1 and H3 subregions. MATs greater than 12 °C have been recorded in the last decade in the H2, H3, I2, I1, J1, J2 and K1 subregions.

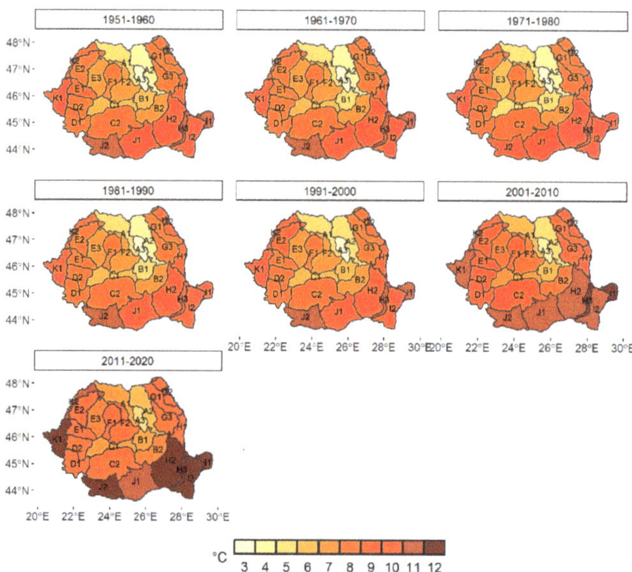

Figure 2. The variation of mean annual temperature by decade for the provenance subregions.

A comparative analysis of the MAT for 1991–2020, as the climatological norm, with the MAT for 1951–1960, as the reference decade, shows an increase by an average of 0.67 °C over the last 30 years throughout the country. At the subregion level, the increase in MAT varied from 0.33 °C, as registered in the B2 subregion, to 1.11 °C, as recorded in the G2 subregion (Figure 3). The lowest increases in MAT over the last 30 years, compared to the reference decade, were registered in the B1, B2, D2 and A3 subregions. The highest increases in temperature were recorded in the G1, G2, G3 subregions, followed by I2, I1, E3, A1, A2, H1, H3, H2, F1 and J1. If we compare the last decade (2011–2020) with the reference decade (1951–1960), the differences in MAT were much higher, varying between 0.95 °C in B2 and 1.79 °C in G2.

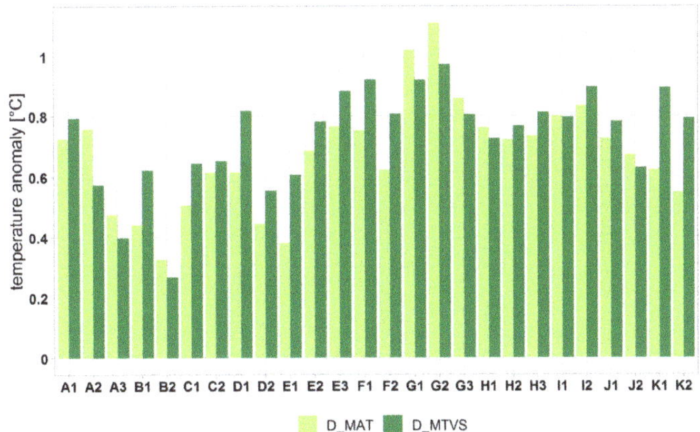

Figure 3. The increases in MAT and MT$_{VS}$ over the last 30 years compared to 1951–1960 decade for the provenance subregions.

In terms of AAP, there was an average increase of 47 mm in the period 1991–2020 compared to 1951–1960 at the provenance subregion level (Figure 4). The differences in the AAP between the analyzed decades varied between −44 mm and +146 mm (Figure 5). The subregions where there was a deficit of precipitation over the last 30 years were I1, C1, K1, K2, B1 and F2. An excess of AAP was registered in the rest of the subregions, with the highest values recorded in E3, A2 and D1. Comparing the last decade (2011–2020) with the reference decade (1951–1960), the differences in AAP varied between −57 mm in K2 and 118 mm in E3.

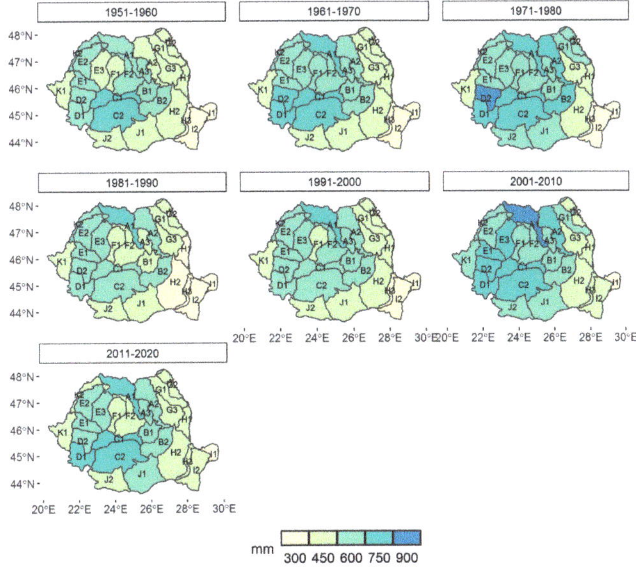

Figure 4. The variation of average annual precipitation on the decades for the provenance subregions.

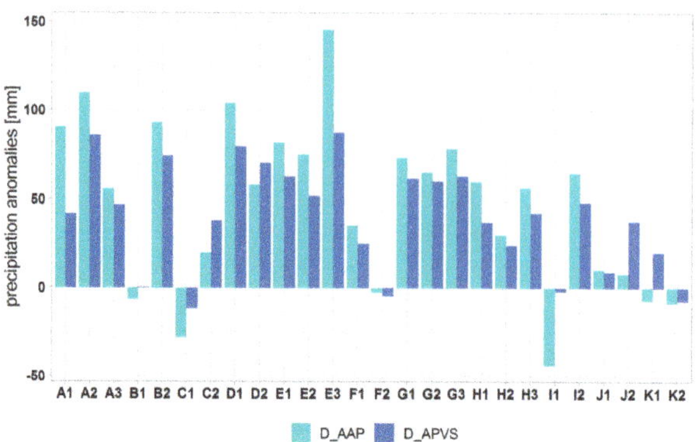

Figure 5. The increases in AAP and APVS over the last 30 years compared to 1951–1960 decade for the provenance subregions.

Analyzing the MT_{VS} at the provenance subregion level, the variation was very similar to that of the MAT (Figure 6). The differences in MT_{VS} between 1991–2020 and 1951–1960 varied between 0.27 °C (in B2) and 0.97 °C (in G2). The subregions most affected by rising MT_{VS} over the last 30 years were G2, G1, G3, F1, F2, I2, K1, E3 and D1 (Figure 3). If we compare the last decade (2011–2020) with the reference decade (1951–1960), the differences in MT_{VS} varied between 0.93 °C in B2 and 1.78 °C in G2. MT_{SV} values close to 20 °C were recorded in the last decade in the H2, H3, I2, I1, J1, J2 and K1 subregions.

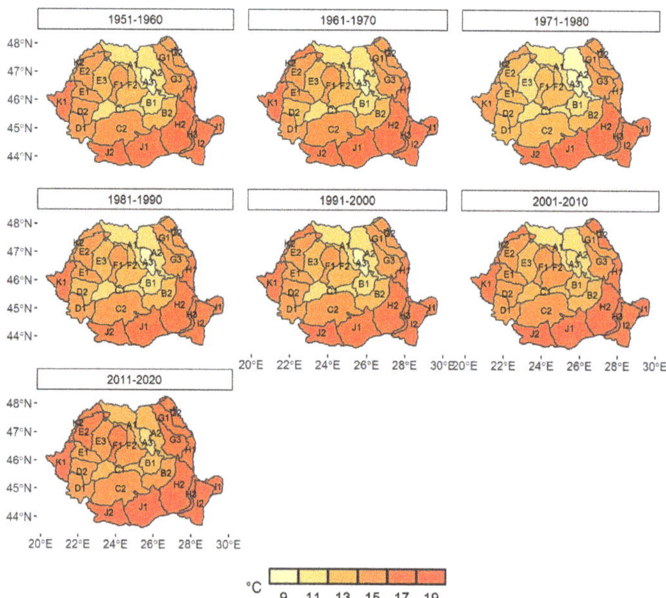

Figure 6. The variation of MT_{VS} on the decades for the provenance subregions.

In terms of the AP_{VS}, in addition to the subregions presented above, a slight excess in humidity was registered in the last few decades in B1 and K1 (Figures 5 and 7).

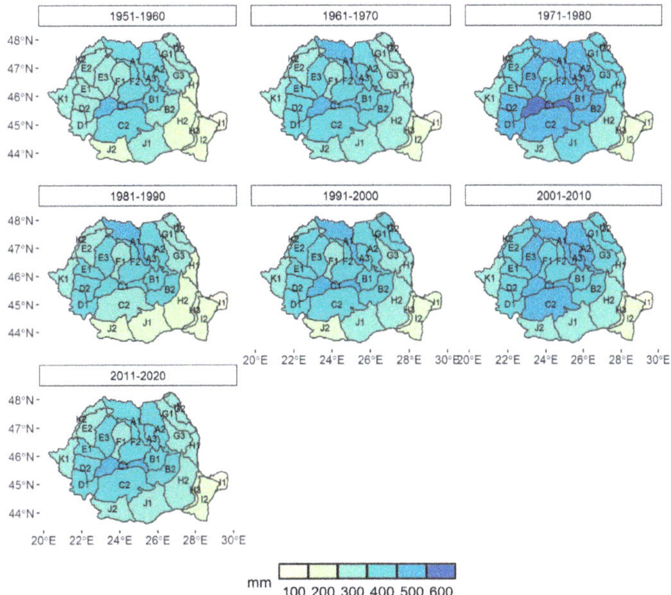

Figure 7. The variation of AP$_{VS}$ on the decades for the provenance subregions.

3.2. Climate Change at the Forest-Type Level under Normal Ecological Conditions

3.2.1. Variation in the Main Climatic Parameters at the Forest-Type Level for 1951–2020

At the forest-type level, as an average throughout the country, the MAT has increased over the last 70 years from a multiannual average of 7.34 °C in 1951–1960 to 8.65 °C in 2011–2020, and to 7.98 °C in 1991–2020.

The increase in MAT has varied greatly over the analyzed period at the forest-type level (Figure 8). Thus, in the last decade, the MAT has varied between 3.02 °C in the 1—Swiss pine and mixed with Norway spruce forests to 12.57 °C in 15—poplar and willow forests. MATs greater than or close to 12 °C have also been recorded in 13—Turkey oak and Hungarian oak forests and 14—meadow forests over the last decade.

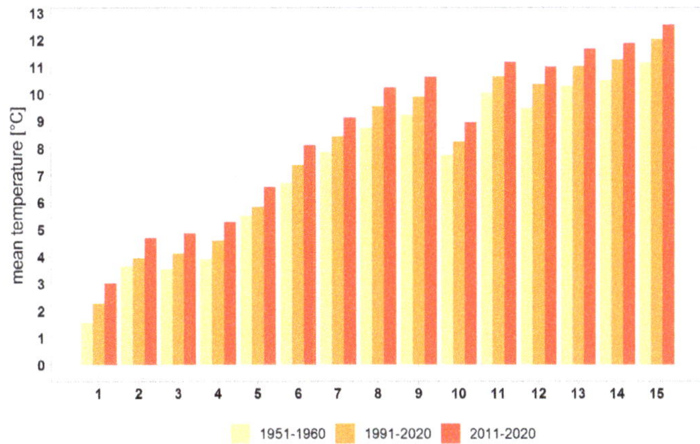

Figure 8. The variation of mean annual temperature over the analyzed period for the main forest types.

Analyzing the differences in MATs between 1991–2020 as the climatological norm and 1951–1960 as the reference decade, it can be seen that there has been an increase by an average of 0.64 °C over the last 30 years across the forest ecosystems. The highest temperature increases were recorded in 12—thermophile oak forests (0.88 °C), 15—poplar and willow forests (0.86 °C) and 9—pedunculate oak forests (0.80 °C) (Figure 9). The lowest MAT increases over the last 30 years, compared to the reference decade, were registered in 5—mixed beech and coniferous species (0.33 °C) and 10—pedunculate oaks in depressions (0.50 °C). If we compare the last decade (2011–2020) with the reference decade (1951–1960), the differences in the MAT were much higher, varying between 1.51 °C in 12—thermophile oak forests and 1.49 °C in 8—pure sessile oak to 1.01 °C in 2—larch and mixed larch and Norway spruce forests.

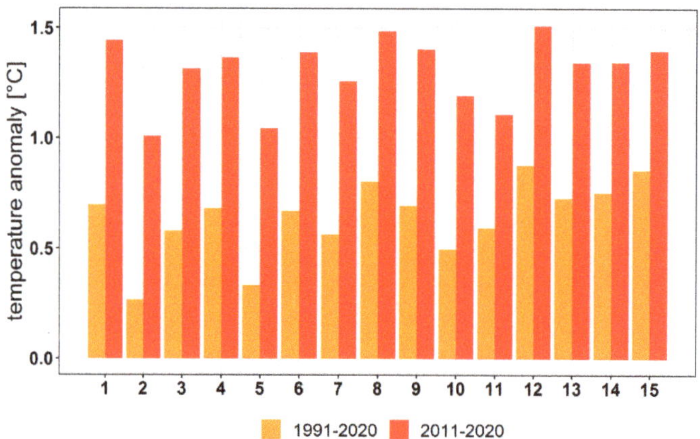

Figure 9. The differences in mean annual temperature over the last 30 years compared to 1951–1960 decade for the main forest types.

In terms of AAP, the increase was, on average, 40 mm in 1991–2020 compared to 1951–1960 at the forest-type level (Figure 10). The differences in AAP between the analyzed periods varied between −40 mm (10—pedunculate oak forests in depressions) to 83 mm (3—Norway spruce forests) (Figure 11). The forest types in which the AAP was very low over the last 30 years were 15—poplar and willow forests, 14—meadow forests, 13—Turkey oak and Hungarian oak forests and 12—thermophile oak species. Comparing the last decade (2011–2020) with the reference decade (1951–1960), the differences in AAP varied between −64 mm in 10—pedunculate oak forests in depressions to 105 mm in 2—larch and mixed larch with Norway spruce forests. In addition to the types of forest already mentioned, affected by the reduction of rainfall in the last decade are also 10—pedunculate oak forests in depressions and 9—pure pedunculate oak and mixed with other oaks.

In analyzing the MT_{VS}, we found an increase of 0.71 °C, on average, for the period 1991–2020, and 1.38 °C for 2011–2020, compared to the reference decade. The amplitude of variation was from 0.16 °C (in 2—larch and mixed larch and Norway spruce forests) to 0.93 °C (in 8—sessile oak forests), and from 0.90 °C (in 2—larch and mixed larch and Norway spruce forests) to 1.61 °C (in 8—sessile oak forests) (Figures 12 and 13). The forest types most affected by the increase in the MT_{VS} in the last 30 years were 8—sessile oak, 15—poplar and willow, 12—thermophile oak, 13—Turkey oak and Hungarian oak, and 14—meadow forests. In the last decade, 9—pedunculate oak, 6—mountain beech, and 1—Swiss pine and mixed Swiss pine with Norway spruce forests can be added to that list.

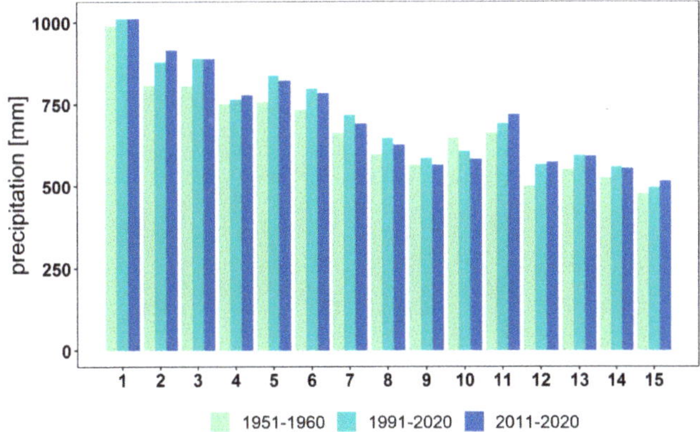

Figure 10. The variation of average annual precipitation over the analyzed period for the main forest types.

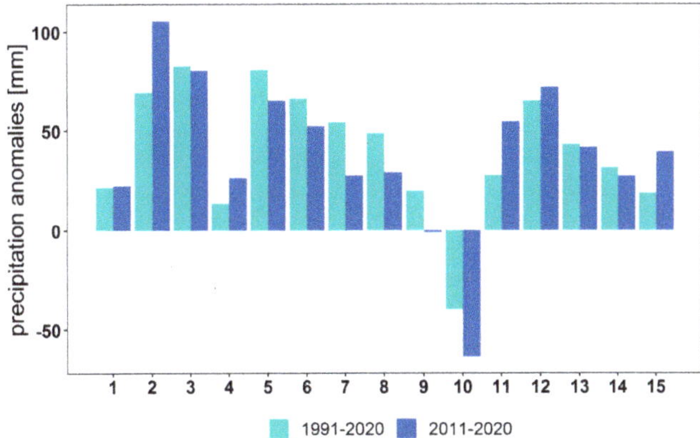

Figure 11. The differences in average annual precipitation over the last 30 years compared to 1951–1960 decade for the main forest types.

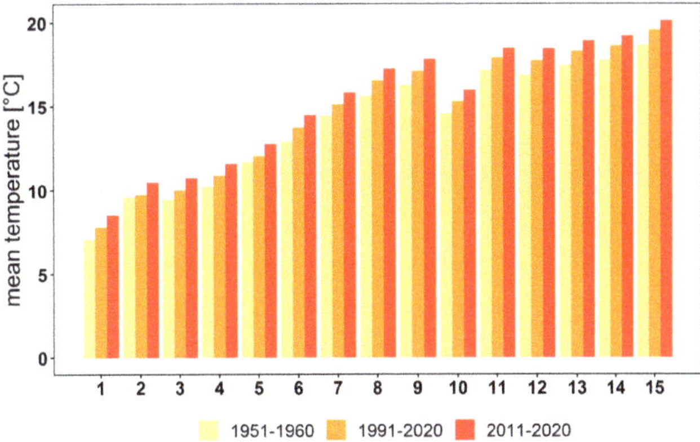

Figure 12. The variation of MT_{VS} over the analyzed period for the main forest types.

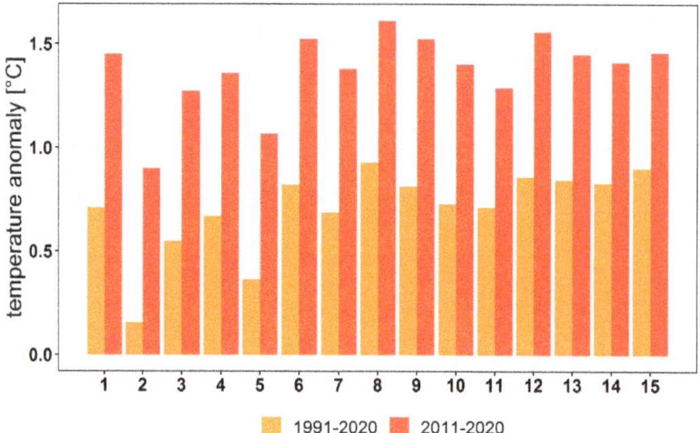

Figure 13. The differences in MT_{VS} over the last 30 years compared to 1951–1960 decade for the main forest types.

Regarding the AP_{VS}, there was an average increase of 33 mm in 1991–2020 compared to 1951–1960 at the forest-type level. The differences in the AP_{VS} between the analyzed periods varied between −33 mm (10—pedunculate oak forests in depressions) to +72 mm (2—larch and mixed larch with Norway spruce forests) (Figures 14 and 15). The forest types in which the AAP was very low over the last 30 years were 1—Swiss pine and mixed Swiss pine with Norway spruce, 9—pedunculate oak, 11—pedunculate oak on piedmont, and 15—poplar and willow forests. In comparing the last decade (2011–2020) with the reference decade (1951–1960), the differences in AP_{VS} have varied between −48 mm in 10—pedunculate oak forests in depressions to +95 mm in 2—larch and mixed larch with Norway spruce forests.

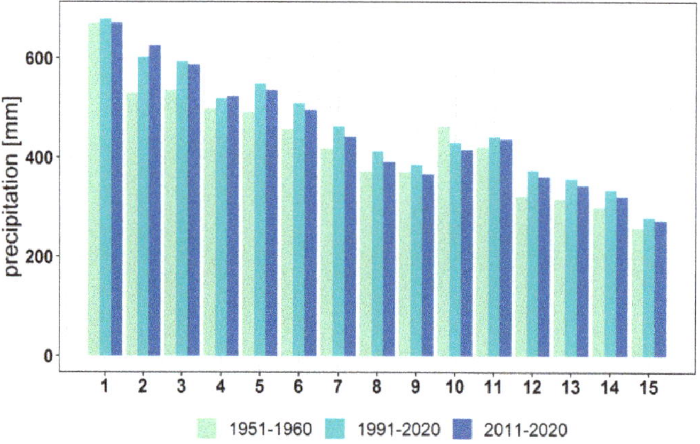

Figure 14. The variation of AP_{VS} over the analyzed period for the main forest types.

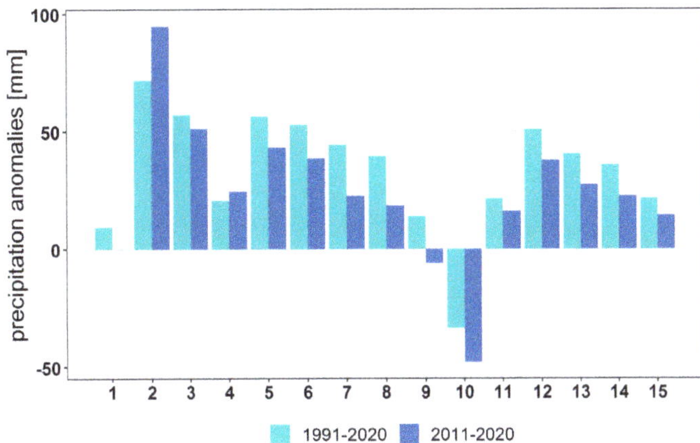

Figure 15. The differences in AP$_{VS}$ over the last 30 years compared to 1951–1960 decade for the main forest types.

Analyzing the forest types in each provenance subregion resulted in 117 ecological sectors. Comparing the MAT over the last three decades (1991–2020) with that of the reference decade, it can be seen that there was a temperature increase that varied between −1.22 °C (in D120) to +1.58 °C (in B110). The highest temperature increases were recorded in: 3—Norway spruce forests from B1, A1 and A2; 5—mixed forests from A1, B2 and E3; 9—pedunculate oak (pure and mixed) forests in F2, G1, G3, E1, E2, E3, B1; 13—Turkey oak and Hungarian oak forests in E3 and E2; 14—meadow forests in G2; 12—thermophile oak forests in G2; and 7—hilly beech forests in A2 and G3.

Comparing the last decade with 1951–1960, the differences in MAT varied between −0.49 °C (in D120) and 2.32 °C (in A120). The highest temperature increases were recorded in: 5—mixed forests in A1, B2 and E2; 9—pedunculate oak (pure and mixed) forests in F2, G1, G3, E1, E2, E3 and B1; 3—Norway spruce forests in B1, A1, A2 and E2; 13—Turkey oak and Hungarian oak forests in E3 and E2; 14—meadow forests in G2; 12—thermophile oak forests in G2; and 7—hilly and mountain beech forests in A1, A2, E3, E2 and G3.

Of all the forest sectors, 86% experienced a temperature increase of >1.0 °C, while in 44% there was an increase of >1.5 °C.

In terms of the AAP, the differences in precipitation between 1991–2020 and 2011–2020 compared to the reference decade varied from −73 mm (in I190) to 252 mm (in E310), and from −114 mm (in C16B) to 279 mm (in E210), respectively. The forest types in which there was a deficit of precipitation over the last three decades were the: 14—meadow forests in K1, J1 and I1; 10—pedunculate oak forests in depressions in B1 and C1; 7—hilly beech forests in F2, B1 and C1; 1—Swiss pine and Norway spruce at altitude in C2; 15—poplar and willow forests in J2 and I1; 6—mountain beech forests in C1; 13—Turkey oak and Hungarian oak forests in J2, C1 and K2; and 9—pedunculate oak forests in F2, E3 and F1.

In analyzing the MT$_{VS}$, the amplitude of variation was from −1.47 °C (in B210) to +1.81 °C (in B220), when compared to the last 30 years and the reference decade, and from −0.71 °C (in B210) to +2.50 °C (in B220) when compared to the last decade and the reference decade. The forest types most affected by the increase in temperature during the growing season were similar to those identified in the case of MAT.

Regarding AP$_{VS}$, the amplitude of variation was from −64 mm to +144 mm when comparing the last 30 years with the reference decade, and from −94 mm to +157 mm when comparing the last decade with the reference decade. The forest types in which there has been a deficit in precipitation over the last three decades are similar to those identified in the case of AAP.

3.2.2. Ecoclimatic Index Calculated for Forest Types under Normal Ecological Conditions

The values of the ecoclimatic indices (DMAI, LRI and EC) by forest types and ecological sectors are presented in Table S1.

The DMAI shows the degree of aridity of the ecological sector, thus highlighting the degree of vulnerability of the forest types to climate change. The DMAI values calculated for 1991–2020 varied between 110 in 1—Swiss pine and mixed with Norway spruce from C21E ecological sector, with values of 14 in 14—meadow forests from I190 and 16 in 15—poplar and willow forests from I19A ecological sector. Corresponding to this aridity index, a very high vulnerability was identified for: 15—poplar and willow forests, 14—meadow forests and 12—thermophile oaks in all subregions of provenance; 9—pure or mixed oaks, other than those in E260 and F260; 8—sessile oak forests from G350, H150 and I250; and 13—Turkey oak and Hungarian oak forests from C170, E270, H180, H280, I270, I280, J170, J180, J270, J280 and K270. The hilly beech forests in the G340 and A240 subregions were also identified as vulnerable.

In analyzing the DMAI values for the last decade, it was found that the degree of vulnerability at the ecological-sector level was high and the trend is increasing. Calculating this index for the vegetation season (DMAI$_{VS}$), it is found that, out of the total number of 117 ecological sectors presented in Table S1, 11 fell into the arid category, 90 fell into the semiarid category and 7 fell into the moderately arid category, which indicates a very high degree of vulnerability. Corresponding to the classification of the DMAI, values between 5 to 10 (arid) indicate desert climatic conditions, 11 to 30 (semiarid and moderate-arid) indicate the presence of steppe climatic conditions, while values between 31 to 35 (semihumid) indicate forest steppe.

The LRI also highlights significant climate changes and the tendency for some sectors to dry up. Thus, steppe climatic conditions occurred in last decade in 15—poplar and willow forests from I1 and H3, in 14—meadow forests from I1, in 13—Turkey oak and Hungarian oak forests from I2 and in 8—sessile oak forests from I2. Forest steppe climatic conditions occurred in all pure or mixed pedunculate oaks forests (9), thermophile oak species (12), meadow forests (14), poplar and willow (15), and Turkey oak and Hungarian oak forests (13) except those in C2. The degree of vulnerability increased in the case of forest types located along the low altitudinal limit of the distribution area, such as 7—hilly beech forests in G3, 8—sessile oak forests in I2, H1, E2, G1 and G3, 9—pure or mixed pedunculate oak forests in E3.

The EC indicates the level of compatibility between the distribution of forest types and the current zonal climate. The EC values for 1991–2020 clearly show the process of global warming and indicate which species are suitable for the current climate. It can be seen that this warming process is more accentuated at the altitudinal limits (both upper and lower) of the species distributions, especially in the following forest types: 1—Swiss pine and mixed with Norway spruce, 2—larch and mixed with Norway spruce, 3—Norway spruce, 4—Norway spruce in mountain depression, 5—mixed beech with coniferous species, 14—meadow forests, 15—poplar and willow. In addition, the process of aridity occurred in 7—hilly beech forests in F2 and G3 ecological sectors, 8—sessile oak forests in H1 and I2 ecological sectors and 9—pedunculate oak forests in E3 and G1 ecological sectors.

3.3. Climate Change at the Forest-Type Level under Extreme Ecological Conditions

3.3.1. Variation in the Main Climatic Parameters at the Forest-Type Level for 1951–2020

At the forest-type level under extreme vegetation conditions, the MAT has increased over the last 70 years from a multiannual average of 7.98 °C in 1951–1960 to 9.40 °C in 2011–2020, and 8.73 °C in 1991–2020. In the last decade, the increase in the MAT has varied between 3.97 °C in Norway spruce forests at the upper altitudinal limit and 12.29 °C in *Quercus pedunculiflora* forests on sand. MAT values >12 °C have been recorded in 10E—steppe island forests, in the last decade.

In analyzing the differences in MAT between 1991–2020, as the climatological norm, and 1951–1960, as the reference decade, an average increase of 0.75 °C can be seen over the last 30 years at these forest types. The highest temperature increases were recorded in 10E—steppe island forests and 2E—Norway spruce on marshland. The lowest increases over the last 30 years, compared to the reference decade, were recorded in 9E—Turkey oak and Hungarian oak forests on pseudo-gleyic soils (0.60 °C), and in 6E—pedunculate oak on marshland (0.62 °C). If we compare the last decade (2011–2020) with the reference decade (1951–1960), the differences in MAT were much greater, ranging between 1.68 °C in 2E—Norway spruces forests on marshland and 1.18 °C in 9E—Turkey oak and Hungarian oak forests on pseudo-gleyic soils.

Regarding the AAP, the increase was, on average, 45 mm in 1991–2020 compared to 1951–1960 at the level of forest types. The differences in AAP between the analyzed periods varied between −37 mm (7E—thermophile pedunculate oak on sand) up to 104 mm (3E—beech at the upper altitudinal limit). The forest types in which the amount of precipitation has been very low over the last 30 years were: 10E—steppe island forests, 8E—*Quercus pedunculiflora* on sand and 5E—thermophile sessile oak and mixed with other deciduous species. Comparing the last decade (2011–2020) with the reference decade (1951–1960), the differences in AAP varied between −67 mm in 7E—thermophile pedunculate oak on sand and 118 mm in 3E—beech at the upper altitudinal limit.

In analyzing the MT_{VS}, an increase of 0.83 °C, on average, was found for 1991–2020, and 1.50 °C for 2011–2020, compared to the reference decade. The temperature differences ranged from 1.04 °C (in 2E—Norway spruce on marshland) and 1.05 °C (in 7E-Thermophile pedunculate oak on sand) to 0.57 °C (in 3E—beech at the upper altitudinal limit), and from 1.78 °C (in 2E—Norway spruces on marshland soil) to 1.21 °C (in 9E—Turkey oak and Hungarian oak forests on pseudo-gleyic soils), respectively. The forest types most affected by the increase in MT_{VS} over the last 30 years were 2E—Norway spruce on marshland, 7E—thermophile pedunculate oak on sand and 10E—steppe island forests.

Regarding the AP_{VS}, there was an average increase of 34 mm in 1991–2020 compared to 1951–1960. The differences in AP_{VS} between the analyzed periods varied between −19 mm (7E—thermophile pedunculate oak on sand) up to 77 mm (3E—beech at the upper altitudinal limit). The forest types in which the AP_{VS} has been very low over the last 30 years are: 10E—steppe island forests, 9E—Turkey oak and Hungarian oak forests on pseudo-gleyic soils, 8E—*Quercus pedunculiflora* on sand, 7E—thermophile pedunculate oak on sand and 5E—thermophile sessile oak. Comparing the last decade (2011–2020) with the reference decade (1951–1960), the differences in AP_{VS} varied between −53 mm in 7E—thermophile pedunculate oak on sand and +73 mm in 3E—beech at the upper altitudinal limit.

The EC indicates a mismatch between the distributions of forest species and current climatic conditions. The forest types which are the most affected by climate warming are the following: 1E—Norway spruce or mixed with other specie at upper altitudinal limit, 2E—Norway spruce on marshland and 8E—*Quercus pedunculiflora* on sand from all ecological sectors, 4E—sessile and pedunculate oaks on pseudo-gleyic soils in F25A, 5E—thermophile sessile oak and mixed with broadleaf species in H15B and I25B.

3.3.2. Ecoclimatic Index Calculated for Forest Types under Extreme Ecological Conditions

The values of the ecoclimatic indices for forest types and ecological sectors under extreme ecological conditions are presented in Table S2.

The DMAI values calculated for 1991–2020 varied between 84 in 1E—Norway spruce forests at the upper altitudinal limit from C21B to 22 in 10E—steppe island forests from I28A. Calculating this index for the vegetation season (DMAIVS) during 2011–2020, out of the total number of 38 ecological sectors presented in Table S2, 3 fell into the arid category, 19 fell into the semiarid category and 5 into the moderately arid category, which indicates that 71% of ecological sectors from extreme ecological conditions present a very high degree of vulnerability.

The LRI also highlights that steppe climatic conditions intensified in the last decade in 10E—steppe island forests. Forest steppe climatic conditions occurred in 12 ecological sectors and especially in the case of the following forest types: 5E—thermophile sessile oak, 7E—thermophile pedunculate oak on sand, 8E—*Quercus pedunculiflora* on sand and 9E—Turkey oak and Hungarian oak forests on pseudo-gleyic soils.

3.4. Changes in the Distribution of Climate Envelope of the Forest Species

The shifts in distribution of the climate envelope of the forest species are presented in Figure 16. For analyzed species, the climatic envelope will withdraw in some regions while expanding in others. The most obvious changes in the optimum climatic of species are along an altitudinal gradient. In addition, the shifts occur in the southern regions (J2, J1 and H2), along the latitude, and in the eastern regions (G1, G2, G3 and H1) where climate habitat for many species will withdraw toward the west.

Thus, for Norway spruce and silver fir, whose distribution areas are partially or entirely in mountainous areas, the shifts will be towards higher altitudes. The same tendency would be in the case of European beech and sessile oak, in the future. The oak species from southern plain regions will expand substantially at more northerly latitudes. Climate habitat for Hungarian oak and Turkey oak will expand substantially in many regions, replacing current climate envelopes of pedunculate oak and sessile oak ecosystems. Pedunculate oak will withdraw from southern, western and eastern regions towards the extra-Carpathian hills and the intra-Carpathian regions.

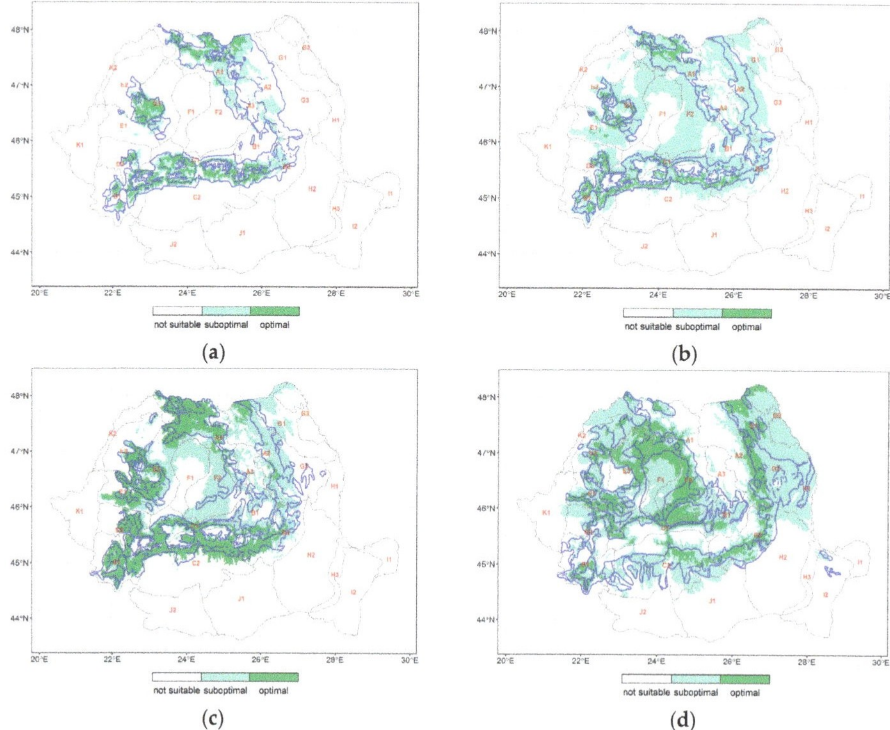

Figure 16. *Cont.*

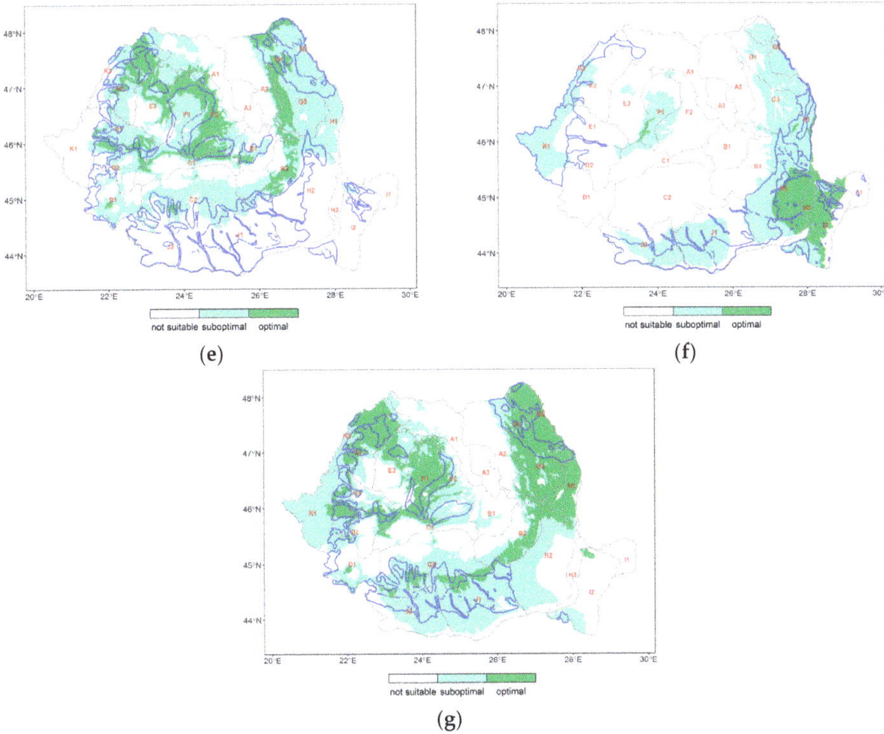

Figure 16. Climate envelope of: (**a**) Norway spruce; (**b**) silver fir; (**c**) European beech; (**d**) sessile oak; (**e**) pedunculate oak; (**f**) *Quercus pedunculiflora* and pubescent oak; (**g**) Hungarian oak and Turkey oak. The blue polygons show the current spatial distribution.

Warming climate will create suitable growth conditions for expansion of xerophilous oak species in many regions from the south, east and west of the country, replacing current climate envelopes of mesophilic oak species. Furthermore, for species such as *Quercus pedunculiflora*, pubescent oak and Hungarian oak it can be seen that there is almost no spatial overlap with the zone's current climatic distribution.

The largest climate habitat changes are the expansion of the climatic envelope of xerophilous (*Quercus pedunculiflora*, pubescent oak) or semi-xerophilous species (Hungarian oak and Turkey oak), typical for the forest steppe region and decrease the habitat for the coniferous forests.

4. Discussion

In this study, four climatic parameters and ecoclimatic index have been calculated and analyzed at the level of provenance regions, subregions and ecological sectors (forest types) in Romania, during the period 1951–2020.

The results show a general shift towards warmer and drier conditions in the last 30 years compared to the reference decade. MAT recording increases between 0.33 °C to 1.11 °C across provenance subregions, and MT_{VS} between 0.27 °C to 0.97 °C. The provenance subregions most affected by rising temperature are G1, G2, G3, H2, H3, I1 and I2 in the east of the country; J1 in the south; K1, K2, E3 in the west; and F1, F2 in the intra-Carpathian area. Precipitation has recorded a decline in the following subregions: I1, C1, K1, K2, B1 and F2, and a slight excess in the rest of the subregions.

Evident climate changes also occurred at the ecological-sector level (forest types) during the last three decades. Compared to the reference decade, there have been increases by an average of 0.64 °C in MAT and 0.71 °C in MT_{VS} at the forest-type level growing under normal ecological conditions and by an average of 0.75 °C in MAT and 0.83 °C in MT_{VS} at the forest types under extreme ecological conditions. The results show that climate change is not uniform throughout provenance regions, and each forest ecosystem might be affected slightly differently. Currently, the most affected by the climate warming are the following forest types: 12—thermophile oak, 15—poplar and willow, 8—pure sessile oak and mixed with broadleaf species, but also 14—meadow forests, 13—Turkey and Hungarian oak and 6—mountain beech forests if we consider the temperature values recorded during growing season. Considering both temperature increases and precipitation deficit recorded in the last decade, 9—pedunculate oak and 1—Swiss pine and mixed Swiss pine with Norway spruce forests should be added to this list. In case of the forests growing in extreme vegetation conditions, the most sensitive to climate change will be the following: 6E—pedunculate oak on marshland, 7E—thermophile pedunculate oak on sand, 4E—sessile and pedunculate oak forests on pseudo-gleyic soils and 10E—steppe island forests.

Another phenomenon associated with climate change is the tendency for some ecological sectors to dry up, as evidenced by the DMAI and LRI. The values of DMAI for the vegetation season show that 86% of the ecological sectors under normal ecological conditions fell into the arid and semiarid categories, which indicates a very high degree of vulnerability for forest species. On the LRI values, forest steppe climatic conditions occurred in all pure or mixed pedunculate oaks forests, thermophile oak species, meadow forests except those in I1, poplar and willow except those in I1 and H3, and Turkey oak and Hungarian oak forests. There are steppe climatic conditions in the case of sessile oak forests from I2—Dobrogea Plateau, meadow forests from I1—Danube Delta and poplar and willow forests from I1—Danube Delta and H3—Danube water holes. Regarding ecological sectors under extreme ecological conditions 71% fell into arid, semiarid and moderately arid categories.

EC values highlight that the warming process is more evident along the altitude, and that the degree of vulnerability increases at lower altitudes or at the edge of species distribution because the amount of water available is reduced. That is the case for hilly beech forests in the south of the Moldavian Plateau (G3), sessile oak forests in the eastern regions (G1, G3, H1, I2) and in the Western Apuseni Mountains (E2), and pure or mixed pedunculate oak forests in Eastern Apuseni Mountains (E3).

Our findings show that the climate in eastern Europe is changing, and according to the projections of the climate scenarios the change will continue at an even faster rate. For Romania an increase in the average annual temperature by 1.2 °C is forecast in the period 2021–2050 compared to the period 1991–2020 [49] and by over 2 °C in the next 100 years, respectively, in 2061–2090 vs. 1961–1990 [62].

Climate change in the last 30 years has produced a spatial mismatch between the current distribution of species and their optimal climate envelope. Considering that the aridity process will be emphasized in some regions from Romania, the implications for tree growth are that some species will continue to grow well and will expand, while other species would disappear from some areas currently located at the edge of the distribution range. The main feature of climate change will be expansion of the distribution area of many species towards higher altitudes and retractions at low-latitude and low-elevation limits.

The climate envelopes for the main forest species have already shifted to another ecosystem's climate. The maps of climate envelopes reveal the following general trends: (1) some of the most important conifer species, such as Norway spruce and silver fir, will expand to higher altitudes but will significantly decrease in frequency and lose their habitat, particularly in the eastern Carpathians; (2) oak species that currently have a more southern distribution are expected to gain suitable habitat toward the north; (3) European beech and sessile oak will expand to higher altitudes but will lose substantial habitat in eastern regions; (4) pedunculate oak will withdraw from southern, western and eastern regions towards the extra-Carpathian hills and the intra-Carpathian regions; (5) xerophilous or semi-xerophilous species will expand in many regions, replacing current climate envelopes of pedunculate oak and sessile oak ecosystems.

Therefore, the impact on the forest ecosystems and reforestation practices might be drastic if climate changes as predicted. In order to improve adaptability of forest species, urgent actions are needed based on the sustainable use and deployment of forest reproductive material. In this context, regions of provenance have a crucial role ensuring a match between the ecological requirements of the species to planting conditions and increasing the capacity of forest species to cope with climate changes and extreme events. However, the system of provenance regions, according to the OECD Scheme and EU directive, was thought to be encouraging the use of the local seed sources, under the concept 'local is the best', because it was considered that these sources are most adapted to local and regional environmental conditions [63,64]. The results of numerous provenance experiments, in Romania and abroad, revealed small local adaptation for growth traits among provenance regions. In Romania, local provenances are, generally, less performing than some provenances from other regions [65]. Furthermore, a significant site effect was detected, suggesting that phenotypic plasticity rather than local adaptation explains phenotypic differences among provenances [66–68]. Similar results were also observed in other geographical regions of Europe [69–71].

Genetic adaptation is the microevolutionary process that enhances the fitness of a population in accordance with the environmental conditions [72]. Therefore, minimizing regional maladaptation involves optimizing provenance region boundaries. Our findings do not validate current provenance region delineation in terms of ecological criteria. The re-delimitation of current provenance regions according to climatic criteria rather than the geographic or administrative ones is required in order to promote adaptation. For this reason, provenance regions must be wide enough to provide a high level of genetic diversity within species, so that the future forest stands can survive and remain productive [73,74]. The delineation of too-small provenance regions, as is currently the case of xerophilous oak species with fragmented populations in the southern, eastern and western plains of Romania, may decrease levels of genetic diversity and increase levels of inbreeding [75,76].

However, to ensure genetic adaptation, it is not enough to know where a species will be suitable in the future, but also which populations of a species will perform well in areas of potential new habitat. In the establishment of new forests through reforestation, a gradual adaptation may be achieved through moderate transfer of forest reproductive material from latitudinal–adjacent regions or from lower altitudinal distances and by selection and transfer of high-productive and resilient forest reproductive material-assisted migration [54]. Assisted migration is considered as part of a forest climate change adaptation strategy because it can prevent species extinction, minimize economic loss and sustain ecosystem services and biodiversity. The authors of [77–79] summarized the adaptive actions for forest management into three categories: societal adaptation, adaptation of the forest (e.g., species selection, tree breeding) and adaptation to the forest (e.g., changing rotation age, modifying wood-processing technology).

The present study argues that the delineation of the provenance regions of reproductive materials are fundamental for an adaptive forest management in Europe. New approaches regarding the delineation of provenance regions according to species climate envelope, future climate projections and intraspecific genetic variation are requested.

5. Conclusions

The results highlight significant changes in the variation of climatic variable in the last decades at both the provenance-region and forest-type levels.

Based on the results of this study, we can state that the aridity process of some regions in Romania will increase in future and will considerably alter growing conditions of the forest species, particularly the growth and survival of stand regeneration and new plantations.

Our results stress that climate changes will occur within the lifetime of a single tree generation and will be faster than species can adapt or migrate. The risks of climatic changes will be very high in forestry; consequently, major changes in forest management will become necessary. In this regard, some measures for an adaptive management would be identifying the suitable sources of forest reproductive material; replacing sensitive species with others better adapted, particularly on sites exposed to aridization; admixing better-adapted provenances with local sources; enriching the species mixtures; and using a forest reproductive material that holds a high level of genetic diversity, such as that from seed orchards.

Therefore, delineation of the provenance regions for forest species together with selection of the seed sources and transfer of forest reproductive material (assisted migration) could be fundamental tools for an adaptive forest management.

Supplementary Materials: The following supporting information can be downloaded at: https://www.mdpi.com/article/10.3390/f13081203/s1, Figure S1: The geographical position of Romania in the European continent. The country topography is represented in the lower right box; Table S1: Ecoclimatic indices (DMAI, LRI and EC) for each forest type and ecological sector under normal ecological conditions; Table S2: Ecoclimatic indices (DMAI, LRI and EC) for each forest type and ecological sector under extreme ecological conditions.

Author Contributions: Conceptualization, G.M. and M.-V.B.; methodology, G.M. and M.-V.B.; climate data analysis and processing, A.-M.A., I.-A.N. and M.-V.B.; maps, I.-A.N.; writing, all authors. All authors have read and agreed to the published version of the manuscript.

Funding: This study was carried out within the framework of the project PN 19070303 (Revision of the provenance regions for production and deployment of the forest reproductive materials in Romania in order to increase the adaptability of forest ecosystems to climate change). This work was supported by the Ministry of Research, Innovation and Digitalization in Romania, in BIOSERV Nucleu Program.

Institutional Review Board Statement: Not applicable.

Informed Consent Statement: Not applicable.

Data Availability Statement: Data can be transmitted to anyone interested via email request.

Acknowledgments: The authors would like to thank the anonymous reviewers for their comments and suggestions that contributed positively to this paper.

Conflicts of Interest: The authors declare no conflict of interest.

References

1. O'Neill, G.A.; Aitken, S.N. Area-Based Breeding Zones to Minimize Maladaptation. *Can. J. For. Res.* **2004**, *34*, 695–704. [CrossRef]
2. Ying, C.C.; Yanchuk, A.D. The Development of British Columbia's Tree Seed Transfer Guidelines: Purpose, Concept, Methodology, and Implementation. *For. Ecol. Manag.* **2006**, *227*, 1–13. [CrossRef]
3. Buiteveld, J.; de Vries, S.M.G.; Kranenborgen, K.G. *Delimitations of Regions of Provenance for Forest Reproductive Material in Western Europe*; Alterra-Report; Alterra, Research Instituut voor de Groene Ruimte: Wageningen, The Netherlands, 2000; 45p.
4. Ducci, F.; Vannuccini, M.; Carone, G.; Vedele, S.; Cili, S.; Apuzzo, S. Delimitations of Regions of Provenance for Forest Reproductive Material in Western Europe. *Ann. C.R.A.-SEL* **2008**, *35*, 133–142.
5. Alía, R.; del Barrio, J.M.G.; Iglesias, S.; Mancha, J.A.; De Miguel, J.; Nicolás, J.; Pérez, F.; de Ron, D.S. *Regiones de Procedencia de Especies Forestales En España*; MARM: Madrid, Spain, 2009.
6. Auñon, F.J.; del Barrio, J.M.G.; Mancha, J.A.; de Vries, S.M.G.; Alía, R. Regions of Provenance of European Beech (*Fagus Sylvatica* L.) in Europe. In *Genetic Resources of European Beech* (Fagus sylvatica L.) *for Sustainable Forestry*; Ministerio de Ciencia e Innovacion: Madrid, Spain, 2011; pp. 141–148.
7. Matyas, C. Modeling Climate Change Effects with Provenance Test Data. *Tree Physiol.* **1994**, *14*, 797–804. [CrossRef]
8. Rehfeldt, G.E.; Ying, C.C.; Spittlehouse, D.L.; Hamilton, D.A. Genetic Responses to Climate in Pinus Contorta: Niche Breadth, Climate Change, and Reforestation. *Ecol. Monogr.* **1999**, *69*, 375–407. [CrossRef]
9. Hamann, A.; Wang, T. Potential Effects of Climate Change on Ecosystem and Tree Species Distribution in British Columbia. *Ecology* **2006**, *87*, 2773–2786. [CrossRef]
10. Wang, T.; O'Neill, G.; Aitken, S.N. Integrating Environmental and Genetic Effects to Predict Responses of Tree Populations To climate. *Ecol. Appl.* **2010**, *20*, 153–163. [CrossRef] [PubMed]

11. Hamann, A.; Gylander, T.; Chen, P.Y. Developing Seed Zones and Transfer Guidelines with Multivariate Regression Trees. *Tree Genet. Genomes* **2011**, *7*, 399–408. [CrossRef]
12. Davis, M.; Shaw, R. Range Shifts and Adaptive Responses to Quaternary Climate Change. *Science* **2001**, *292*, 673–679. [CrossRef]
13. Jobbágy, E.G.; Jackson, R.B. Global Controls of Forest Line Elevation in the Northern and Southern Hemispheres. *Glob. Ecol. Biogeogr.* **2000**, *9*, 253–268. [CrossRef]
14. Parmesan, C.; Yohe, G. A Globally Coherent Fingerprint of Climate Change Impacts across Natural Systems. *Nature* **2003**, *421*, 37–42. [CrossRef] [PubMed]
15. Pretzsch, H. Diversity and Productivity in Forests: Evidence from Long-Term Experimental Plots. In *Forest Diversity and Function. Ecological Studies*; Scherer-Lorenzen, M., Körner, C., Schulze, E.D., Eds.; Springer: Berlin, Germany, 2005; Volume 176. [CrossRef]
16. Christensen, J.H.; Hewitson, B.; Busuioc, A.; Chen, A.; Gao, X.; Held, I.; Jones, R.; Kolli, R.K.; Kwon, W.T.; Laprise, R.; et al. Regional Climate Projections. In *Climate Change 2007: The Physical Science Basis. The Fourth Assessment Report of the Intergovernmental Panel on Climate Change*; Solomon, S.D., Qin, M., Manning, Z., Chen, M., Marquis, K., Averyt, K., Tignor, M., Miller, H., Eds.; Cambridge University Press: New York, NY, USA, 2007.
17. Hamrick, J.L. Response of Forest Trees to Global Environmental Changes. *For. Ecol. Manag.* **2004**, *197*, 323–335. [CrossRef]
18. Lindner, M.; Maroschek, M.; Netherer, S.; Kremer, A.; Barbati, A.; Garcia-Gonzalo, J.; Seidl, R.; Delzon, S.; Corona, P.; Kolström, M.; et al. Climate Change Impacts, Adaptive Capacity, and Vulnerability of European Forest Ecosystems. *For. Ecol. Manag.* **2010**, *259*, 698–709. [CrossRef]
19. Pulido, F.; Berthold, P. Microevolutionary Response to Climatic Change. *Adv. Ecol. Res.* **2004**, *35*, 151–183. [CrossRef]
20. Kremer, A.; Ronce, O.; Robledo-Arnuncio, J.J.; Guillaume, F.; Bohrer, G.; Nathan, R.; Bridle, J.R.; Gomulkiewicz, R.; Klein, E.K.; Ritland, K.; et al. Long-Distance Gene Flow and Adaptation of Forest Trees to Rapid Climate Change. *Ecol. Lett.* **2012**, *15*, 378–392. [CrossRef] [PubMed]
21. Chen, I.C.; Hill, J.K.; Ohlemüller, R.; Roy, D.B.; Thomas, C.D. Rapid Range Shifts of Species Associated with High Levels of Climate Warming. *Science* **2011**, *333*, 1024–1026. [CrossRef] [PubMed]
22. Aitken, S.N.; Yeaman, S.; Holliday, J.A.; Wang, T.; Curtis-McLane, S. Adaptation, Migration or Extirpation: Climate Change Outcomes for Tree Populations. *Evol. Appl.* **2008**, *1*, 95–111. [CrossRef]
23. Vranckx, G.; Jacquemyn, M.; Muys, B.; Honnay, O. Meta Analysis of Susceptibility of Woody Plants to Loss of Genetic Diversity through Habitat Fragmentation. Conservation Biology? *J. Soc. Conserv. Biol.* **2012**, *26*, 228–237. [CrossRef]
24. Aitken, S.N.; Bemmels, J.B. Time to Get Moving: Assisted Gene Flow of Forest Trees. *Evol. Appl.* **2016**, *9*, 271–290. [CrossRef]
25. Keenan, R.J. Climate Change Impacts and Adaptation in Forest Management: A Review. *Ann. For. Sci.* **2015**, *72*, 145–167. [CrossRef]
26. Lindner, M.; Fitzgerald, J.B.; Zimmermann, N.E.; Reyer, C.; Delzon, S.; van der Maaten, E.; Schelhaas, M.J.; Lasch, P.; Eggers, J.; van der Maaten-Theunissen, M.; et al. Climate Change and European Forests: What Do We Know, What Are the Uncertainties, and What Are the Implications for Forest Management? *J. Environ. Manag.* **2014**, *146*, 69–83. [CrossRef]
27. Alistair, S.J.; Josep, P. Running to Stand Still: Adaptation and the Response of Plants to Rapid Climate Change. *Ecol. Lett.* **2005**, *8*, 1010–1020.
28. Jones, T.A. When Local Isn't Best. *Evol. Appl.* **2013**, *6*, 1109–1118. [CrossRef] [PubMed]
29. De Kort, H.; Mergeay, J.; Vander Mijnsbrugge, K.; Decocq, G.; Maccherini, S.; Kehlet Bruun, H.H.; Honnay, O.; Vandepitte, K. An Evaluation of Seed Zone Delineation Using Phenotypic and Population Genomic Data on Black Alder Alnus Glutinosa. *J. Appl. Ecol.* **2014**, *51*, 1218–1227. [CrossRef]
30. Parnuta, G.; Lorent, A.; Teodoroiu, M.; Petrila, M. *Regiuni de Provenienta Pentru Materialele de Baza Din Care Se Obtin Materialele Forestiere de Reproducere Din Romania (Provenance Regions for Basic Materials to Produce Forest Reproductive Material in Romania)*; Forest Publishing: Bucharest, Romania, 2010.
31. Enescu, V.; Donita, N.; Bindiu, C.; Contescu, L. *Zonele de Recoltare a Semintelor Forestiere in R.S. Romania (Provenance Zones for Harvesting Forest Seeds in Romania)*; Forest Research and Management Institute: Bucharest, Romania, 1988.
32. Cheval, S.; Birsan, M.V.; Dumitrescu, A. Climate Variability in the Carpathian Mountains Region over 1961–2010. *Glob. Planet. Change* **2014**, *118*, 85–96. [CrossRef]
33. Meinshausen, M.; Smith, S.J.; Calvin, K.; Thomson, A.; Daniel, J.S.; Kainuma, M.L.T.; Matsumoto, K.; Lamarque, J.; Raper, S.C.B.; Riahi, K.; et al. The RCP Greenhouse Gas Concentrations and Their Extensions from 1765 to 2300. *Clim. Change* **2011**, *109*, 213–241. [CrossRef]
34. Spinoni, J.; Szalai, S.; Szentimrey, T.; Lakatos, M.; Bihari, Z.; Nagy, A.; Németh, Á.; Kovács, T.; Mihic, D.; Dacic, M.; et al. Climate of the Carpathian Region in the Period 1961–2010: Climatologies and Trends of 10 Variables. *Int. J. Climatol.* **2015**, *35*, 1322–1341. [CrossRef]
35. Ionita, M.; Scholz, P.; Chelcea, S. Assessment of Droughts in Romania Using the Standardized Precipitation Index. *Nat. Hazards* **2016**, *81*, 1483–1498. [CrossRef]
36. Birsan, M.V.; Dumitrescu, A.; Micu, D.M.; Cheval, S. Changes in Annual Temperature Extremes in the Carpathians since AD 1961. *Nat. Hazards* **2014**, *74*, 1899–1910. [CrossRef]
37. Birsan, M.-V.; Micu, D.-M.; Niță, I.-A.; Mateescu, E.; Szép, R.; Keresztesi, Á. Spatio-Temporal Changes in Annual Temperature Extremes over Romania (1961–2013). *Rom. J. Phys.* **2019**, *64*, 816.

38. Busuioc, A.; Dobrinescu, A.; Birsan, M.V.; Dumitrescu, A.; Orzan, A. Spatial and Temporal Variability of Climate Extremes in Romania and Associated Large-Scale Mechanisms. *Int. J. Climatol.* **2015**, *35*, 1278–1300. [CrossRef]
39. Birsan, M.V.; Dumitrescu, A. Snow Variability in Romania in Connection to Large-Scale Atmospheric Circulation. *Int. J. Climatol.* **2014**, *34*, 134–144. [CrossRef]
40. Micu, D.M.; Dumitrescu, A.; Cheval, S.; Nita, I.A.; Birsan, M.V. Temperature Changes and Elevation-Warming Relationships in the Carpathian Mountains. *Int. J. Climatol.* **2021**, *41*, 2154–2172. [CrossRef]
41. Birsan, M.V.; Marin, L.; Dumitrescu, A. Seasonal Changes in Wind Speed in Romania. *Rom. Rep. Phys.* **2013**, *65*, 1479–1484.
42. Birsan, M.V.; Nita, I.-A.; Craciun, A.; Sfica, L.; Keresztesi, Á.; Szep, R.; Micheu, M. Observed Changes in Mean and Maximum Monthly Wind Speed over Romania since Ad 1961. *Rom. Rep. Phys.* **2020**, *72*, 702.
43. Busuioc, A.; Birsan, M.V.; Carbunaru, D.; Baciu, M.; Orzan, A. Changes in the Large-Scale Thermodynamic Instability and Connection with Rain Shower Frequency over Romania: Verification of the Clausius-Clapeyron Scaling. *Int. J. Climatol.* **2016**, *36*, 2015–2034. [CrossRef]
44. Manea, A.; Birsan, M.V.; Tudorache, G.; Cărbunaru, F. Changes in the Type of Precipitation and Associated Cloud Types in Eastern Romania (1961–2008). *Atmos. Res.* **2016**, *169*, 357–365. [CrossRef]
45. Nita, I.A.; Sfîcă, L.; Apostol, L.; Radu, C.; Birsan, M.V.; Szep, R.; Keresztesi, Á. Changes in Cyclone Intensity over Romania According to 12 Tracking Methods. *Rom. Rep. Phys.* **2020**, *72*, 706.
46. Nita, I.-A.; Apostol, L.; Patriche, C.; Sfica, L.; Bojariu, R.; Birsan, M.-V. Frequency of Atmospheric Circulation Types over Romania According to Jenkinson-Collison Method Based on Two Long-Term Reanalysis Datasets. *Rom. J. Phys.* **2022**, *67*, 812.
47. Tîmpu, S.; Sfîcă, L.; Dobri, R.-V.; Cazacu, M.-M.; Nita, A.-I.; Birsan, M.-V. Tropospheric Dust and Associated Atmospheric Circulations over the Mediterranean Region with Focus on Romania's Territory. *Atmosphere* **2020**, *11*, 349. [CrossRef]
48. Cheval, S.; Busuioc, A.; Dumitrescu, A.; Birsan, M.V. Spatiotemporal Variability of Meteorological Drought in Romania Using the Standardized Precipitation Index (SPI). *Clim. Res.* **2014**, *60*, 235–248. [CrossRef]
49. Cheval, S.; Dumitrescu, A.; Birsan, M.V. Variability of the Aridity in the South-Eastern Europe over 1961–2050. *Catena* **2017**, *151*, 74–86. [CrossRef]
50. Birsan, M.V. Trends in Monthly Natural Streamflow in Romania and Linkages to Atmospheric Circulation in the North Atlantic. *Water Resour. Manag.* **2015**, *29*, 3305–3313. [CrossRef]
51. Micheu, M.M.; Birsan, M.V.; Szép, R.; Keresztesi, Á.; Nita, I.A. From Air Pollution to Cardiovascular Diseases: The Emerging Role of Epigenetics. *Mol. Biol. Rep.* **2020**, *47*, 5559–5567. [CrossRef]
52. Micheu, M.M.; Birsan, M.V.; Nita, I.A.; Andrei, M.D.; Nebunu, D.; Acatrinei, C.; Sfîcă, L.; Szép, R.; Keresztesi, Á.; Hernáez, P.F.D.A.; et al. Influence of Meteorological Variables on People with Cardiovascular Diseases in Bucharest, Romania (2011–2012). *Rom. Rep. Phys.* **2021**, *73*, 707.
53. Mihai, G.; Birsan, M.V.; Dumitrescu, A.; Alexandru, A.; Mirancea, I.; Ivanov, P.; Stuparu, E.; Teodosiu, M.; Daia, M. Adaptive Genetic Potential of European Silver Fir in Romania in the Context of Climate Change. *Ann. For. Res.* **2018**, *61*, 95–108. [CrossRef]
54. Mihai, G.; Alexandru, A.M.; Stoica, E.; Birsan, M.V. Intraspecific Growth Response to Drought of Abies Alba in the Southeastern Carpathians. *Forests* **2021**, *12*, 387. [CrossRef]
55. Donita, N. *Elaborarea Hartii Forestiere a Romaniei La Scara 1:500,000 (Development of the Romanian Forest Map at 1:500,000 Scale)*; Forest Research and Management Institute: Bucharest, Romania, 1996.
56. Dumitrescu, A.; Birsan, M.V. ROCADA: A Gridded Daily Climatic Dataset over Romania (1961–2013) for Nine Meteorological Variables. *Nat. Hazards* **2015**, *78*, 1045–1063. [CrossRef]
57. Vlăduţ, A.Ş.; Nikolova, N.; Licurici, M. Influence of Climatic Conditions on the Territorial Distribution of the Main Vegetation Zones within Oltenia Region, Romania. *Olten. Stud. Comun. Stiintele Nat.* **2017**, *33*, 154–164.
58. WMO/GWP Integrated Drought Management Programme (IDMP). *Handbook of Drought Indicators and Indices*; WMO-No. 1173; World Meteorological Organization: Geneva, Switzerland, 2016.
59. Ellenberg, H. *Vegetation Mitteleuropas Mit Den Alpen*; Eugen Ulmer: Stuttgart, Germany, 1963.
60. Sofletea, N.; Curtu, L. *Dendrologie*; Editura Pentru Viata: Brasov, Romania, 2001.
61. IPCC. *Climate Change 2001: Mitigation. Contribution of Working Group III to the Third Assessment Report of the Intergovernmental Panel on Climate Change*; Cambridge University Press: Cambridge, UK, 2001.
62. Dumitrescu, A.; Bojariu, R.; Birsan, M.V.; Marin, L.; Manea, A. Recent Climatic Changes in Romania from Observational Data (1961–2013). *Theor. Appl. Climatol.* **2015**, *122*, 111–119. [CrossRef]
63. Savolainen, O.; Pyhäjärvi, T.; Knürr, T. Gene Flow and Local Adaptation in Trees. *Annu. Rev. Ecol. Evol. Syst.* **2007**, *38*, 595–619. [CrossRef]
64. Pluess, A.R.; Frank, A.; Heiri, C.; Lalagüe, H.; Vendramin, G.G.; Oddou-Muratorio, S. Genome-Environment Association Study Suggests Local Adaptation to Climate at the Regional Scale in Fagus Sylvatica. *New Phytol.* **2016**, *210*, 589–601. [CrossRef] [PubMed]
65. Mihai, G. *Surse de Seminţe Testate Pentru Principalele Specii de Arbori Forestieri Din România [Tested Seed Sources for the Main Forest Tree Species from Romania]*; Editura Silvică: Bucharest, Romania, 2009.
66. Şofletea, N.; Curtu, A.L.; Daia, M.L.; Budeanu, M. The Dynamics and Variability of Radial Growth in Provenance Trials of Norway Spruce (Picea Abies (L.) Karst.) within and beyond the Hot Margins of Its Natural Range. *Not. Bot. Horti Agrobot. Cluj-Napoca* **2015**, *43*, 265–271. [CrossRef]

67. Mihai, G.; Mirancea, I.; Birsan, M.V.; Dumitrescu, A. Patterns of Genetic Variation in Bud Flushing of Abies Alba Populations. *IForest* **2018**, *11*, 284–290. [CrossRef]
68. Mihai, G.; Teodosiu, M.; Birsan, M.V.; Alexandru, A.M.; Mirancea, I.; Apostol, E.N.; Garbacea, P.; Ionita, L. Impact of Climate Change and Adaptive Genetic Potential of Norway Spruce at the South–Eastern Range of Species Distribution. *Agric. For. Meteorol.* **2020**, *291*, 108040. [CrossRef]
69. Worrell, R. A Comparison between European Continental and British Provenances of Some British Native Trees: Growth, Survival and Stem Form. *Forestry* **1992**, *65*, 253–280. [CrossRef]
70. Pâques, L.E. *Forest Tree Breeding in Europe*; Springer: Dordrecht, The Netherlands, 2009.
71. Whittet, R.; Cavers, S.; Ennos, R.; Cottrell, J. *Genetic Considerations for Provenance Choice of Native Trees under Climate Change in England*; Forestry Commission: Edinburgh, UK, 2019; pp. 1–56.
72. Matyas, C. *Guidelines for the Choice of Forest Reproductive Material in the Face of Climate Change*; FORGER Guidelines; Bioversity International: Rome, Italy, 2016; 8p.
73. Reed, D.H.; Frankham, R. Correlation between Fitness and Genetic Diversity. *Conserv. Biol.* **2003**, *17*, 230–237. [CrossRef]
74. Leimu, R.; Mutikainen, P.; Koricheva, J.; Fischer, M. How General Are Positive Relationships between Plant Population Size, Fitness and Genetic Variation? *J. Ecol.* **2006**, *94*, 942–952. [CrossRef]
75. Young, A.; Boyle, T.; Brown, T. The Population Genetic Consequences of Habitat Fragmentation for Plants. *Trends Ecol. Evol.* **1996**, *11*, 413–418. [CrossRef]
76. O'Neill, G.A.; Hamann, A.; Wang, T. Accounting for Population Variation Improves Estimates of the Impact of Climate Change on Species' Growth and Distribution. *J. Appl. Ecol.* **2008**, *45*, 1040–1049. [CrossRef]
77. Pedlar, J.H.; McKenney, D.W.; Aubin, I.; Beardmore, T.; Beaulieu, J.; Iverson, L.; O'Neill, G.A.; Winder, R.S.; Ste-Marie, C. Placing Forestry in the Assisted Migration Debate. *Bioscience* **2012**, *62*, 835–842. [CrossRef]
78. Williams, M.I.; Dumroese, R.K. Preparing for Climate Change: Forestry and Assisted Migration. *J. For.* **2013**, *111*, 287–297. [CrossRef]
79. Spittlehouse, D.L.; Stewart, R.B. Adaptation to Climate Change in Forest Management. *BC J. Ecosyst. Manag.* **2003**, *4*, 1–7.

Article

Predicting Distribution and Range Dynamics of Three Threatened *Cypripedium* Species under Climate Change Scenario in Western Himalaya

Naveen Chandra [1], Gajendra Singh [1], Ishwari Datt Rai [2], Arun Pratap Mishra [3], Mohd. Yahya Kazmi [1], Arvind Pandey [4], Jeewan Singh Jalal [5], Romulus Costache [6,7,8], Hussein Almohamad [9,*], Motrih Al-Mutiry [10] and Hazem Ghassan Abdo [11,12,13]

[1] Department of Forestry and Climate Change, Uttarakhand Space Application Centre, Upper Aamwala, Nalapani, Dehradun 248008, India
[2] Department of Forestry and Ecology, Indian Institute of Remote Sensing, Indian Space Research Organization (ISRO), Kalidas Road, Dehradun 248001, India
[3] Department of Habitat Ecology, Wildlife Institute of India, Chandrabani, Dehradun 248001, India
[4] Department of Remote Sensing and GIS, SSJ Campus, Soban Singh Jeena University, Almora 263601, India
[5] Botanical Survey of India, Headquarters, CGO Complex, 3rd MSO Building, Block F, Salt Lake City, Kolkata 700064, India
[6] National Institute of Hydrology and Water Management, București-Ploiești Road, 97E, 1st District, 013686 Bucharest, Romania
[7] Department of Civil Engineering, Transilvania University of Brasov, 5, Turnului Str, 500152 Brasov, Romania
[8] Danube Delta National Institute for Research and Development, 165 Babadag Street, 820112 Tulcea, Romania
[9] Department of Geography, College of Arabic Language and Social Studies, Qassim University, Buraydah 51452, Saudi Arabia
[10] Department of Geography, College of Arts, Princess Nourah bint Abdulrahman University, Riyadh 11671, Saudi Arabia
[11] Geography Department, Faculty of Arts and Humanities, University of Tartous, Tartous P.O. Box 2147, Syria
[12] Geography Department, Faculty of Arts and Humanities, Damascus University, Damascus P.O. Box 30621, Syria
[13] Geography Department, Arts and Humanities Faculty, Tishreen University, Lattakia P.O. Box 30621, Syria
[*] Correspondence: h.almohamad@qu.edu.sa

Citation: Chandra, N.; Singh, G.; Rai, I.D.; Mishra, A.P.; Kazmi, M.Y.; Pandey, A.; Jalal, J.S.; Costache, R.; Almohamad, H.; Al-Mutiry, M.; et al. Predicting Distribution and Range Dynamics of Three Threatened *Cypripedium* Species under Climate Change Scenario in Western Himalaya. *Forests* **2023**, *14*, 633. https://doi.org/10.3390/f14030633

Academic Editors: Any Mary Petritan and Mirela Beloiu

Received: 2 February 2023
Revised: 8 March 2023
Accepted: 13 March 2023
Published: 21 March 2023

Abstract: Climate change and anthropogenic pressure have significantly contributed to the decline of biodiversity worldwide, particularly in mountain ecosystems such as the Himalaya. In addition to being relatively sensitive to disturbances, orchids may also respond more quickly to climate change impacts than other plant species. Because of their complex biology and anthropogenic pressures on their habitat in the Himalayan region, lady's slipper orchids are considered to be a highly vulnerable group of orchids. In the present study, we examine the effect of climate change on the distribution of three threatened *Cypripedium* species (*Cypripedium cordigerum*, *Cypripedium elegans, and Cypripedium himalaicum*), utilizing ecological niche modeling for present and future climatic scenarios to identify key environmental determinants and population parameters. A community climate system model (CCSM ver. 4) was used to identify suitable distribution areas for future scenarios. Based on the least correlated characteristics of the species bioclimatic, topographical, and physiological characteristics, the species' climatic niche was determined. According to the results, the true skill statistic (TSS), area under the receiver operating characteristic curve (AUC), and Cohen's kappa provide more reliable predictions. Precipitation during the wettest month and precipitation during the coldest quarter are the primary climatic variables that influence the distribution of suitable areas. A total of 192 km^2 of the area was estimated to be suitable for all three species under current climate conditions. Under future climate conditions, the model predicts a trivial increase in suitable habitat areas with a shift toward the northwest. However, highly suitable habitat areas will be severely diminished. There are currently highly suitable habitats in Tungnath and the Valley of Flowers, but due to climatic factors, the habitats will become unsuitable in the future. Additionally, under future climatic scenarios, viable habitats will be identified for priority conservation to cope with the effects of climate change and anthropogenic activities. In light of these findings, conservation methods for the target species may be designed that will be successful and have the potential to prevent local extinctions.

Keywords: potential distribution; range expansion; anthropogenic pressure; climate change

1. Introduction

As one of the world's Global Biodiversity Hotspots [1,2], the Himalaya is home to more than 10,500 flowering plant species that inhabit a diverse range of eco-climatic zones. Although climate change and other anthropogenic disturbances such as habitat fragmentation, invasion by alien species, and intensive livestock grazing have contributed to the continued disruption the Himalayan ecosystems' structural and functional reliability has been altered [3,4]. Due to the vulnerability of the Himalayan ecosystems to climate change and their socio-ecological importance [5], the National Action Plan on Climate Change (NAPCC) has established sustainable ecosystem utilization to ensure that ecosystem services continue, and biodiversity is conserved [6]. It is particularly important to pay attention and take measures to conserve the climate-sensitive, endangered, threatened, and endemic plant species in the Himalaya.

Due to their specialized reproductive strategies and mycorrhizal selectivity, orchids are considered among the most ecologically sensitive vascular plants. With over 736 genera and 28,000 species, the Orchidaceae family is the second largest and most diverse family of flowering plants in the world [7]. As a result, it is currently facing an unusual process of extinction [8]. The majority of vulnerable orchid species in the world are terrestrial orchids, despite making up a minor percentage of the family. Of the 244 orchid species reported in Uttarakhand, Western Himalaya, 129 species are terrestrial in nature. In the Western Himalayas, there are 14 endemic orchid species, of which 13 are endemic to Uttarakhand [9].

Cypripedium L. is a perennial herb commonly known as Lady's Slipper. This genus consists of about 59 known species (TPL, 2017) and two varieties. According to Jalal et al. [10], there are three species of *Cypripedium* found: *C. cordigerum* D. Don, *C. elegans* Reichb.f., and *C. himalaicum* Rolfe in the Western Himalaya. Considering the scale, the dispersion of these species is affected by a number of biotic and abiotic factors, including territory size, light, and soil conditions [11]. This genus is widely distributed, with populations ranging from 2400 to 3900 m in altitude [12]. It has been reported by the International Union for Conservation of Nature and Natural Resources (IUCN) that the overall population of orchids has decreased. As a result of various factors, such as overgrazing, collection, deforestation, and climate change, suitable habitat has been lost. The three *Cypripedium* species require specific habitat conditions in order to germinate and grow, including the presence of mycorrhizal parasites, nutrient availability, and sufficient sunlight [13]. In light of the significant threat, the Convention on International Trade in Endangered Species of Wild Fauna and Flora (CITES) included all *Cypripedium* species in Appendix II. *C. cordigerum* is categorized as Vulnerable (VU), *C. elegans*, and *C. himalaicum* as Endangered (EN) due to declining populations.

An assessment of the potential distribution of threatened species and their habitats can be made with the aid of species distribution models (SDMs). A predictive map of suitable habitats for the species can be created using environmental data and known species occurrences [14]. In addition to guiding targeted surveys, SDMs can help prioritize areas for conservation action and inform land-use planning decisions by identifying areas where threatened species are likely to occur. Furthermore, SDMs can assist in predicting the impacts of climate change on species distributions, enabling proactive management of these impacts to be accomplished.

Characterizing a species' ecological niche gives the essential knowledge required to identify crucial areas that may require prioritization of conservation actions [14] and gives an idea to constantly monitor the status of the growth parameters of various plant species in their natural habitats [15]. Terrestrial endangered species are negatively affected by climate change, and more information about their distribution is required in order to identify and rehabilitate them. Western Himalayan orchids are subject to very scanty information regarding ecological niche dynamics, distribution patterns, and the impact of climate change. *Cypripedium* species have previously been studied primarily in terms of distribution and biology [16]. Keeping in view above, it is imperative to assess their availability and potential zone in the region. Accordingly, the study attempted to clarify the present and future distribution and impact of climate change on *Cypripedium* species in the western Himalaya. The three objectives of our study are (i) to provide present geographical distribution and possible future habitats, (ii) to elucidate the effect of each independent variable in the creation of a model, and (iii) to detect suitable areas for species bestowing to global climate change projections for the years 2050 and 2070. In general, our study will provide valuable insights into climate change's potential impact on the distribution and range dynamics of *Cypripedium* species in the Western Himalayas and assist in the development of conservation strategies that will protect these threatened species for the foreseeable future.

2. Materials and Methods

2.1. Study Area

In the Indian Himalayan region, the Western Himalayas extend across Uttarakhand and Himachal Pradesh [17]. This study examined the vegetation of the cool temperate to alpine regions of Uttarakhand state (Figure 1). Approximately 8990 km^2 of this region is covered by broadleaved forest and sub-alpine coniferous forest, as well as alpine herbaceous/grassy slopes. Extremely varied topography in the alpine zone includes steep slopes terminating at famed summits, deep gorges, extensive moraines, ridges, spurs, upland plateaus, and lakes. Temperatures and rainfall are extreme in the inner dry valleys. As the Trans-Himalayan zones transition to the Greater Himalayas, the aridity increases. Also, the vegetation is sparsely distributed with an increasing cold-dry climate. There are four broad classes of vegetation in the study area: Himalayan moist temperate forests, sub-alpine forests, alpine moist scrubs, and moist alpine meadows. In addition to being habitats for various native species, these formations are unique assemblages of specialized growth forms adapted to particular environmental conditions. Poets, saints, tourists, trackers, mountaineers, researchers, pilgrims, and people from every walk of life are drawn to the region because of its panoramic beauty.

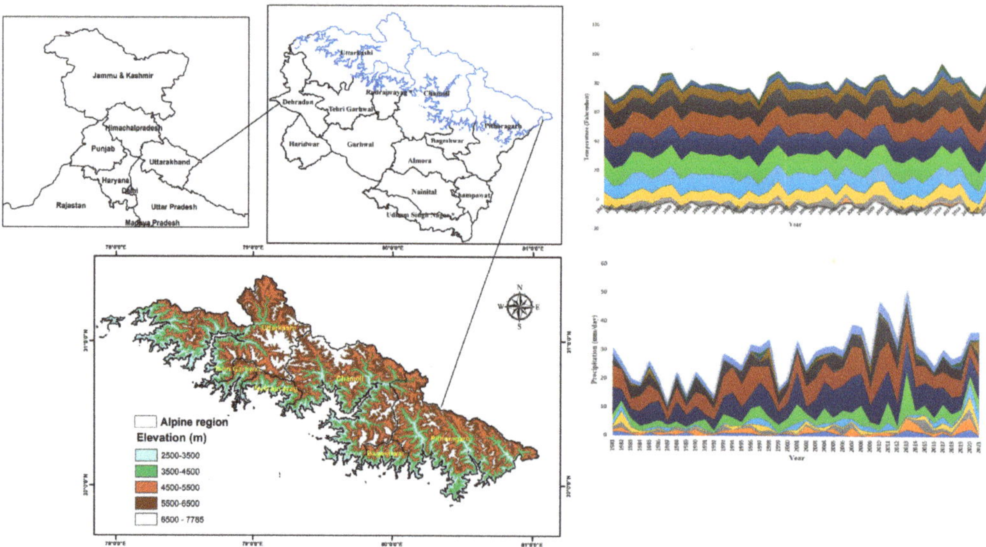

Figure 1. Study area map with decadal climatic data (Temperature, Precipitation).

2.2. Study Species

This study prioritized the conservation of three threatened terrestrial orchid species commonly known as Lady's Slipper orchids, *Cypripedium himalaicum* (EN), *Cypripedium elegans* (EN), and *Cypripedium cordigerum* (VU). Continuous harvesting, habitat destruction, over-grazing, and emerging climate change scenarios are deteriorating their areas of occurrence and leading to a rapid population decline, which reflects its endangered status. In the western Himalayas, the species have a very restricted distribution, and in the state, these species are represented by a handful of populations [10].

2.3. Data Collection

The field surveys were conducted in the regions of high-elevation forests and alpine regions of Uttarakhand during the period 2019–2021. A Global Positioning System (Garmin eTrexH) was used to determine the locations of the target species in the field. A GeoCAT assessment was conducted based on population abundance and coordinates in order to assess the IUCN threat status of the species in the study area [18]. Additionally, distribution data were compiled from global sources such as the Global Biodiversity Information Facility (GBIF.org) and local herbaria such as the Forest Research Institute (DD), Botanical Survey of India (BSD), and Wildlife Institute of India (WII).

2.4. Predictor Variables

In order to determine the impact of climate on species dispersal, several climatic factors were considered. Therefore, gridded climatic data with a spatial resolution of 1 km was acquired from the WorldClim database [19]. A Digital Elevation Model (ASTER, DEM) with a 30 m spatial resolution was used to determine topographic variables, such as slope, aspect, and elevation. Based on Sentinel-2 satellite images and the Random Forest (RF) classification algorithm, a detailed thematic vegetation map of the area was prepared in order to establish the relationship between environmental factors and habitats. Using the nearest neighbor sampling method, all variables' layers were resampled at 1 km grids. The variable with the highest correlation (Pearson correlation coefficient, 0.75) was removed from the analysis by performing a multi-collinearity analysis. Final prediction models were based on only ten variables (from a total of 12).

The future scenarios for *C. himalaicum*, *C. cordigerum*, and *C. elegans* in the region were assessed using environmental predictors (current, 2050, and 2070). The probability distribution modeling was conducted using a total of 22 environmental variables for each period. For the future climatic scenario, 19 bioclimatic variables with a spatial resolution of 1 km were also used. According to the IPCC 5th assessment report, the climatic scenarios for the 2050s and 2070s have been derived from the global climate model (GCM). In order to predict the probable impact of future climate change on the species of interest, we used the community climate system model version 4 (CCSM 4). As a means of assessing potential suitability ranges for species habitation in the 2050s (2041–2060) and the 2070s (2061–2080), the scenarios RCP 2.6 (the lowest GHG emissions scenario) and RCP 8.5 (the highest GHC emissions scenario) were used.

2.5. Predictive Modeling

We created the model using the Maximum Entropy Modeling (MaxEnt) machine learning methodology. For *C. himalaicum*, *C. elegans*, and *C. cordigerum*, 20, 12, and 10 presence records were used. The Jack-knife test was utilized to understand the relative influence of each predictor variable [20]. In addition, the predictive model allows for the creation of replicated runs which are used for cross-ratification, bootstrapping, and repeated sub-sampling. To prevent over-prediction and reduce model over-fitting, the regularization multiplier was set at 0.1 [20]. Therefore, the presence data is randomly divided into 'training' and 'test' sets [21] in the ratio of 75:25 while maintaining the other values as defaults. A Jack-knife test involves the prediction of a single location that is excluded from the 'training' dataset [22]. The predictive performance of the model was evaluated by (ROC) analysis, true skill statistics (TSS), Kappa statistics, sensitivity (true positive rate (TPR)) and false alarm rate (FAR) (1—specificity). The model's robustness was also assessed by the AUC value. AUC value 0.5 shows a low fit model, 0.5–0.7 indicates lower accuracy of the model, 0.7–0.9 denotes the reasonable performance of the model, and values > 0.9 demonstrate the high performance of the selected model [23].

3. Results

3.1. Habitat Preferences of the Species

Cypripedium populations were recorded between 2500 and 4000 m in different microhabitat conditions: cool temperate forests, sub-alpine forests, grassy slopes, and alpine moist scrubs. Slopes ranging from 25 to 40 degrees and which are exposed to the south and north characterize the microhabitat condition. There is a significant difference in the distribution of species between different types of habitats. It was found that mixed herbaceous meadows (short forbs), mat-forming shrubs, and tall forbs were the most preferred types of habitats. In warm temperate sub-alpine forests and on the upper edges of alpine meadows, *C. cordigerum* is found sporadically in birch (Betula utilis) forests.

As determined by the Jackknife test, the mean diurnal range (Bio-2), soil type, and bio-3 (isothermality) were the variables that did the best job of predicting suitable habitats when used separately. Cumulatively, these variables contributed 70%. Additionally, altitude, slope, aspect, and temperature seasonality (Bio-4), as well as vegetation type, did not contribute significantly to the weight. As a consequence, they are likely to have a limited impact on the distribution of *Cypripedium* species. Mean daylight range (Bio-2) and aspect were the strongest predictors for *C. himalaicum* distribution, with 44.9% and 25.2%, respectively. Soil type and bio 11 (mean temperature of the coldest quarter) are considered strong predictors for *C. cordigerum* with 30% and 16%, respectively, while isothermally (Bio-3), altitude, and aspect structured the habitat suitability of *C. elegans* with 48.4%, 22.1%, and 8.4%, respectively (Table 1). The combination of all climate and topographic elements structured the distribution pattern of all the *Cypripedium* species.

Table 1. Environmental variable contributions showing their respective percent contribution.

Variables	*C. himalaicum*	*C. cordigerum*	*C. elegans*
Bio-2	44.9	2.5	1.4
Aspect	25.2	14.5	8.4
Bio-12	11.4	3.4	–
Bio-11	9.3	16.6	6.2
Altitude	7.7	1.3	22.1
Bio-3	–	4.9	48.4
Slope	1.1	4.1	–
Vegetation type	–	2.4	–
Bio-4	–	0.4	7.9
Soil type	0.5	30.1	–

3.2. Potential Distribution and Habitat Preferences

Habitat suitability analysis and field surveys revealed that *Cypripedium* species prefer the understory of *Betula* forest, alpine grassy slopes, and moist rocky habitats. *C. himalaicum* showed high suitability in the moist grassy slopes and herbaceous meadows in association with *Danthonia cachemyriana, Anemone tetrasepala, Aconitum heterophyllum, Rhododendron lepidotum,* and *Nardostachys jatamansi.* It is occasionally found under the canopy of *Betula utilis.* Similarly, *C. cordigerum* is mainly distributed in the forested area of *Quercus semecarpifolia, Abies pindrow,* and *A. spectabilis,* and *Betula utilis* having gentle to steeper slopes (>35°) in the broken rocky areas and amidst open scrubs. *C. elegans* is showed distribution in moist slopes along 26–34°, and humus-rich soils. It was found to be associated with *Smilacina purpurea, Viola biflora, Fragaria nubicola, Gerbera* sp., *Caltha palustris, Clematis* sp.,

Ranunculus hirtellus, and *Selinum vaginatum.* In some places, it forms an association with *Bergenia stracheyi* and *Rhododendron* thickets.

3.3. Habitat Suitability under Current Climatic Conditions

Habitat suitability analysis under current climatic conditions reveals that *C. himalaicum* has a highly suitable habitat of about 209 km^2, whereas the less suitable habitat of 363 km^2. *C. cordigerum* has a highly suitable habitat of about 199 km^2 and a less suitable of about 600 km^2 area, and C. *elegans* showed an area of 215 km^2 as a highly suitable habitat and 315 km^2 as the less suited. All three species showed maximum suitable habitats on grassy slopes (55%) dominated by *D. cachemyriana* and sub-alpine (*Betula utilis, Q. semicarpifolia*) forests (40%) and moist herbaceous meadows (5%) (Figure 2).

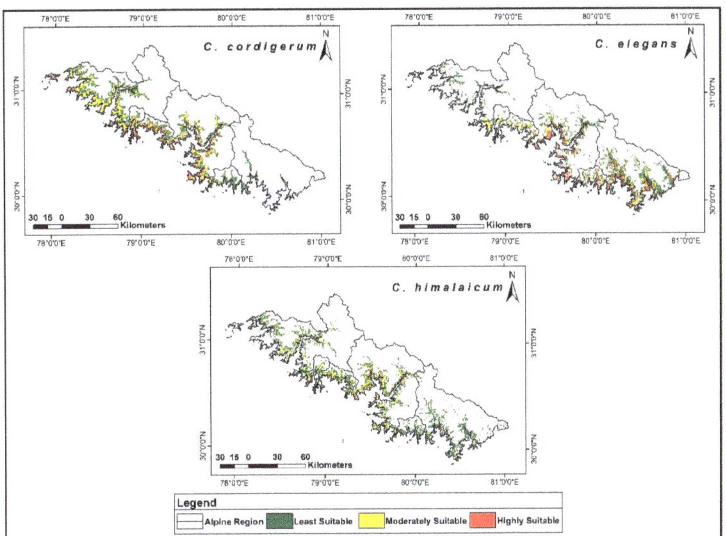

Figure 2. Current habitat suitability of *C. cordigerum, C. elegans,* and *C. himalaicum.*

3.4. Distribution Prediction in Future Climate Scenarios

According to RCP 4.5 and RCP 8.5, Figures 3–5 illustrate the projected distributions of each species for 2050 and 2070. The AUC value ranged from 0.70 to 0.92 for predicting the suitability of *Cypripedium* species for future distribution, which indicates that the models performed well. Furthermore, the range of TSS values (0.75 to 0.90) indicated the robustness of the model. As shown in Table 2, Figure 5, the probable area of occurrence for *C. himalaicum* is highest at 2.6% (190.23 km^2) in RCP 2.6 for 2050, while the lowest at 107 km^2) in RCP 8.5 for 2070. The maximum probable zone (97 km^2) for *C. cordigerum* in 2050 was predicted based on RCP 2.6, and the lowest zone (51 km^2) was predicted based on RCP 8.5 in 2070 (Table 3 and Figure 3). Maximum future climatic suitability (199 km^2) for *C. elegans* was in 2070 of RCP 2.6, while the lowest (102 km^2) was in the year 2050 RCP of 2.6 (Table 4, Figure 4).

Annual mean temperature (Bio-1) and annual precipitation (Bio-12) have been highly influential on *C. himalaicum* distributions in 2050. By 2070, the coldest quarter (Bio-19) and the wettest month (Bio-13) will play a significant role. Under RCP 2.6 and 8.6, *C. cordigerum* will be influenced by elevation and annual mean temperature (Bio-1), whereas in 2070, the mean temperature of the coldest month (Bio-6) and precipitation of the wettest month (Bio-13) will determine its distribution. During the 2050 time period, *C. elegans* distribution will be influenced by annual precipitation and precipitation of the wettest month, while the distribution will be influenced by the coldest quarter's precipitation (Bio-19) and annual precipitation (Bio-12).

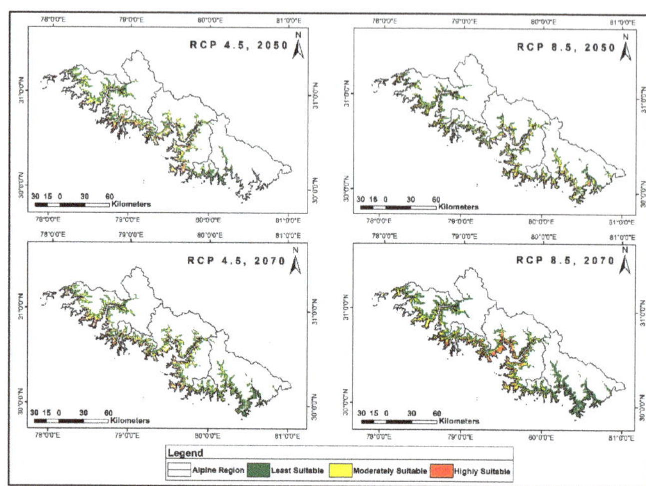

Figure 3. Predicted future habitat suitability for *C. cordigerum*.

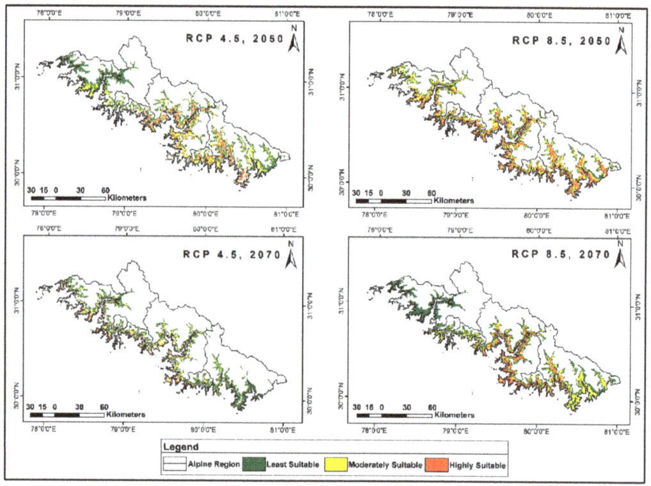

Figure 4. Predicted future habitat suitability for *C. elegans*.

Table 2. Prediction accuracy of *C. himalaicum* species under future climatic scenarios.

	Year 2050				Year 2070			
	RCP 2.6		**RCP 8.6**		**RCP 2.6**		**RCP 8.6**	
AUC value	0.84		0.75		0.92		0.87	
TSS value	0.75		0.81		0.88		0.74	
Percentage of contribution	Bio-1	Bio-19	Bio-12	Bio-6	Bio-19	Bio-1	Bio-13	Bio-6
Value (%)	34	27	41	25	37	24	31	30
Area in km² (10th percentile training presence threshold rule)	190		124		140		107	
Percentage of Area	2.6		1.5		1.7		1.3	
Highest probability of species occurrence	0.74		0.65		0.71		0.81	

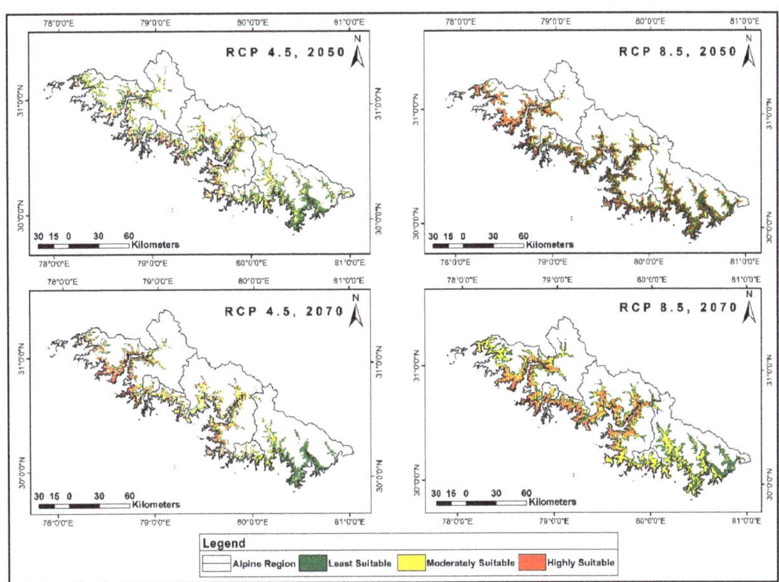

Figure 5. Predicted future habitat suitability for *C. himalaicum*.

Table 3. Prediction accuracy of *C. cordigerum* under future climatic scenarios.

	2050				Year 2070			
	RCP 2.6		RCP 8.6		RCP 2.6		RCP 8.6	
AUC value	0.71		0.71		0.81		0.91	
TSS value	0.75		0.81		0.79		0.69	
Percentage of contribution	Altitude	Bio-19	Bio-1	Bio-12	Bio-12	Bio-6	Bio-13	Bio-19
Value (%)	40	21	35	27	23	31	34	29
Area in km² ((10th percentile training presence threshold rule)	97		55		86		51	
Percentage of Area	1.2		0.7		1.09		0.6	
Highest probability of species occurrence	0.65		0.71		0.68		0.7	

Table 4. Prediction accuracy of *C. elegans* under future climatic scenarios.

	2050				Year 2070			
	RCP 2.6		RCP 8.6		RCP 2.6		RCP 8.6	
AUC value	0.78		0.80		0.89		0.70	
TSS value	0.80		0.82		0.87		0.90	
Percentage of contribution	Bio-12	Bio-19	Bio-13	Slope	Bio-19	Bio-1	Bio-12	Bio-19
value	37	25	34	26	37	24	37	32
Area in km² ((10th percentile training presence threshold rule)	102		189		199		154	
Percentage of Area	1.3		2.4		2.5		1.9	
Highest probability of species occurrence	0.61		0.62		0.65		0.59	

3.5. Range Dynamics under Future Climatic Scenarios

According to future scenarios, Tungnath, Panwalikantha, Garbyang, Badrinath, Deodi (Rishiganga valley), Janki Chatti, Tungnath, Yamnotri are predicted to become unsuitable for *C. himalaicum* growth by 2050. Additionally, the Ralam valley, Tungnath, Bajmora, and Dibrugheta areas will no longer be suitable for *C. elegans*. In the future, however, forest-dwelling species of *C. cordigerum* that have a limited distribution in the alpine region will be less affected. In future scenarios, the following areas are likely to be more suitable for all *Cypripedium* species: Kedarnath Wildlife Sanctuary, Valley of Flowers National Park, Kandara, and Har-Ki-Don (Figures 6–8).

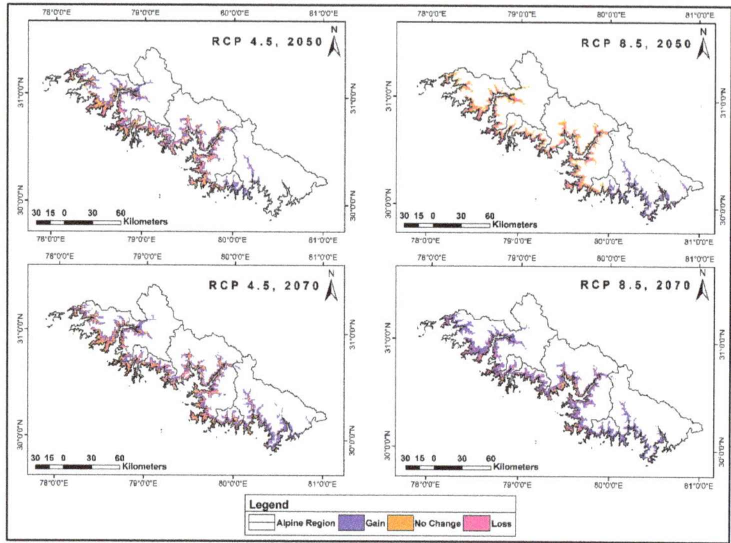

Figure 6. Predicted range contraction for *C. cordigerum*.

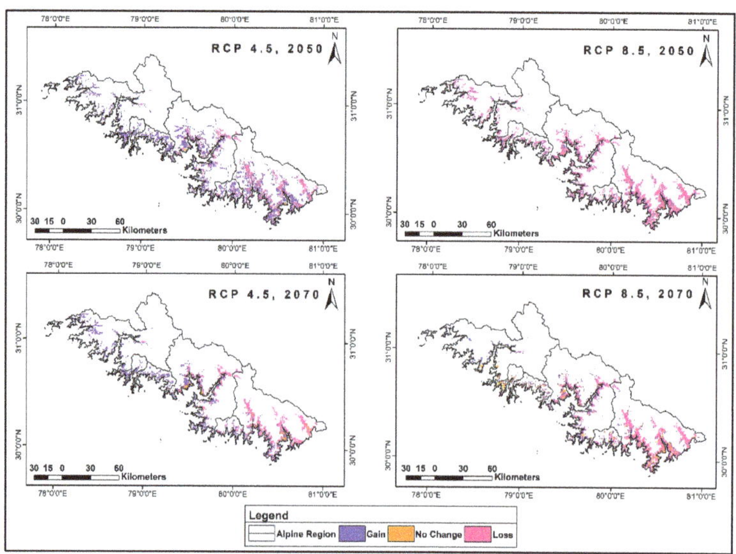

Figure 7. Predicted range contraction for *C. elegans*.

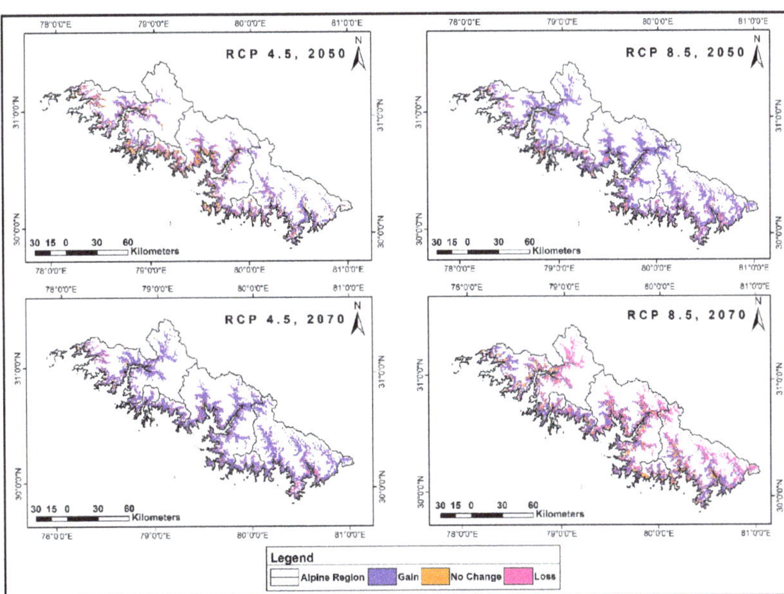

Figure 8. Predicted range contraction for *C. himalaicum*.

On the basis of the distribution trends for two time periods (2050 and 2070), as well as future climate scenarios (RCPs 4.5 and 8.5), a significant decline in the overall habitat is predicted. RCP8.5 (2070) showed moderate habitat declines for all species, with RCP 4.5 (2050) showing an overall decline. According to these future scenarios, 60% of suitable habitats will be damaged and no longer suitable for these species.

4. Discussion

Climate and precipitation play a crucial role in defining species tolerance ranges of distribution, among other environmental factors. A change in global temperatures and precipitation patterns has a detrimental impact on vulnerable species' habitats and distributions [24]. Various methods can be used to evaluate the probable range of species occurrence and population demographics. A gridded dataset of environmental parameters as well as the presence and absence of specific species, is used to develop these methods [25]. An important correlative approach to assessing a species' potential ecological niche and predicting its future distribution is to model its ecological niche using information about its environmental conditions. As a result of correlation niche modeling, conservation and management of rare taxa have been supported and have been encouraged to identify suitable habitats for the long-term maintenance of species populations [26]. They are spatially explicit and can be applied to a wide range of species at a scale that is suitable for conservation purposes [27].

For centuries, scientists have observed and documented consistent relationships between the distribution of species and their physical environment. Since the advent of numerical models, quantitative algorithms are increasingly being used for both the description of patterns and the prediction of future patterns [27]. Mathematical techniques can be applied to a wide range of applications with varying degrees of success. A number of published studies have been used in this study to demonstrate that species distribution models (SDMs) are capable of accurately predicting the natural distribution of species. With the help of high-quality survey data as well as relevant predictors, in addition to being functionally viable, these specific models are capable of providing genuine distribution ranges. They provide both ecological insights as well as a high degree of predictability. As a tool for predicting the existing and future distribution ranges of species, species distribution modeling (SDM) is essential for the development of various strategic management approaches for habitat conservation and management. Technological advancements in SDMs have enabled ensemble modeling to become a viable method due to growing data availability [28].

This study aims to model the current and future distributions of three *Cypripedium* species using a machine-learning approach. This study examines the dynamics of the species niches, both in the present and in the future, and estimates the size and pace of possible range expansions and contractions. Data with a 1 km^2 spatial resolution were used because the current study area is topographically complex. Many applications require data with a high spatial resolution (1 km^2) in hilly and other regions with severe climate gradients in order to capture environmental variation that would otherwise be lost at lower spatial resolutions [19,29].

Climate may still play a significant role in determining how quickly natural populations regenerate and disperse. Many climate prediction models indicate that climate change will be a dominant stressor in the second half of the twenty-first century. It is likely that species with major human threats, a small distribution range, a more compact population structure, and high habitat distinctiveness will be particularly vulnerable to habitat loss and changes in distribution [14,30]. Through the use of species distribution models, it has been possible to find potentially suitable habitats for vulnerable species. Climate change has been shown to have a significant impact on species distributions [31–33] by expanding, changing, or reducing ranges.

Overlaying the modal outputs on satellite images, a moist temperate forest, open *R. anthopogon* scrub, moist-shady areas under Himalayan Birch, Kharsu Oak, and Salix forests had high habitat suitability, while Abies forest in temperate and less disturbed herbaceous meadows had medium to low habitat suitability.

Satellite images showed high habitat suitability of moist temperate forests, open *R. anthopogon* scrub, and moist-shady areas under Himalayan Birch, Kharsu Oak, and *Salix* trees. Abies forests in temperate herbaceous meadows had medium to low habitat suitability.

The climate is one of several abiotic factors that define a species' range limits and determine its ecological niche, which is used in SDM models [34]. It is crucial for the management and protection of vulnerable and endemic species to be able to predict appropriate ecological niches under current and future climatic conditions [35]. Because climatic factors are not the primary determinant of habitat suitability, the real niche of species is often smaller than that suggested by model-based forecasts [36]. In terms of topographic variables, the aspect contributed the highest weight when used alone. Climate predictors such as diurnal range and isothermality (Bio-2 and Bio-3) provide considerable information on the distribution of all *Cypripedium* species. Alpine distribution patterns and community structure are greatly influenced by topographical (elevation, aspect) and bioclimatic factors [37]. There are some species that prefer moist and marshy habitats over open grasslands or rocky terrain, which are crucial factors in alpine biodiversity [38]. There has been rapid climate change in the Himalayas [39]. Species in alpine habitats are more vulnerable to extinction because they have a limited geographic range. Species migrating to high latitudes or elevations are disproportionately affected by climate change [40].

Based on model output, climate scenarios RCP 4.5 and RCP 8.5 indicate climatically acceptable habitats move northerly or northeasterly and become less suitable as elevation decreases. As a result, we hypothesize that north-facing mountain slopes have greater biomass, coverage, height, species diversity, and soil nutrient content than south-facing slopes [41]. Earlier studies in the Himalayan region have also reported species moving north and northeast due to climate change [42–44]. Changing temperatures will displace cold-adapted species with warm-adapted ones, which may result in habitat loss [45]. Northwestern Indian Himalayan rainfall declined similarly, according to Hasan et al. [46]. Changes in climatic conditions contribute to significant declines in precipitation for the driest quartile while increasing precipitation and temperature for the wettest quartile [47]. In the Western Himalayas, winter precipitation has declined, and dry days have increased continuously [48].

Future climate scenarios will influence the distribution of *Cypripedium* species based on average yearly temperatures (Bio-1), precipitation (Bio-12), precipitation of the coldest quartile (Bio-19), precipitation of the wettest month (Bio-13), and topography. RCP 4.5 (2050) would result in fewer suitable habitats for this species than RCP 8.5 (2050 and 2070), indicating that climatic changes and changes in land use patterns due to increasing emission rates would harm habitat suitability. The model predicts that climatic changes will result in new habitat suitability classes (referred to as habitat gains) at lower emission rates (such as RCP 4.5 vs. RCP 8.5), with the biggest gains occurring at lower emission rates. Cypripedium species are likely to benefit more from the RCP 4.5 scenario than from the RCP 8.5 scenario in the northern part of Uttarakhand. In addition, future climatic niche loss may be explained by the warming of lower altitudes or locations where the species are now found.

Based on our ensemble model, we predicted that habitat suitability for these three orchid species would drastically decline under future climate change scenarios, peaking at RCP 8.5 by 2070. Although many of the currently suitable ecosystems will become unsuitable in the future, certain areas with unsuitable climates will be able to adapt to changing climatic conditions. Conservation areas for rewilding and restoration may be established in regions that create appropriate habitats. According to observations, the rapid warming of the climate in the western Himalayas might be one of the major reasons for the significant decline in the niches of species, which is observed to be faster than the projected increases in temperature in other similar ecosystems around the world. Due to lower GHG emissions in RCP 4.5, most of the regions are expected to remain acceptable habitats for both species. According to RCP 8.5, the number of suitable habitats is expected to decline by approximately 55%. It is expected that the Valley of Flowers National Park, Kedarnath Wildlife Sanctuary, Kandara sites, and Har ki Doon will remain suitable for all species. A limited range expansion is also observed at RCPs 4.5 and 8.5 for *C. himalaicum*, *C. cordigerum*, and *C. elegans* in the eastern parts of the Uttarakhand. As a result, the range expansion in western Uttarakhand is greater. The results of this study confirm predictions that many Himalayan plant species will lose habitat as a result of climate change [49]. Under the influence of climate change, increased temperatures and earlier snowmelt may lead to enzymatic dysfunction, inhibiting plant growth and placing limitations on their developmental route [50]. Changing temperature affects plant phenology, which may reduce future habitat appropriateness [51]. There have been significant increases in the yearly average temperature and the average temperature of the wettest quartile of the north-western Himalaya [52].

As well as climate change, habitat loss resulting from excessive grazing, livestock trampling, and various development activities (road construction) have significantly affected the species. In addition to being medicinal, orchids are also horticultural [53,54]. Uncontrolled wild collecting and alien species invasion have put many of these species at risk [16]. A CITES listing of the Orchidaceae family has been imposed due to the impending threat (Appendices II). Western Himalaya showed a contiguous distribution of all three Cypripedium species owing to habitat specificity [55]. Orchids seem to prefer moist, shady, and humus-rich soils, as many of them grow in moist, shady, and humus-rich areas [56]. It has been reported that Pindari valley had the highest density of *C. cordigerum*, *C. elegans*, and *C. himalaicum* [57], respectively, of 0.8, 5.5, and 0.9 individuals/m^2, respectively.

The results of predictive modeling indicate that the species of *Cypripedium* are likely sensitive to changes in Annual mean temperature (Bio-1), annual precipitation (Bio-12), precipitation of wettest month (Bio-13), precipitation of coldest quarter (Bio-19), altitude and aspect. In general, the species' optimal bioclimatic requirements are defined as warm quartile temperatures of 6–7 °C, total precipitation of 1500–1700 mm per month, overall precipitation of 1200–1500 mm, altitude between 2700–3800 m on an east-facing slope.

Substantial alterations in these typical bioclimatic and topographic conditions could have a significant impact on the phenological behavior and the future spread of species. The annual total precipitation is useful in assessing how significant water availability is to a species range because it closely equates to total water intake. When the potential range of a species is impacted by extreme precipitation conditions throughout the year, the wettest month can be useful. While precipitation during the coldest quarter of the year gives total precipitation during the three coldest months of the year, this information can be useful for understanding how such climatic circumstances may affect species' seasonal distributions. The current study gives an overview of habitat suitability and expected changes in response to future climatic scenarios, as well as changes in the species' geomorphological profile. Several conservation activities on a local and regional scale in the Western Himalayas could benefit from the findings of this study. It is important to note that the model was built to predict the fundamental niche of the species rather than its understood niche, which is one of the study's limitations. The species' actual niche might not match what our model predicted. Another obstacle is that biotic elements like plant-mycorrhizal association or plant-pollinator interaction were not taken into account while modeling the habitat appropriateness of *Cypripedium* at different time scales (2050 and 2070). The potential range of the Orchidaceae species might be overestimated due to biotic interactions, especially mycorrhizal interactions [58].

5. Conclusions

In addition to being habitat specific, each of the three threatened *Cypripedium* species are suffering from severe stress and requires immediate attention. Anthropogenic stress and environmental perturbations contribute to additional strain on the species. Consequently, the present study evaluates the habitat suitability of three *Cypripedium* species and predicts the likely effects of climatic warming on their spread in the Western Himalayas in the future. The major predictors were the temperature average over the year (Bio- 1), precipitation average over the year (Bio-12), precipitation during the coldest quarter (Bio-19), and precipitation during the wettest month (Bio-13), as well as topography. There is an ecological niche for the species of approximately 192 km^2, which corresponds to approximately 1.5% of the total geographic area of the alpine region in the state. Climate change is likely to result in a greater loss of habitat for species in the eastern portion of the state. Climate change scenarios (RCP 4.5 & 8.5) will lead to a substantial decline in species habitat suitability compared to current predictions. The most important priority should be the selection and maintenance of a stable ecosystem that is resistant to the effects of climate change based on the predictions of habitat contraction. A number of additional preventive actions may also be undertaken in sensitive areas of the state in order to protect and restore the species. A significant threat to the population arrangement of the selected species is the disruption and alteration of land use/land cover caused by human activities such as the expansion of human habitats, agricultural operations, and animal husbandry. There are a variety of phytosociological features that demonstrate this.

In addition to providing useful information for conservation planning and participatory management, these distribution maps will also enable the identification of other sites not previously documented. Thus, the Valley of Flowers area is crucial for protecting Cypripedium's wild germplasm resources since it is one of the best-preserved tracts of extremely favorable habitat in the world for *Cypripedium* in the future. Several policymaking agencies may benefit from the information generated, including the Department of Forestry, the Board of Biodiversity, the Board of Medicinal Plants (SMPB), and the Ministry of Environment, Forest, and Climate Change (MOEF & CC). Through the incorporation of a Biodiversity Management Committee (BMC), the restoration of existing habitats would be one step forward.

Author Contributions: Conceptualization, N.C., G.S., M.Y.K., I.D.R., A.P.M., A.P., J.S.J. and R.C.; methodology, N.C., G.S., M.Y.K., I.D.R., A.P.M., A.P., J.S.J. and R.C.; software, N.C., G.S., M.Y.K., I.D.R., A.P.M., A.P., J.S.J. and R.C.; validation, N.C., G.S., M.Y.K., I.D.R., A.P.M., A.P., J.S.J. and R.C.; formal analysis, N.C., G.S., M.Y.K., I.D.R., A.P.M., A.P., J.S.J. and R.C.; investigation, N.C., G.S., M.Y.K., I.D.R., A.P.M., A.P., J.S.J. and R.C.; resources, N.C., G.S., M.Y.K., I.D.R. and R.C.; data curation, N.C., G.S., A.P., J.S.J. and R.C.; writing—original draft preparation, N.C., G.S., M.Y.K., I.D.R., A.P.M., A.P., J.S.J. and R.C.; writing—review and editing, N.C., G.S., M.Y.K., I.D.R., A.P.M., A.P., J.S.J. and R.C.; visualization, N.C., G.S., M.Y.K., I.D.R., A.P.M., A.P., J.S.J. and R.C.; supervision, N.C., G.S., M.Y.K., I.D.R., A.P.M., H.G.A., A.P., J.S.J. and R.C.; project administration, N.C., G.S., M.Y.K., I.D.R., A.P.M., A.P., J.S.J. and R.C.; funding acquisition, H.A., N.C., M.A.-M., G.S., M.Y.K., I.D.R., A.P.M., A.P., J.S.J. and R.C. All authors have read and agreed to the published version of the manuscript.

Funding: This project was funded by Princess Nourah bint Abdulrahman University Research Supporting Project Number PNURSP2022R24, Princess Nourah bint Abdulrahman University, Riyad, Saudi Arabia. The article processing charge was funded by the Deanship of Scientific Research at Qassim University.

Data Availability Statement: Data will be made available on request.

Acknowledgments: The researchers would like to thank the Deanship of Scientific Research, Qassim University, for funding the publication of this project. We thank Director Uttarakhand Space Application Centre (USAC), Dehradun, for providing the required facilities. The financial support from the National Mission on Himalayan Studies (NMHS) of the Ministry of Environment, Forest and Climate Change (MoEF & CC), Govt. of India under "Himalayan Alpine Biodiversity Characterization and Information System-Network" (Grant Number: GBPNI/NMHS-2019-20/MG-rev/399) is duly acknowledged. We also thank the officials of the State Forest Department of Uttarakhand for permission to undertake field studies in the area.

Conflicts of Interest: The authors declare no conflict of interest.

References

1. Myers, N. Threatened biotas: "hot spots" in tropical forests. *Environmentalist* **1988**, *8*, 187–208. [CrossRef] [PubMed]
2. Rana, S.K.; Rana, H.K.; Ranjitkar, S.; Ghimire, S.K.; Gurmachhan, C.M.; O'Neill, A.R.; Sun, H. Climate-change threats to distribution, habitats, sustainability and conservation of highly traded medicinal and aromatic plants in Nepal. *Ecol. Indic.* **2020**, *115*, 106435. [CrossRef]
3. Brandt, J.S.; Haynes, M.A.; Kuemmerle, T.; Waller, D.M.; Radeloff, V.C. Regime shift on the roof of the world: Alpine meadows converting to shrublands in the southern Himalayas. *Biol. Conserv.* **2013**, *158*, 116–127. [CrossRef]
4. Singh, L.; Tariq, M.; Sekar, K.C.; Bhatt, I.D.; Nandi, S.K. Ecological niche modelling: An important tool for predicting suitable habitat and conservation of the Himalayan medicinal herbs. *ENVIS Bull. Himal. Ecol.* **2017**, *25*, 154–155.
5. Thakur, S.; Negi, V.S.; Dhyani, R.; Satish, K.V.; Bhatt, I.D. Vulnerability assessments of mountain forest ecosystems: A global synthesis. *Trees For. People* **2021**, *6*, 100156. [CrossRef]
6. Negi, V.S.; Pathak, R.; Rawal, R.S.; Bhatt, I.D.; Sharma, S. Long-term ecological monitoring on forest ecosystems in Indian Himalayan Region: Criteria and indicator approach. *Ecol. Indic.* **2019**, *102*, 374–381. [CrossRef]
7. Christenhusz, M.J.M.; Byng, J.W. The number of known plants species in the world and its annual increase. *Phytotaxa* **2016**, *261*, 201–217. [CrossRef]
8. Cribb, P. *The Genus Cypripedium*; Timber Press Inc.: Portland, OR, USA, 1997.
9. Jalal, J.S.; Jayanthi, J. An annonated checklist of the the orchid of western Himalaya, India. *Lankesteriana* **2015**, *15*, 7–50. [CrossRef]
10. Jalal, J.S.; Kumar, P.; Rawat, G.S.; Pangtey, Y.P.S. Orchidaceae, Uttarakhand, Western Himalaya, India. *Check List* **2008**, *4*, 304–320. [CrossRef]
11. Diez, J.M.; Pulliam, H.R. Hierarchical analysis of species distributions and abundance across environmental gradients. *Ecol. Soc. Am.* **2007**, *88*, 3144–3152. [CrossRef]
12. Rai, I.D.; Adhikari, B.S.; Rawat, G.S. A unique patch of timberline ecotone with three species of Lady's slipper orchids in Garhwal Himalaya, India. *J. Threat Taxa* **2010**, *2*, 766–769. [CrossRef]
13. Kull, T. *Cypripedium calceolus* L. *J. Ecol.* **1999**, *87*, 913–924. [CrossRef]
14. Warren, J.; McLaughlin, M.; Bardsley, J.; Eich, J.; Esche, C.A.; Kropkowski, L.; Risch, S. The Strengths and Challenges of Implementing EBP in Healthcare Systems. *Worldviews Evid. Based Nurs.* **2016**, *13*, 15–24. [CrossRef] [PubMed]
15. Polak, T.; Saltz, D. Reintroduction as an Ecosystem Restoration Technique. *Conserv. Biol.* **2011**, *25*, 424–425. [CrossRef] [PubMed]
16. Suyal, R.; Bhatt, D.; Rawal, R.S.; Tewari, L.M. Status of two threatened astavarga herbs, Polygonatumcirrhifolium and Malaxismuscifera, in West Himalaya: Conservation implications. *Proc. Natl. Acad. Sci. India Sect. B Biol. Sci.* **2020**, *90*, 695–704. [CrossRef]
17. Hooker, J.D. *A Sketch of the Flora of British India*; Eyre and Spottiswoode: London, UK, 1907; Volume 1, pp. 157–212.

18. Bachman, S.; Moat, J.; Hill, A.W.; de la Torre, J.; Scott, B. Supporting Red List threat assessments with GeoCAT: Geospatial conservation assessment tool. *ZooKeys* **2011**, *150*, 117–126. [CrossRef]

19. Fick, S.E.; Hijmans, R.J. WorldClim2: New 1-km spatial resolution climate surfaces for global land areas. *Int. J. Climatol.* **2017**, *37*, 4302–4315. [CrossRef]

20. Phillips, S.J.; Anderson, R.P.; Schapire, R.E. Maximum entropy modelling of species geographic distributions. *Ecol. Model.* **2006**, *190*, 231–259. [CrossRef]

21. Fielding, A.H.; Bell, J.F. A review of methods for the assessment of prediction errors in conservation presence/absence models. *Environ. Conserv.* **1997**, *24*, 38–49. [CrossRef]

22. Pearson, R.G.; Raxworthy, C.J.; Nakamura, M.; Peterson, A.T. Predicting species distributions from small numbers of occurrence records: A test case using cryptic geckos in Madagascar. *J. Biogeogr.* **2007**, *34*, 102–117. [CrossRef]

23. Peterson, A.T.; Soberon, R.G.; Pearson, R.P.; Anderson, E.; Martinez-Meyer, M.; Araujo, M.B. *Ecological Niches and Geographic Distributions*; Princeton University Press: Princeton, NJ, USA, 2011.

24. Thuiller, W.; Araujo, M.B.; Lavorel, S. Do we need land-cover data to model species distributions in Europe? *J. Biogeogr.* **2004**, *31*, 353–361. [CrossRef]

25. Elith, J.; Graham, C.H.; Anderson, R.P.; Dudík, M.; Ferrier, S.; Guisan, A.; Hijmans, R.; Huettmann, F.; Leathwick, J.R.; Lehmann, A.; et al. Novel methods improve prediction of species' distributions from occurrence data. *Ecography* **2006**, *29*, 129–151. [CrossRef]

26. Guisan, A.; Tingley, R.; Baumgartner, J.B.; Naujokaitis-Lewis, I.; Sutcliffe, P.R.; Tulloch, A.I.; Regan, T.J.; Brotons, L.; McDonald-Madden, E.; Mantyka-Pringle, C.; et al. Predicting species distributions for conservation decisions. *Ecol. Lett.* **2013**, *16*, 1424–1435. [CrossRef]

27. Elith, J.; Leathwick, J.R. Species distribution models: Ecological explanation and prediction across space and time. *Annu. Rev. Ecol. Evol. Syst.* **2009**, *40*, 677–697. [CrossRef]

28. Araujo, M.; New, M. Ensemble forecasting of species distributions. *Trends Ecol. Evol.* **2007**, *22*, 42–47. [CrossRef]

29. Zhang, X.Q.; Li, G.Q.; Du, S. Simulating the potential distribution of *Elaeagnus angustifolia* L. based on climatic constraints in China. *Ecol. Eng.* **2018**, *113*, 27–34. [CrossRef]

30. Santiz, E.C.; Lorenzo, C.; Carrillo-Reyes, A.; Navarrete, D.A.; Islebe, G. Effect of climate change on the distribution of a critically threatened species. *Therya* **2016**, *7*, 147–159. [CrossRef]

31. Thomas, C.D.; Cameron, A.; Green, R.E.; Bakkenes, M.; Beaumont, L.J.; Collingham, Y.C.; Erasmus, B.F.N.; De Siqeira, M.F.; Grainger, A.; Hannah, L.; et al. Extinction risk from climate change. *Nature* **2004**, *427*, 145–148. [CrossRef]

32. Yuan, H.; Wei, Y.; Wang, X. Maxent modeling for predicting the potential distribution of Sanghuang, an important group of medicinal fungi in China. *Fungal Ecol.* **2015**, *17*, 140–145. [CrossRef]

33. Li, R.; Xu, M.; Wong, M.H.G.; Qiu, S.; Sheng, Q.; Li, X.; Song, Z. Climate change-induced decline in bamboo habitats and species diversity: Implications for giant panda conservation. *Divers. Distrib.* **2015**, *21*, 379–391. [CrossRef]

34. Lenoir, J.; Ge'gout, J.C.; Marquet, P.; De Ruffray, P.; Brisse, H.J.S. A significant upward shift in plant species optimum elevation during the 20th century. *Science* **2008**, *320*, 1768–1771. [CrossRef] [PubMed]

35. Qin, A.; Liu, B.; Guo, Q.; Bussmann, R.W.; Ma, F.; Jian, Z.; Xu, G.; Pei, S. Maxent modeling for predicting impacts of climate change on the potential distribution of *Thuja sutchuenensis* Franch., an extremely endangered conifer from south western China. *Glob. Ecol. Conserv.* **2017**, *10*, 139–146. [CrossRef]

36. Buisson, L.; Thuiller, W.; Casajus, N.; Lek, S.; Grenouillet, G. Uncertainty in ensemble forecasting of species distribution. *Glob. Chang. Biol.* **2010**, *16*, 1145–1157. [CrossRef]

37. Chitale, V.S.; Behera, M.D.; Roy, P.S. Future of endemic flora of biodiversity hotspots in India. *PLoS ONE* **2014**, *9*, e115264. [CrossRef] [PubMed]

38. Singh, L.; Bhatt, I.D.; Negi, V.S.; Nandi, S.K.; Rawal, R.S.; Bisht, A.K. Population status, threats, and conservation options of the orchid *Dactylorhiza hatagirea* in Indian Western Himalaya. *Reg. Environ. Chang.* **2021**, *21*, 40. [CrossRef]

39. Chen, I.C.; Hill, J.K.; Ohlemuller, R.; Roy, D.B.; Thomas, C.D. Rapid range shifts of species associated with high levels of climate warming. *Science* **2011**, *333*, 1024–1026. [CrossRef] [PubMed]

40. Bertrand, R.; Lenoir, J.; Piedallu, C.; Riofrı´o-Dillon, G.; De Ruffray, P.; Vidal, C.; Pierrat, J.C.; Gégout, J.C. Changes in plant community composition lag behind climate warming in lowland forests. *Nature* **2011**, *479*, 517–520. [CrossRef] [PubMed]

41. Kutiel, P.; Lavee, H. Effect of slope aspect on soil and vegetation properties along an aridity transect. *Isr. J. Plant Sci.* **1999**, *47*, 169. [CrossRef]

42. Boisvert-Marsh, L.; Perie, C.; de Blois, S. Shifting with climate? Evidence for recent changes in tree species distribution at high latitudes. *Ecosphere* **2014**, *5*, 1–33. [CrossRef]

43. Yu, F.; Wang, T.; Groen, T.A.; Skidmore, A.K.; Yang, X.; Ma, K.; Wu, Z. Climate and land use changes will degrade the distribution of Rhododendrons in China. *Sci. Total Environ.* **2019**, *659*, 515–528. [CrossRef] [PubMed]

44. Rana, S.K.; Rawat, G.S. Database of Himalayan plants based on published floras during a century. *Data* **2017**, *2*, 36. [CrossRef]

45. Sobrino, E.; Gonzalez, A.; Sanz-Elorza, M.; Dana, E.; Sanchez-Mata, D.; Gavilan, R. The expansion of thermophilic plants in Iberian peninsula as a sign of climate change. In *Fingerprints of Climate Change. Adaptive Behaviour and Shifting Species Range*; Walther, G.R., Burga, C.A., Edwards, P.J., Eds.; Kulwer Publishers: Dordrecht, The Netherlands, 2001; pp. 163–184.

46. Hassan, T.; Hamid, M.; Wani, S.A.; Malik, A.H.; Waza, S.A.; Khuroo, A.A. Substantial shifts in flowering phenology of Sternbergiavernalis in the Himalaya: Supplementing decadal field records with historical and experimental evidences. *Sci. Total Environ.* **2021**, *795*, 148811. [CrossRef] [PubMed]

47. Bhutiyani, M.R.; Kale, V.S.; Pawar, N.J. Climate change and the precipitation variations in the northwestern Himalaya: 1866–2006. *Int. J. Climatol.* **2010**, *30*, 535–548. [CrossRef]

48. Livensperger, C.; Steltzer, H.; Darrouzet-Nardi, A.; Sullivan, P.F.; Wallenstein, M.; Weintraub, M.N. Earlier snowmelt and warming lead to earlier but not necessarily more plant growth. *Ann. Bot.* **2016**, *8*, plw021. [CrossRef] [PubMed]

49. Manish, K.; Telwala, Y.; Nautiyal, D.C.; Pandit, M.K. Modelling the impacts of future climate change on plant communities in the Himalaya: A case study from Eastern Himalaya, India. *Model Earth Syst. Environ.* **2016**, *2*, 92. [CrossRef]

50. Chuine, I. Why does phenology drive species distribution? *Philos. Trans. R. Soc. B* **2010**, *365*, 3149–3160. [CrossRef]

51. Dash, S.K.; Jenamani, R.K.; Kalsi, S.R.; Panda, S.K. Some evidences of climate change in twentieth-century India. *Clim. Chang.* **2007**, *85*, 299–321. [CrossRef]

52. Keller, F.; Goyette, S.; Beniston, M. Sensitivity analysis of snow cover to climate change scenarios and their impact on plant habitats in alpine terrain. *Clim. Chang.* **2005**, *72*, 299–319. [CrossRef]

53. Hossain, M.M. Therapeutic orchids: Traditional uses and recent advances—An overview. *Fitoterapia* **2011**, *82*, 102–140. [CrossRef]

54. Deb, C.R.; Imchen, T. Orchids of horticultural importance from Nagaland, India. *Pleione* **2011**, *5*, 44–48.

55. Suyal, R.; Joshi, P.; Bahukhandi, A.; Bhandari, S. Diversity and Distribution Pattern of Orchids Along an Altitudinal Gradient: Pindari Valley, West Himalaya. *Proc. Natl. Acad. Sci. India Sect. B Biol. Sci.* **2022**, *92*, 817–824. [CrossRef]

56. Jalal, J.S.; Rawat, G.S.; Pankaj, K. Status, distribution and habitats of orchids in Uttarakhand. *J. Orchid. Soc. India* **2010**, *24*, 35–41.

57. Suyal, R.; Rawal, R.S.; Jalal, J.S. Noteworthy additions to the orchids of Kumaun Himalaya, India. *Indian For.* **2018**, *144*, 778–780.

58. Tsiftsis, S.; Djordjević, V. Modelling sexually deceptive orchid species distributions under future climates: The importance of plant–pollinator interactions. *Sci. Rep.* **2020**, *10*, 10623. [CrossRef] [PubMed]

Article

Impact of Environmental Gradients on Phenometrics of Major Forest Types of Kumaon Region of the Western Himalaya

Vikas Dugesar [1,2], Koppineedi V. Satish [2], Manish K. Pandey [2,3], Prashant K. Srivastava [2,*], George P. Petropoulos [4], Akash Anand [2,5] and Mukunda Dev Behera [6]

[1] Department of Geography, Institute of Science, Banaras Hindu University, Varanasi 221005, India
[2] Remote Sensing Laboratory, Institute of Environment and Sustainable Development, Banaras Hindu University, Varanasi 221005, India
[3] Centre for Quantitative Economics and Data Science, Birla Institute of Technology, Mesra 835215, India
[4] Department of Geography, Harokopio University of Athens, 17671 Athens, Greece
[5] Department of Forestry and Wildlife Ecology, University of Wisconsin-Madison, Madison, WI 53706, USA
[6] Centre for Oceans, Rivers, Atmosphere and Land Sciences (CORAL), Indian Institute of Technology, Kharagpur 721302, India
* Correspondence: prashant.iesd@bhu.ac.in

Citation: Dugesar, V.; Satish, K.V.; Pandey, M.K.; Srivastava, P.K.; Petropoulos, G.P.; Anand, A.; Behera, M.D. Impact of Environmental Gradients on Phenometrics of Major Forest Types of Kumaon Region of the Western Himalaya. *Forests* **2022**, *13*, 1973. https://doi.org/10.3390/f13121973

Academic Editors: Any Mary Petritan and Mirela Beloiu

Received: 30 August 2022
Accepted: 10 November 2022
Published: 22 November 2022

Publisher's Note: MDPI stays neutral with regard to jurisdictional claims in published maps and institutional affiliations.

Abstract: Understanding ecosystem functional behaviour and its response to climate change necessitates a detailed understanding of vegetation phenology. The present study investigates the effect of an elevational gradient, temperature, and precipitation on the start of the season (SOS) and end of the season (EOS), in major forest types of the Kumaon region of the western Himalaya. The analysis made use of the Normalised Difference Vegetation Index (NDVI) time series that was observed by the optical datasets between the years 2001 and 2019. The relationship between vegetation growth stages (phenophases) and climatic variables was investigated as an interannual variation, variation along the elevation, and variation with latitude. The SOS indicates a delayed trend along the elevational gradient (EG) till mid-latitude and shows an advancing pattern thereafter. The highest rate of change for the SOS and EOS is 3.3 and 2.9 days per year in grassland (GL). The lowest rate of temporal change for SOS is 0.9 days per year in mixed forests and for EOS it is 1.2 days per year in evergreen needle-leaf forests (ENF). Similarly, the highest rate of change in SOS along the elevation gradient is 2.4 days/100 m in evergreen broadleaf forest (EBF) and the lowest is −0.7 days/100 m in savanna, and for EOS, the highest rate of change is 2.2 days/100 m in EBF and lowest is −0.9 days/100 m in GL. Winter warming and low winter precipitation push EOS days further. In the present study area, due to winter warming and summer dryness, despite a warming trend in springseason or springtime, onset of the vegetation growth cycle shows a delayed trend across the vegetation types. As vegetation phenology responds differently over heterogeneous mountain landscapes to climate change, a detailed local-level observational insight could improve our understanding of climate change mitigation and adaptation policies.

Keywords: NDVI; phenometrics; start of the season; end of the season; elevational gradient; climate change

1. Introduction

Variability in climate models is linked to variations in vegetation dynamics, which can affect ecosystem processes, such as water and carbon exchange, water and energy fluxes, and species interactions. Land surface phenology (LSP) became an increasingly significant field of study in recent years due to its ability to track how land ecosystems adapt to environmental changes on scales ranging from local to global. Phenology is the study of the yearly cycles that plants go through in their many stages of growth [1]. Climate change with significant impact on plant phenology, directly altering carbon flow, dynamic nutrient balance, biodiversity, and related feedback to the climate system [1,2]. Because phenological processes are subject to change in climate, it is important to keep an eye on them in order to mitigate the impact. Due to its substantial impact on the balance of terrestrial ecosystems, the onset of the vegetation growing season is considered the most consistent and effective climate change indicator. Understanding the causes, consequences, and variability of changes in phenology is crucial for predicting the future trajectories of ecological communities [3–8]. Changes in plant phenology associated with climate change are observed around the world. What is little known is whether and how phenological responses to global warming will vary from year to year, season to season, habitat to habitat, and one species to another.

Forest phenology is a systematic process of green-up, maturity, and green-down events. These events are called SOS, and EOS are used to monitor short- and long-term responses to climatic variations [9,10]. For the Himalayan forests ecosystem, an accurate determination of seasonal phenological events is essential to understanding the patterns of response to physiographic and climatic variables [11,12]. Phenological field observations are only made in a few places and over a short period. Using satellite remote sensing data, a different strategy was created that enables explicit spatio-temporal observation on a broad scale [1,13,14], making it an effective tool for the efficient modelling of forest biophysical parameters. NDVI and EVI (enhanced vegetation index), two of the most often used remotely sensed indices, were utilised to track the phenological and seasonal variation in vegetation growth [5,7]. Quantification of forest phenology is widely used to determine forest health and seasonal variation in forest canopy response to climate change. Several studies in recent years demonstrated the sensitivity of different phenological parameters (i.e., SOS and LOS) of vegetation to the climate change response of forest vegetation [15–18]. Global and regional scale forest phenology is primarily derived from remotely sensed satellite data often called LSP that provides wall-to-wall coverage [12,19–21]. LSP studies mainly focus on trend analysis, which represents the gradual changes in phenological metrics, however, there are considerable variabilities, such as the reversal of trends in the long-term phenological dynamics [22,23]. Interannual variations in phenology are greatly affected by topography and climate. Altitudinal gradient has varying effects on the seasonality of dominant tree species and, in turn, affects the productivity and functioning of the ecosystem. Temperature and precipitation directly impact the onset of greening and senescence of the leaf and the length of the season. Precipitation-controlled tropical dry climate can push green-up onset by a month and by half a month in temperature-controlled temperate climate [24,25]. Time series analysis of the phenological patterns is a strong and realistic tool. In order to get usable statistical data features, such as autocorrelation, trend, or biophysical variables, it is necessary to reconstruct, model, and analyse the data using a number of techniques [26].

The LSP is calculated using a variety of logistic functions, with the metrics extraction algorithm, the considered phenological metrics, and the appropriate parameter numbers being the main differences. To extract SOS and EOS dates from vegetation indices (VIs) time series from satellite data, a number of approaches were devised. The number of seasons is calculated using the local maxima and minima of the curve to define the change of state as in prior techniques [27–29]. A season is determined when three successive local minima, maxima, and minimum points are found. This search has to be modified with certain constraints in order to reduce contributions from undesired artifacts at low frequencies. Using a prominent threshold in decomposition and analysis of time series software (DATimeS), these faulty estimations may be minimized. As a consequence, peaks that do not exceed the set prominence value are automatically rejected as noise. DATimeS chooses the largest local maxima/minima when the value is greater than zero for separation. Following the breakdown of the time series, each rising season is investigated using established threshold approaches to detect complicated phenological occurrences (e.g., SOS and EOS) [15,30–32].

Most of the phenological studies in the Himalayas are conducted considering the whole Himalayas as a unit or by taking field plots or small field plots in a particular vegetation type. The present study investigates the upshift of the phenophases over the year and along the elevational gradient for the dominant vegetation types. The present study aims to assess (i) the spatial patterns of phenophases between and within the vegetation types, (ii) quantify the temporal trends of phenometrics, and (iii) understand the climatic attributions of the phenometric changes.

2. Material and Methods

2.1. Study Area

The Kumaon region of the western Himalayas was chosen as the study area, as it offers one of the most diverse biogeographical regions in the Indian Himalayan region. The Kumaun division of Uttarakhand (Figure 1) covers an area of 21,034 km^2, and is located between latitudes 28°44′ N to 30°49′ N and longitudes 78°45′ E to 81°05′ E with elevations ranging from 200 m to 6000 m above mean sea level (amsl).

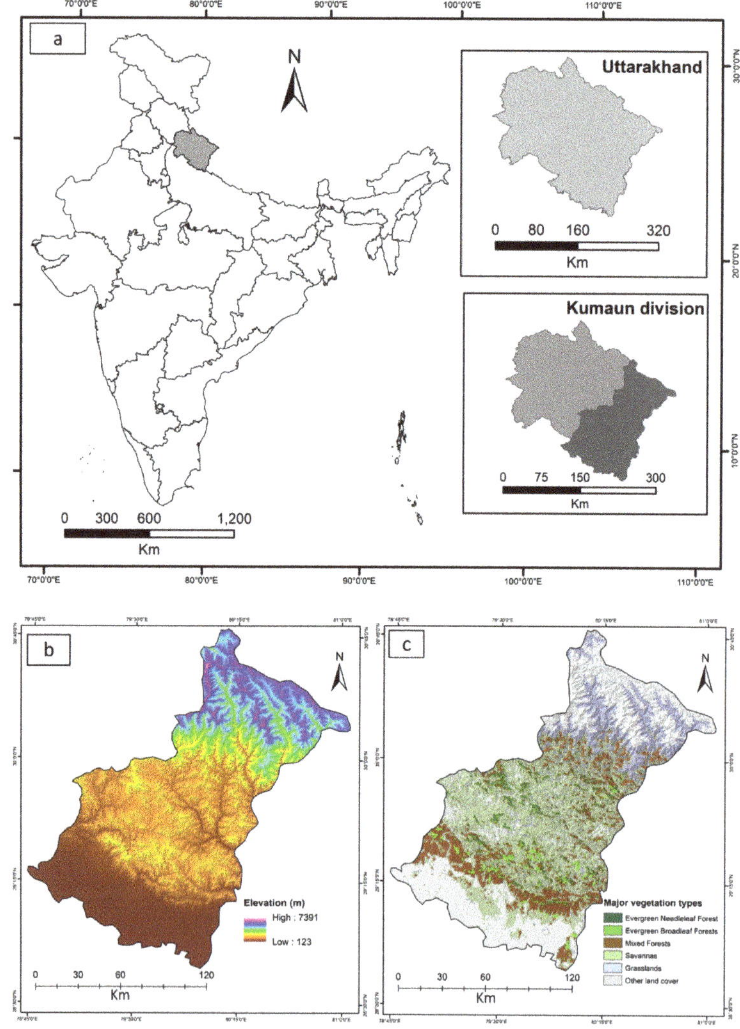

Figure 1. (**a**). Study area location, (**b**) elevation gradient, and (**c**) major vegetation type map.

The area comes under the humid subtropical climate zone of Koppen's classification. As per the International Geosphere–Biosphere Programme (IGBP) classification scheme, the dominant vegetation types of Kumaon Himalaya are need-leaf evergreen, broadleaf evergreen, broadleaf deciduous, and mixed forest along with savanna type vegetation and grassland.

The annual mean temperature remains around 13 °C. The hottest month is June (18.3 °C) and the coldest month is January (6.4 °C). The mean annual rainfall is around 1500 mm, where November is the driest (50 mm) and July remains the wettest (500 mm) month. The monthly average relative humidity ranges from 46 percent (in April) to 93 percent (in September), with the yearly average relative humidity coming at 65.9 percent (in August). The species composition and distribution greatly vary along the altitudinal, temperature, and rainfall gradient.

2.2. Datasets

2.2.1. Earth Observation Datasets

The International Geosphere–Biosphere Programme (IGBP) land cover data of coarse (1 km) spatial resolution are used to identify and delineate the forest types of the study area. ALOS PALSER 12.5 m digital elevation data are used to generate an elevation profile and three vegetation indices products of MODIS (LAI, NDVI, EVI) are used to identify and compare the phenological indicator. The MCD12Q1 version 6 data product provides yearly global land cover type data under six different classification schemes. The land cover classes are obtained using yearly metrics as inputs to the random forest classifier, followed by supervised decision tree classification and post-processing of the map results based on Hidden Markov Models, which significantly reduce interannual variability in the product. In the present study land cover class type, one was used, which has 17 classes, including natural vegetation, human-altered, and non-vegetated. This product also includes layers of seasonal cycles, as the beginning of the vegetation's growth, maturity, and senescence [33]. JAXA provided (ALOS PALSAR) the Digital Elevation Model (DEM) at 12.5 m data that were downloaded from NASA Earth Data's ASF Data search platform. All the tiles were mosaicked and masked using the boundary of the study area.

2.2.2. Meteorological Datasets

The Integrated Multi-Satellite Retrievals for GPM (IMERG) is used as precipitation data. IMERG is a single algorithm that gives precipitation estimates by merging data from all GPM constellation passive microwave devices. GPM-M data were utilised in this work [34,35]. MODIS11A1 version 6 product data, which offers daily per-pixel land surface temperature and emissivity at 1 km spatial resolution, was used to obtain temperature data. The daily data were acquired utilising the Google Earth Engine (GEE) platform, using an elevation zone-wise shape file. In the R environment, monthly mean values are generated from daily mean temperature data.

2.3. Methodology

In the present investigation, we used DATimeS coupled with Savitzky–Golay (SG), and simplified least square fit convolution to extract the phenophases (SOS and LOS) from MODIS datasets for major vegetation types and used SOS and EOS for the final analysis (Figure 2). Afterwards, a correlation between phenophases and climatic variables (temperature and precipitation) was analyzed.

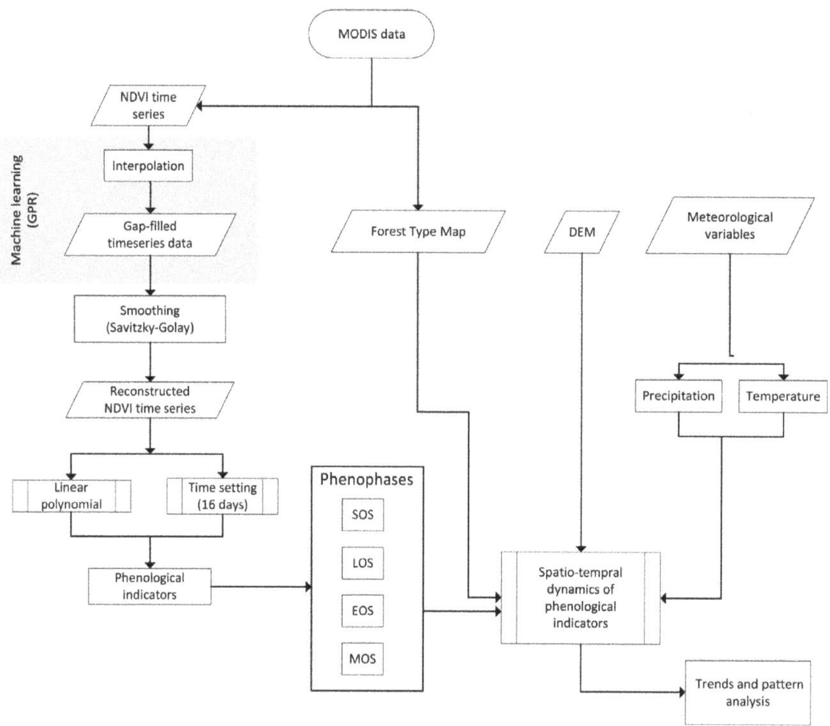

Figure 2. Flowchart of functions used in the present study.

2.3.1. Retrieval of Vegetation Index and Vegetation Map

The MCD12Q1 version 6 data product was used to extract a vegetation type map, which includes ENF, EBF, MF, savanna, and grassland, which are the major vegetation types present in the study area. The elevation data were masked using individual forest-type layers and further divided into four equal elevation gradient zones (H1, H2, H3, and H4). The MOD13A1 version 6 product derived from daily, atmosphere-corrected, bidirectional surface reflectance, provides two vegetation indices (NDVI, EVI) at 16-day intervals, and 500 m × 500 m pixel size; it provides high sensitivity over dense vegetation with a minimum canopy soil variation. Then, the mean NDVI values for each zone of the study vegetation type were obtained from GEE. LSP and phenometrics were then extracted from smoothed MODIS NDVI time series data from 2001 to 2019. NDVI can be calculated by the NIR and red reflectance [36] as:

$$NDVI = \frac{NIR - RED}{NIR + RED} \tag{1}$$

where:

RED = MODIS band 1 surface reflectance
NIR = MODIS band 2 surface reflectance

2.3.2. Time Series Smoothing and Gap Filling

The MODIS NDVI product data are temporarily composited using a prominence threshold approach to filter out the noisy and erroneous points. A prominence of peak is defined as the shortest vertical distance a signal must travel on either side of the peak before either rising to a level higher than the peak again or reaching an endpoint. This distance may be measured in either direction from the peak. Peaks that do not surpass the present prominence value are thus automatically dismissed as noise.

The GPR approach typically depicts the relationship between input samples and output observations, where the use of an additive Gaussian noise with zero mean function encodes that each location has equal probability. The covariance matrix function is to use a kernel function called K to encode the similarity between each combination of the input sample Xi and Xj. The covariance design is extremely important since it needs to consider the key characteristics of the variables to be modelled. Due to its capacity to correctly estimate smoothly changing functions and also take into account asymmetries in the feature space, the asymmetric square exponential (SE) kernel performs well when vegetation indices are to be retrieved from earth observation data [37–39]. The SE kernel is used in the present study and defined as follows:

$$k(x_i, x_j = \sigma_s^2 exp\left(-\frac{1}{2}\sum_{b=1}^{D}\left[\frac{x_i(b) - x_j(b)}{\sigma_b}\right]^2\right) \tag{2}$$

where σ_s^2 denotes the output variance and σ_b is connected to the dispersion of the training information along the input dimension b in such a manner that the inverse of σ_b defines the significance of dataset b in the prediction process. Once the noise variance σ_n^2 and the kernel's free parameters are chosen, the covariance matrix is fully determined. These terms, which are referred to as the hyperparameters of the GPR model, may all be written as $\theta = \{\sigma_s^2, \sigma^2, \sigma_n^2\}$, where $\sigma = [\sigma_1, \ldots, \sigma_D]$.

$$\log p(y|x, f) = -\frac{1}{2}y^T\left(K + \sigma_n^2 I_N\right)^{-1}y - \frac{1}{2}\log\left|K + \sigma_n^2 I_N\right| - \frac{n}{2}\log \pi \tag{3}$$

where N = number of training samples, K = covariance matrix, and y = training output

The first part in Equation (3) is effectively a data-fit term, the second one is a complexity penalty, and the last term is merely a normalizing constant. The maximisation of the log-likelihood method is used to optimise the hyperparameter in the present study. Training the GPR is the common name for this optimisation process [37,38,40,41].

Following that, the data sets were smoothed using the Savitzky–Golay filter (SG filter), and a time series were reconstructed using the DATimeS tool. The SG filter has a balancing ability to decrease noise while keeping the integrity of the NDVI time series. For NDVI time series smoothing, the general equation for the simplified least-squares convolution is:

$$Y_i^* = \frac{\sum_{i=-m}^{i=m} C_i Y_{j+i}}{N} \tag{4}$$

where
    Y = NDVI value (original),
    Y^* = NDVI value (resultant),
    Ci = coefficient for the ith NDVI value of the filter (smoothing window), and
    N = convoluting number of integers.

Finally, phenological events are obtained from the smoothed time series in order to establish the exact timing of the transition between the various phases of plant development. Only SOS and EOS were taken into consideration in the present study. In this study, to obtain phenological events, the use of a polynomial model with a defined degree and the Savitzky–Golay generalised moving average with the filter coefficient being determined by an unweighted linear least-squares regression [39].

2.3.3. Forest Phenological Variables

The geographical area under investigation is very heterogeneous in nature. The elevational gradient for each vegetation type varies differentially. The threshold-based method is adapted to determine the land surface phenology of the Kumaon forest. According to earlier studies [30,42–45] the threshold value varies with the forest type and altitudinal gradients and ranges between 0.2 and 0.3. To make a balanced approach across the altitudinal gradient and forest types, a 25% amplitude is used as the threshold to determine SOS and EOS. The SOS NDVI value for the given vegetation type was quantified using the following equation

$$NDVI_{start} = NDVI_{min} + (NDVI_{max} - NDVI_{min}) * 0.25 \tag{5}$$

$$NDVI_{end} = NDVI_{min} + (NDVI_{max} - NDVI_{min}) * 0.25. \tag{6}$$

The double logistic curve fitting is then used to calculate the SOS and EOS, which can be calculated for any given day using the aforementioned threshold value.

2.3.4. Trend and Correlation Analysis

The trend analysis of the phenological variables is done using the linear regression method. Least square fit algorithm is frequently used to analyse the trend and growth patterns of the vegetation. Further, the correlations between phenometrics and climatic variables are assessed using a linear regression approach. All the statistical analysis is done using the R software package.

The relationship between climatic variables and phenometrics (SOS and EOS) is accomplished using multivariate regression analysis. This step is performed in R statistical software. The equation used in this approach is:

$$Y = \beta_0 + \beta_1 X_1 + \beta_2 X_2 + \in \tag{7}$$

where β_0 = model intercept, β_1 β_2 = coefficients, X_1, X_2 = covariates, and $\in$ = random error.

3. Results

3.1. Spatial Patterns of Phenometrics

Seasonal patterns of plant phenology greatly vary with the vegetation type, latitude, and elevation. The occurrence of SOS varies greatly along the elevation gradient. It even varies considerably within the vegetation type, along the gradient. Both ENF and EBF show a delay (positive) trend in both SOS and EOS. The MF and Savanna show a mixed trend where initially it shows a delayed trend, however, at higher reaches, it starts advancing. GL shows an advancement (negative) trend for both SOS and EOS. Figure 3 exhibits the average SOS and EOS day of the year and its spatial distribution along the elevational gradient. The study shows a consistent pattern in the spatial distribution of phenophases across the vegetation types (Figure 3). The SOS and EOS show a delayed trend for ENF, EBF, and MF. Savanna shows a mixed type of pattern for SOS, where initially it shows a delayed trend up to 2600m and then it becomes reversed and starts advancing (negative) with a rate of 0.7 d/100 m, and the pattern stays similar for the EOS (Table 1). GL shows advancement for SOS, but EOS does not show any significant shift. The highest rate of change in SOS was reported in EBF (2.4 d/100 m) and the lowest rate in the savanna (-0.7 d/100 m). For EOS, the highest rate was in EBF (2.2 d/100 m), but the lowest was in GL (-0.9 d/100 m).

Table 1. Spatial distribution of phenometrics trend within vegetation types.

	ENF (1300–2250 m)	EBF (500–2250 m)	MF (800–3000 m)	Savanna (1500–3300 m)	GL (2000–4000 m)
SOS (days/100 m)	1.5	2.4	1.1	1.6 and −0.7	−1.2
EOS (days/100 m)	1.8	2.2	1.9	−1.4	−0.9

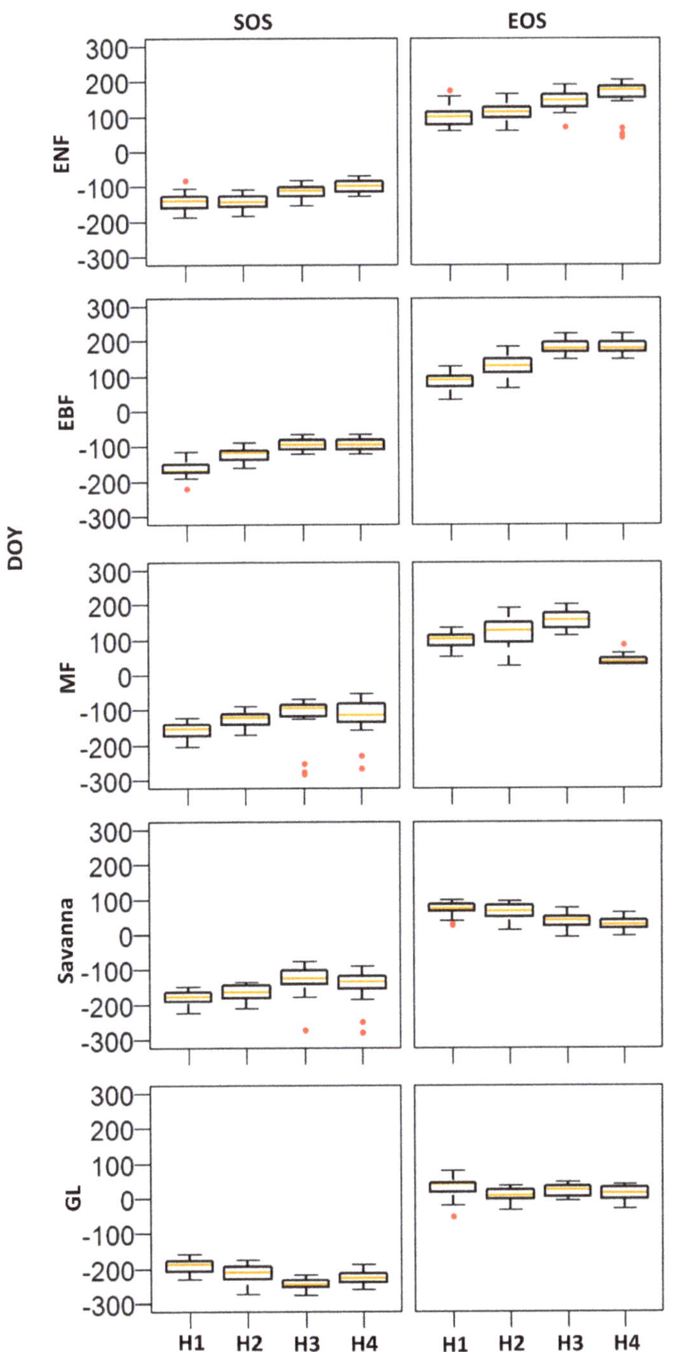

Figure 3. Phenometrics distribution along the elevation gradient across the vegetation types. The positive and negative values indicate starting in the current year and ending in the next year, respectively. The *x*-axis labels are the subzones divided along the altitudinal gradient of the vegetation types.

3.2. Interannual Variations in Phenometrics

SOS shows a delay (positive) trend across the vegetation types, the highest rate of change was reported in the upper reaches of grassland (3.3 DY^{-1}) and the lowest in the upper reaches of mixed forest (0.9 DY^{-1}) (Figure 4). Similarly, the EOS shows a varying degree of change across the vegetation types and along the elevation gradient. The lowest rate was reported in ENF (1.2 DY^{-1}) and the highest in GL (2.9 DY^{-1}). Table 2 shows the rate of change in interannual variations in SOS and EOS. The overall trend in SOS and EOS is not consistent, but shows a decreasing rate of change along with the increasing elevation (Figure 4).

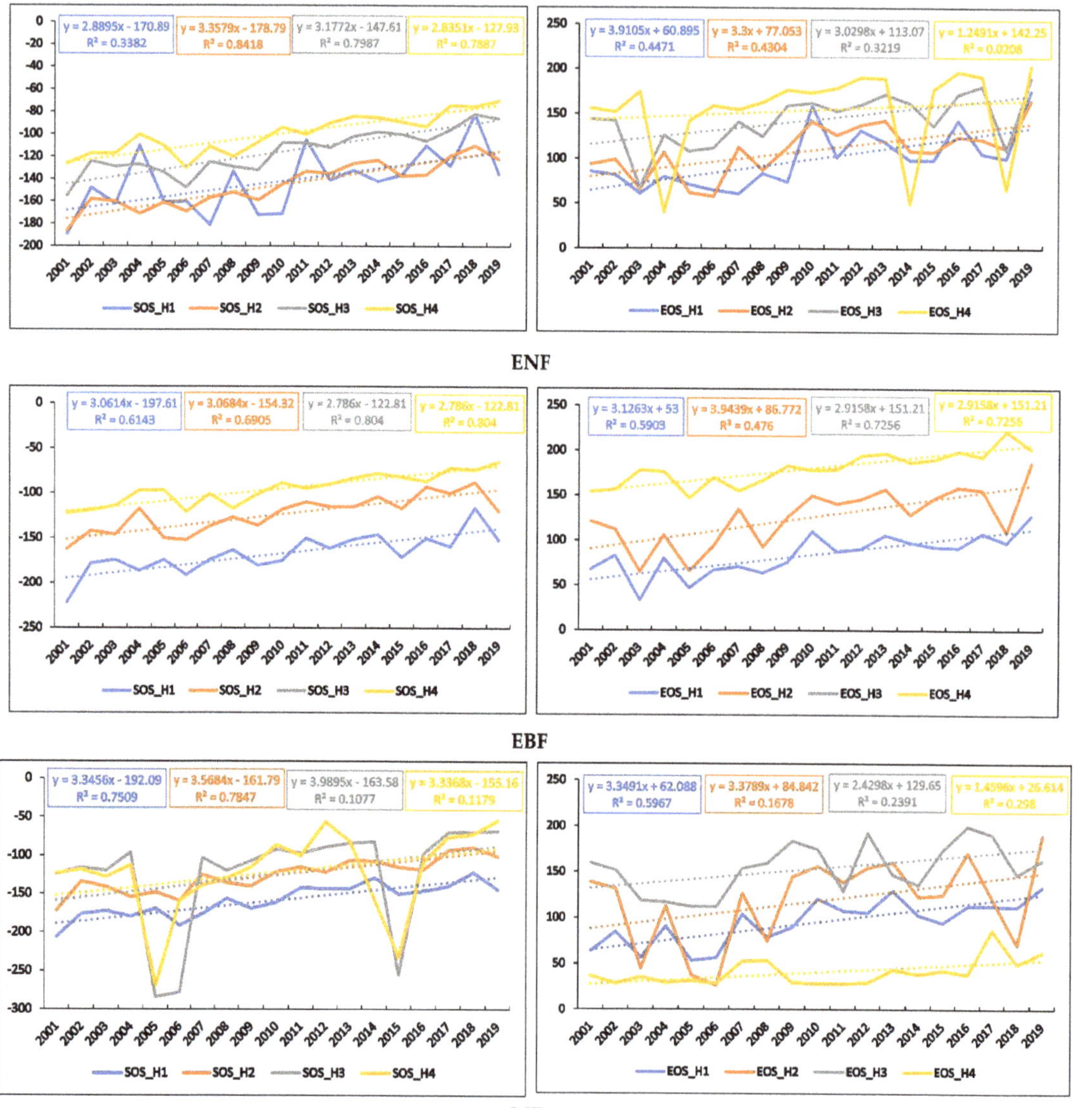

Figure 4. *Cont.*

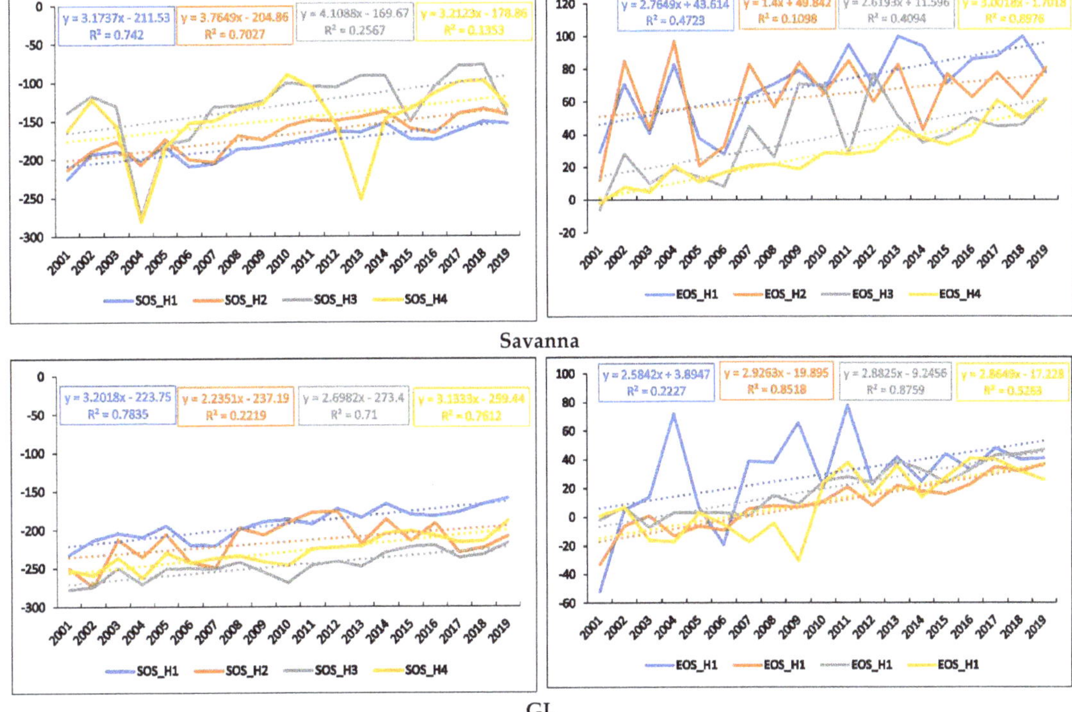

Savanna

GL

Figure 4. Interannual change trend of phenometrics (SOS and EOS) for years from 2001 to 2019 with different vegetation types.

Table 2. Interannual distribution of phenometrics trend in major vegetation types.

	ENF		EBF		MF		Savanna		GL	
Elevation Zones	SOS DY^{-1}	EOS DY^{-1}	SOS DY^{-1}	EOS DY^{-1}	SOS DY^{-1}	EOS DY^{-1}	SOS DY^{-1}	EOS DY^{-1}	SOS DY^{-1}	EOS DY^{-1}
H1	2.6	2.9	2.7	2.1	2.0	2.4	2.5	2.8	2.6	2.6
H2	1.8	2.3	1.2	2.9	1.9	2.4	2.5	2.5	3.1	2.9
H3	1.6	2.0	1.2	1.9	1.3	1.9	2.4	2.1	3.7	2.9
H4	1.2	1.2	1.2	1.9	0.9	2.2	2.1	2.0	3.3	2.9

3.3. Temporal Trends of Climatic Variables and the Relationship between Phenometrics and Climatic Variables

In this work, phenometrics were used to analyse the temporal and spatial relationships between LSP and climatic variables. The association between phenological phases and temperature was investigated with the help of partial correlation analysis and multivariate linear regression. This was performed after the precipitation factor was taken into account. In a similar manner, partial correlation analysis was used in order to evaluate the correlations between phenological phases and precipitation after the temperature was adjusted. The correlations between phenometrics and climatic variables were performed using quarterly (Q1, Q2, Q3 and Q4) data of climatic variables. The relationship between temperature and precipitation in the spring (April, May, and June) season, which also coincides with, or is called the preseason period, and winter + autumn (October to March), was assessed using a multivariate linear regression model. The different vegetation types show variable responses towards climatic variables (Figure 5). SOS of most of the vegetation types shows a negative correlation with the spring temperature, while a positive correlation was observed with the precipitation. Temperature shows a moderate to good negative correlation (R > −0.5, $p < 0.05$) with SOS in preseason months. The SOS of the major vegetation type shows a positive correlation with precipitation. EOS exhibits a significant positive correlation with winter temperature and precipitation and is the most unpredictable at high altitudes and areas with varied species diversity.

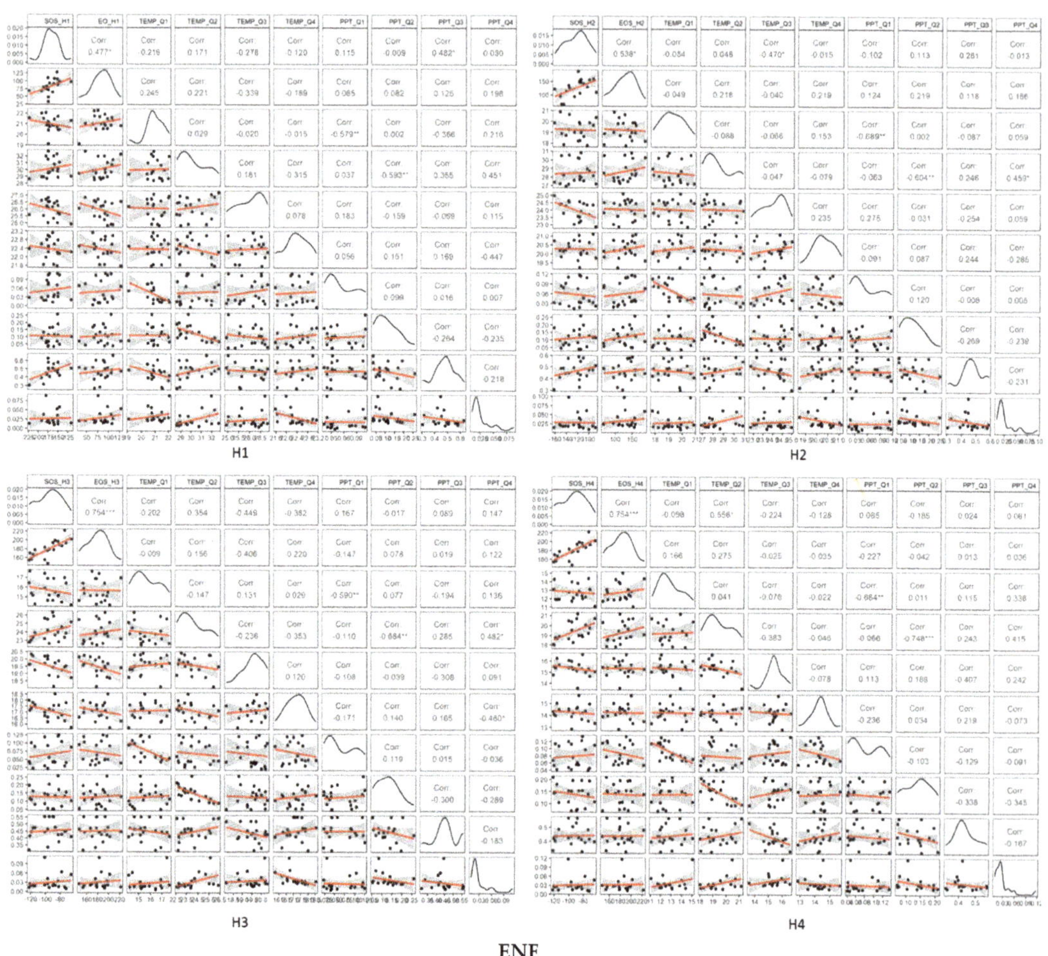

ENF

Figure 5. *Cont.*

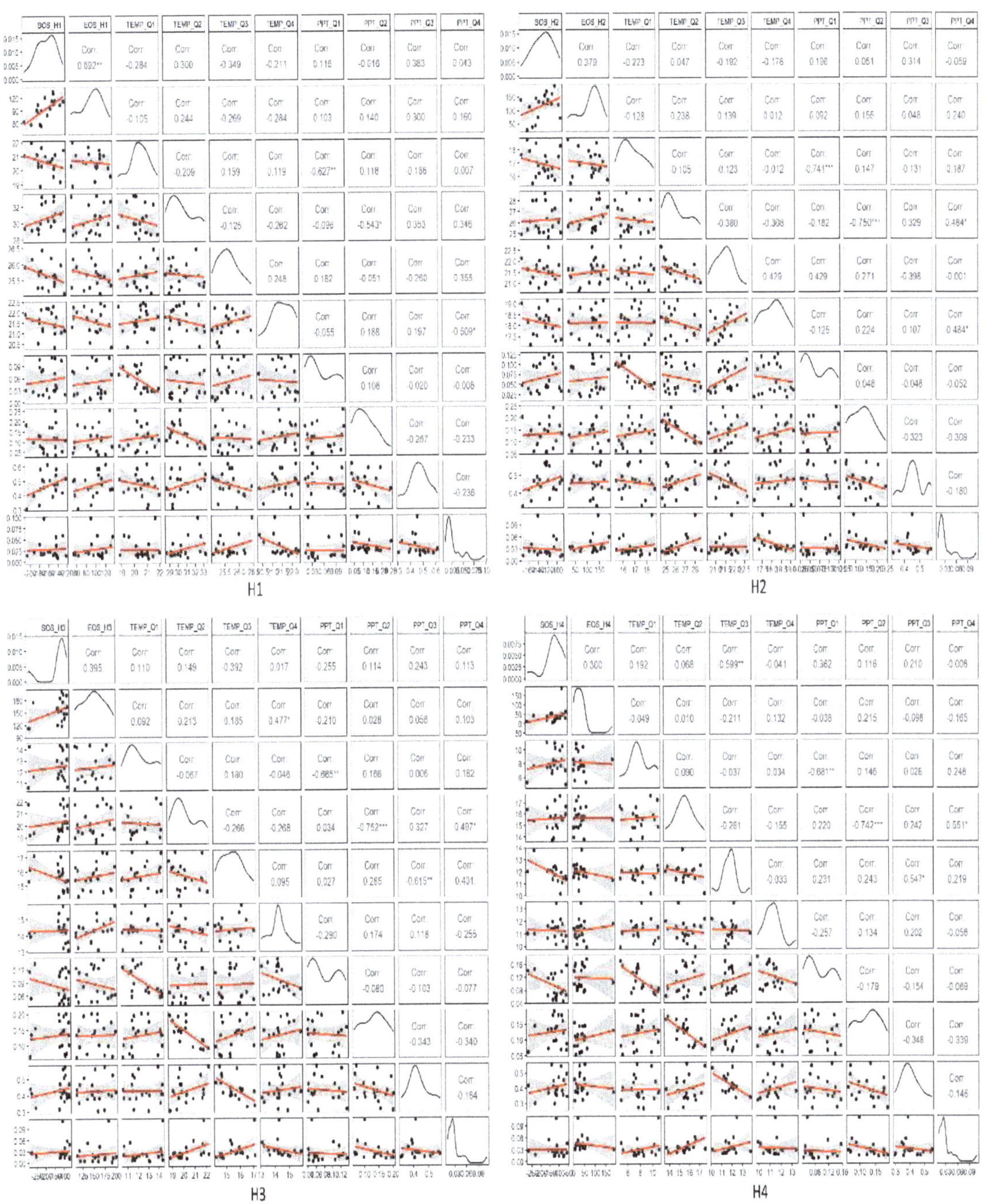

Figure 5. *Cont.*

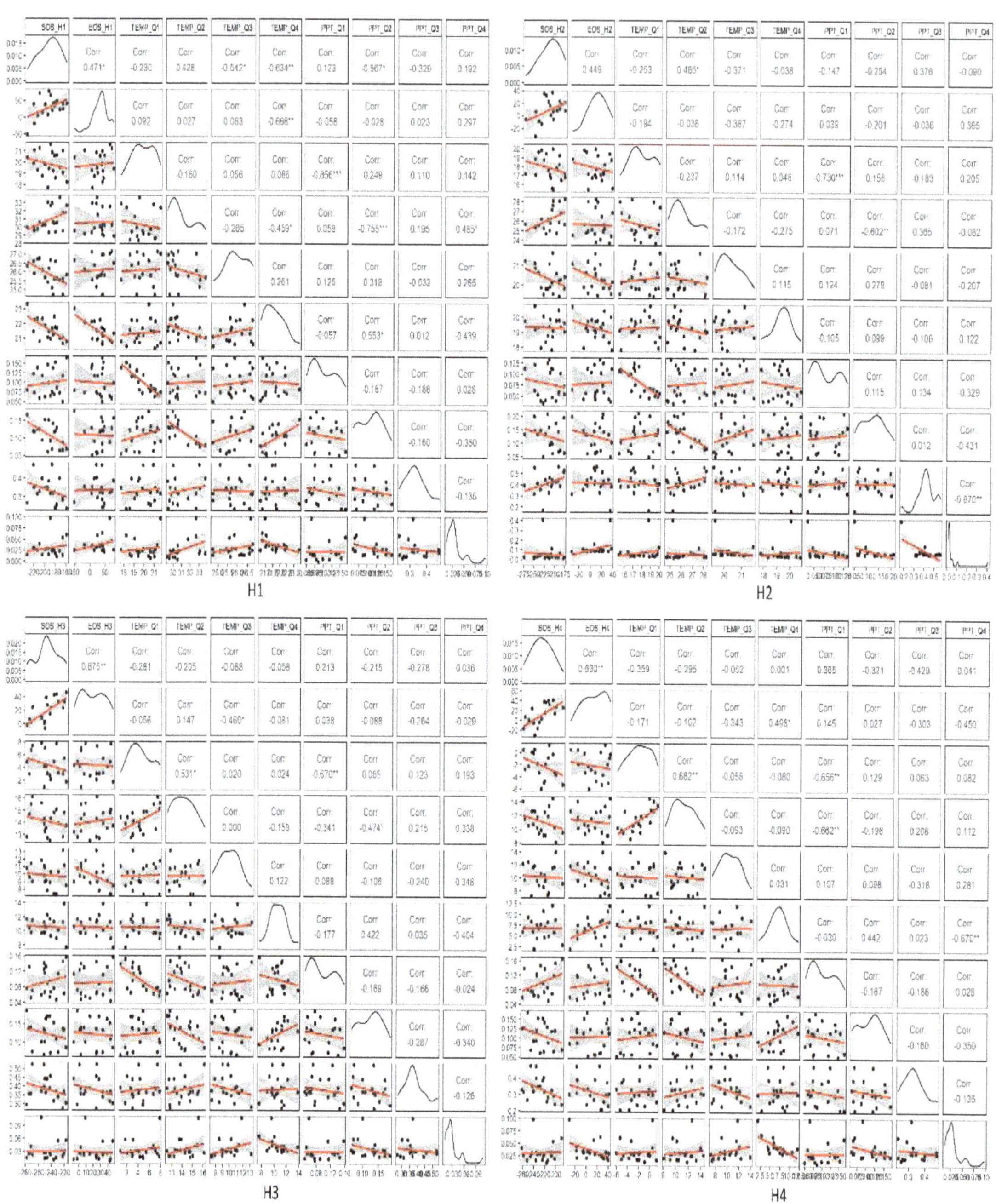

MF

Figure 5. *Cont.*

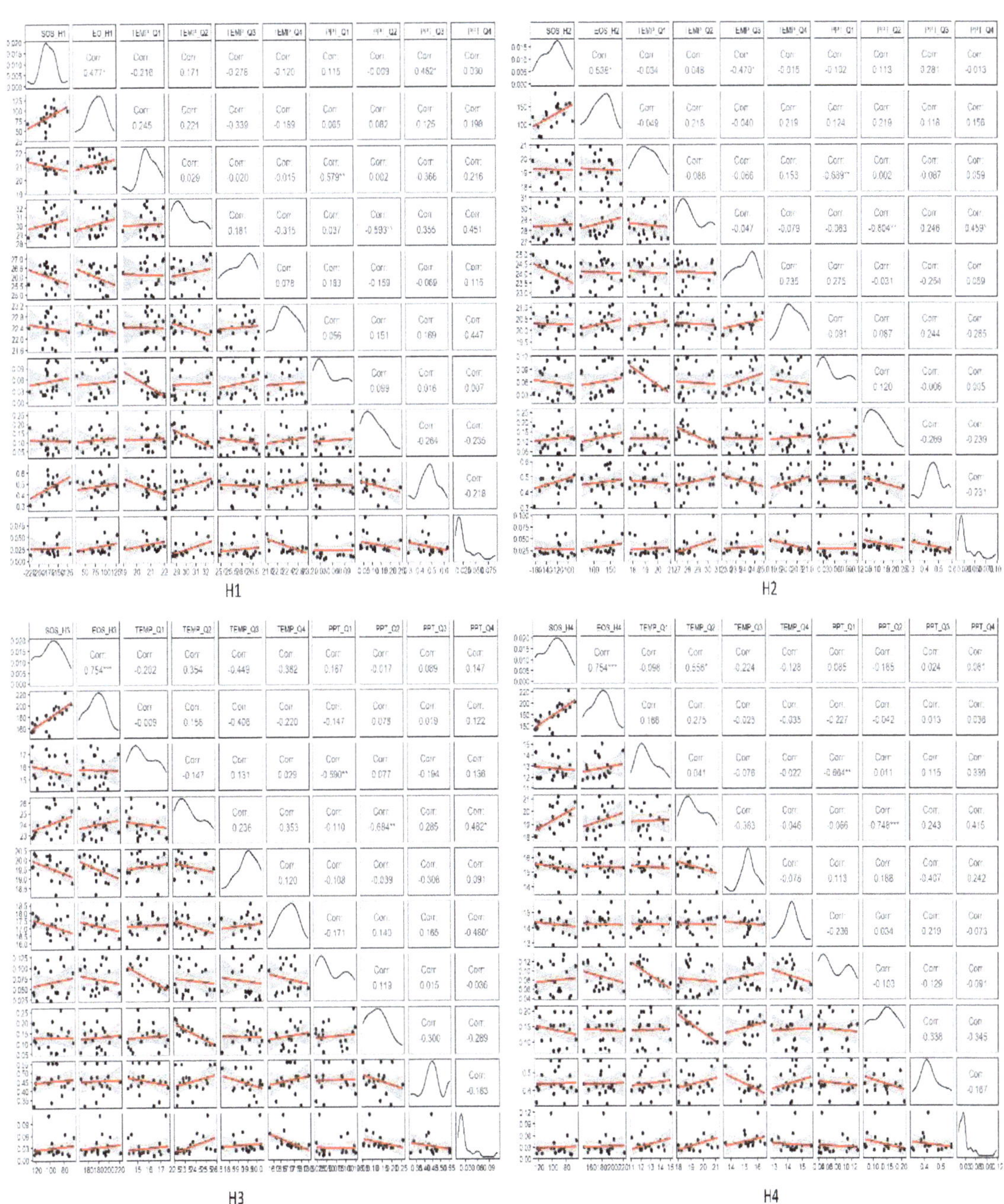

Savanna

Figure 5. *Cont.*

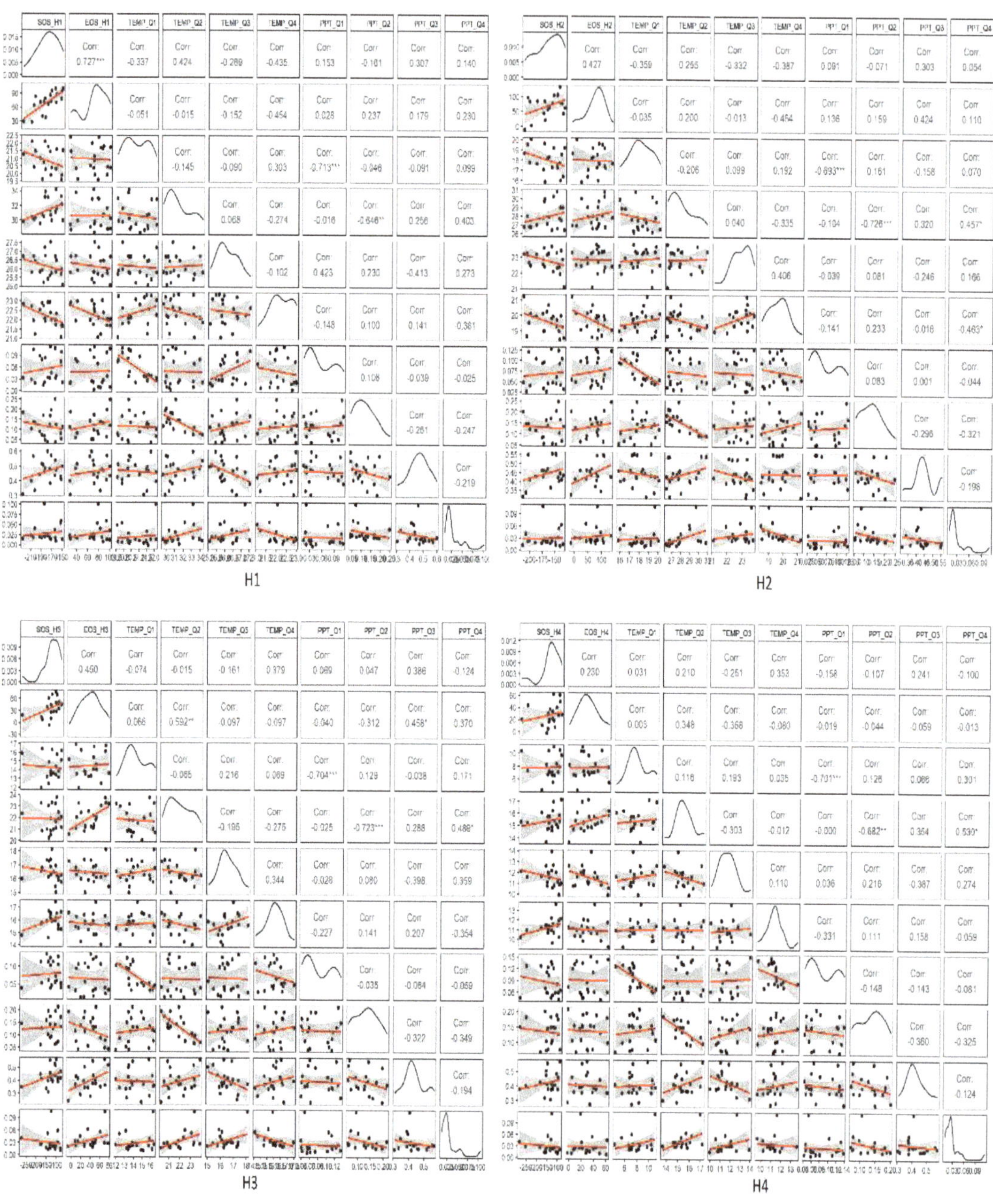

Figure 5. Relationship between phenometrics (SOS and EOS) and climatic variables (temperature and precipitation). The upper triangle consists of correlation value (R2) with significant (*p* value), and diagonal axis shows the density and lower triangle scatter plot with regression fit line. Here H1, H2, H3, and, H4 are the subzones divided along the altitudinal gradient of the vegetation types. The level of significance is depicted by asterisk (* $p \leq 0.05$, ** $p \leq 0.01$, and *** $p \leq 0.001$).

4. Discussion

Climate change altered the dynamics of phenological events in areas from tropical to temperate regions. Many plant species show advancement in their spring phases because of climate change. Strong winter warming, on the other hand, could inhibit required winter chilling, perhaps delaying spring phenology. This phenomenon is particularly noticeable in areas where temperatures are rapidly rising and where vegetation is very temperature-responsive.

4.1. Driving Factors for Phenometrics Changes

Spatial and temporal distribution was evaluated across five major vegetation types of Kumaon Himalaya spanning two decades (2001–2019). The mean SOS and EOS time varies across vegetation types and also with latitude and elevation. The onset of vegetative stages showed a shift from June to July at lower reaches to July to August as we moved to the higher elevation, which shows a negative correlation with the general climate change-induced trend. However, in the upper savanna and alpine region, it shows an advanced onset of the season, a positive correlation with the temperature elevation trend. According to recent studies [46–49], temperature is not the primary governing factor for seasonal behavior along the elevational gradient, but rather the temperature lapse rate due to the elevation gradient, which controls vegetation behavior along the EG in heterogeneous mountainous regions.Our study is also following with these findings and shows a differential behaviour of both SOS and EOS over EG; whereas in MF, Savanna and GL show differential behaviour towards the climatic variable along the EG, even within the vegetation type, because of their large distributional range and topography. As we move upward (lower and middle elevation) it shows a delayed trend in the MF; at higher elevation, it shows a reversal of trend and starts advancing thereafter. Similar results were observed in savanna and GL, they show an advancing trend throughout its EG. These abrupt changes in higher elevation are perhaps influenced by the premature availability of precipitation due to early snowmelt, governed by an increase in soil and ambient temperature. For temporal distribution, all the major vegetation types follow the trends observed in spatial distribution, but the rate of change remains low on the temporal scale in comparison to the spatial scale. This again proves the dominance of lapse rate due to EG across the vegetation types.

4.2. Phenological Changes with Different Gradients

Flowering durations tended to be delayed as elevation increased (−0.07 to 2.2 days/100 m), which is consistent with findings from the Himalayan subalpine environments (0.014 days ml, [46]). The perception that most species show advanced spring phenology was supported by research on phenology changes [47–50]. However, a significant number (~25%) of species in all of these investigations deviates from this pattern and exhibited trends toward delayed spring phenology. To address this, we divided the elevational gradient into four equal vertical zones to know the change patterns along the altitudinal gradient within the vegetation types. Except for mixed forest and savanna, all the vegetation types show a delayed trend that is in agreement with the findings of various studies in the northern hemisphere [51–53]. However, the trends in mixed forest and savanna show the reversal of trends at higher altitudes within the vegetation types. This is because of the weakened temperature effect on higher altitudes. The vertical zones H3 and H4 of mixed forest and savanna, which are overlapped with the grassland show a similar trend as of grassland vegetation. This happens because of the altitude-induced lapse rate, which compensates the climate change-induced warming and reverses the trend at higher altitudes. A few recent studies also observed the trend reversal in grassland and savanna vegetation types at higher altitudes, which supports our findings [52]. These outcomes are not contradictory, but rather show that the change in temperature is generally influenced by the altitudinal gradient in mountainous systems. As the altitude increases the temperature drops constantly and in response, the time to attain accumulated growth temperature for different vegetation types gets altered [53–55].

The spring phenology variations indicate that the responses of plants to the warming are not linear and vary by vegetation type and local climatic behaviour [56]. Most temperate and subtropical plants' spring phases are determined by winter cold and spring heat [57–59] Warming in the spring slows spring phenology, whereas warming in the winter delays chilling needs. The surprising tendency of budburst delay in vegetative events is most likely driven by higher temperatures and low precipitation in winter and the hot dry spring season. Temperature (especially the maximum temperature) increases the rate of evapotranspiration during the driest months of the year, eventually increasing loss of soil water content and further reducing the turgor pressures required to expand developing cells [55], potentially delaying the development of leaves. This is perhaps one of the reasons for delayed SOS in addition to winter warming. If there was more water, an optimal thermal condition would reduce plant heat buildup and enhance budburst, and low rainfall and high temperature would delay the budburst.

In the present study, the phenophases show a delayed trend in ENF, EBF, MF, savanna, and GL SOS despite the spring heating, which does not endorse the general trend shown by most of the studies in various parts of the world. The preseason shows a negative correlation with the SOS, which is in agreement with the general trends reported worldwide. The preseason months show a constant increase in temperature and a decline in precipitation in the study period. The dryer conditions created by low precipitation and high temperature further complicate spring phenology's responses to climate change. In general, the majority of the species vegetative growth and reproductive stages are known to depend on available moisture, whereas blooming and fruiting in plants appear to be related to ambient temperature [57]. The observations in the present study do not confirm the general trend of advancement reported in various studies. Despite the continuous warming of the spring season and phenology-advancing effect [58–61] of rising temperature in spring, this part of the western Himalayas shows a delayed start of the season. A similar trend was also reported by Delbart et al., 2006, [61]. The increasing temperature in October to march and low precipitation in the same time frame [62], coupled with unfavourable hot and dry conditions in spring, pushed forward the onset of the growing season. The temperature of October, November, February, and March shows a warming trend in the last two decades. The delaying impact could be a result of reduced winter chill, causing plants' vernalisation requirements to be met later.

Among all the vegetation types, mixed forest shows a differential trend pattern and does not show agreement with the general trend obtained in the study. Remote sensing modelling of phenology works on several trees and site conditions blended together in a pixel, and that leads to no selection of site characteristics and microclimatic conditions of trees or nearby forest areas. The use of a mix of coniferous and broadleaf species proved to affect phenology data, and this is well investigated. The drop in vegetation indices in pixels with a high percentage of coniferous species is expected to be slower near the end of the season in fall than in pixels with pure deciduous broadleaved species, making estimating the date of phenological events more unclear, as was reported in this study [63,64].

Our findings show a delayed trend in EOS date in the Kumaon Himalaya from 2001 to 2019, which is in agreement with the findings of earlier research in the Northern Hemisphere [65], temperate China [66–68], Europe [69], and North America [66]. The EOS date shows a delayed trend by a rise in preseason temperature, which was similar to findings from prior research using satellite data and field investigations [67]. This is because of the fact that warming in the summer and autumn boosts photosynthetic activity, slows chlorophyll degradation during leaf senescence in the fall, and hence delays leaf senescence. In all the (ENF, EBF, MF, savanna, and GL) regions, preseason precipitation was likewise inversely linked with the EOS. During the preseason, soil moisture will rapidly increase as precipitation increases, enhancing vegetation photosynthesis via altering vegetation carboxylation. The vegetation will develop more quickly and finish the growing season preschedule [68]. However, in the study area under investigation, precipitation shows a negative trend and as result delays the EOS, which is in confirmation with earlier works.

5. Conclusions

The present study unfolded those disparities in phenology changes caused by climate change in major vegetation types of Himalayan Forest ecosystems. We emphasised how different major vegetation types respond to changes in climatic variables at spatial and temporal scales. This study highlighted the necessity for rigorous research into plant phenology in understudied Himalayan forests. Warmer ecosystems control phenology differently than better-studied temperate systems; prediction of spring phenology changes in warmer ecosystems that overlook late summer and early fall temperatures might be misleading. Subtropical plants bloom all year and respond to temperature changes differently than temperate plants. However, subtropical and temperate forests, respond to climate change differently. This study tried to bring some new insight into the field of plant phenology, as there were not many area-specific or vegetation-specific studies on Himalayan ecosystems to study the climate-induced changes along the elevational gradient. For improving our fundamental knowledge of forest ecology and for influencing management and policy choices related to climate and conservation, it is crucial to understand the differences in the mechanisms that govern phenology in forest ecosystems. This research is therefore required to better comprehend climate-driven phenological shifts and forecast future changes in tropical and subtropical environments, which are essential to the environment, ecosystems, and conservation. Awareness of the consequences of climate change on vegetation phenology and ecosystem processes will require an understanding of these changes. The results of this study indicate the importance of the elevation gradient and its relationship with climatic variables among and within the vegetation types. Further research with robust ground truth data and long-term meteorological data from the ground station using better spatial resolution data sets would be helpful in understanding the variations in phenological patterns and complex interaction between climatic variables, elevation, and phenophases.

Author Contributions: Conceptualisation, P.K.S.; data curation, K.V.S. and A.A.; formal analysis, V.D., M.K.P. and A.A.; funding acquisition, P.K.S. and G.P.P.; investigation, V.D., P.K.S. and M.D.B.; methodology, V.D., M.K.P. and P.K.S.; project administration, P.K.S.; resources, P.K.S. and G.P.P.; supervision, P.K.S.; writing–original draft, V.D., K.V.S. and G.P.P.; writing–review and editing, P.K.S. and M.D.B. All authors have read and agreed to the published version of the manuscript.

Funding: This research was funded by the National Mission for Himalayan Studies (NMHS), Ministry of Environment, Forest and Climate Change (P-07/683).

Data Availability Statement: All the datasets presented in this study can be available to the user on request.

Acknowledgments: The authors are thankful to the Director, Institute of Environment and Sustainable Development for the lab facilities and logistic support.

Conflicts of Interest: The authors have declared no conflict of interest.

References

1. Wang, X.; Gao, Q.; Wang, C.; Yu, M. Spatiotemporal patterns of vegetation phenology change and relationships with climate in the two transects of East China. *Glob. Ecol. Conserv.* **2017**, *10*, 206–219. [CrossRef]
2. Wang, X.; Xiao, J.; Li, X.; Cheng, G.; Ma, M.; Che, T.; Dai, L.; Wang, S.; Wu, J. No Consistent Evidence for Advancing or Delaying Trends in Spring Phenology on the Tibetan Plateau. *J. Geophys. Res. Biogeosci.* **2017**, *122*, 3288–3305. [CrossRef]
3. Richardson, A.D.; Hollinger, D.Y.; Dail, D.B.; Lee, J.T.; Munger, J.W.; O'keefe, J. Influence of spring phenology on seasonal and annual carbon balance in two contrasting New England forests. *Tree Physiol.* **2009**, *29*, 321–331. [CrossRef]
4. Piao, S.; Cui, M.; Chen, A.; Wang, X.; Ciais, P.; Liu, J.; Tang, Y. Altitude and temperature dependence of change in the spring vegetation green-up date from 1982 to 2006 in the Qinghai-Xizang Plateau. *Agric. For. Meteorol.* **2011**, *151*, 1599–1608. [CrossRef]
5. Cong, N.; Wang, T.; Nan, H.; Ma, Y.; Wang, X.; Myneni, R.B.; Piao, S. Changes in satellite-derived spring vegetation green-up date and its linkage to climate in China from 1982 to 2010: A multimethod analysis. *Glob. Chang. Biol.* **2013**, *19*, 881–891. [CrossRef]

6. Shen, M.; Piao, S.; Cong, N.; Zhang, G.; Jassens, I.A. Precipitation impacts on vegetation spring phenology on the Tibetan Plateau. *Glob. Chang. Biol.* **2015**, *21*, 3647–3656. [CrossRef]

7. Peng, D.; Wu, C.; Li, C.; Zhang, X.; Liu, Z.; Ye, H.; Luo, S.; Liu, X.; Hu, Y.; Fang, B. Spring green-up phenology products derived from MODIS NDVI and EVI: Intercomparison, interpretation and validation using National Phenology Network and AmeriFlux observations. *Ecol. Indic.* **2017**, *77*, 323–336. [CrossRef]

8. Wang, X.; Zhou, Y.; Wen, R.; Zhou, C.; Xu, L.; Xi, X. Mapping spatiotemporal changes in vegetation growth peak and the response to climate and spring phenology over northeast China. *Remote Sens.* **2020**, *12*, 3977. [CrossRef]

9. Liu, Q.; Fu, Y.H.; Zeng, Z.; Huang, M.; Li, X.; Piao, S. Temperature, precipitation, and insolation effects on autumn vegetation phenology in temperate China. *Glob. Chang. Biol.* **2016**, *22*, 644–655. [CrossRef]

10. Ni, W.; Sun, G.; Pang, Y.; Zhang, Z.; Liu, J.; Yang, A.; Wang, Y.; Zhang, D. Mapping Three-Dimensional Structures of Forest Canopy Using UAV Stereo Imagery: Evaluating Impacts of Forward Overlaps and Image Resolutions with LiDAR Data as Reference. *IEEE J. Sel. Top. Appl. Earth Obs. Remote Sens.* **2018**, *11*, 3578–3589. [CrossRef]

11. Hmimina, G.; Dufrêne, E.; Pontailler, J.Y.; Delpierre, N.; Aubinet, M.; Caquet, B.; de Grandcourt, A.; Burban, B.; Flechard, C.; Granier, A.; et al. Evaluation of the potential of MODIS satellite data to predict vegetation phenology in different biomes: An investigation using ground-based NDVI measurements. *Remote Sens. Environ.* **2013**, *132*, 145–158. [CrossRef]

12. Bórnez, K.; Descals, A.; Verger, A.; Peñuelas, J. Land surface phenology from VEGETATION and PROBA-V data. Assessment over deciduous forests. *Int. J. Appl. Earth Obs. Geoinf.* **2020**, *84*, 101974. [CrossRef]

13. Baldocchi, D.D.; Black, T.A.; Curtis, P.S.; Falge, E.; Fuentes, J.D.; Granier, A.; Gu, L.; Knohl, A.; Pilegaard, K.; Schmid, H.P.; et al. Predicting the onset of net carbon uptake by deciduous forests with soil temperature and climate data: A synthesis of FLUXNET data. *Int. J. Biometeorol.* **2005**, *49*, 377–387. [CrossRef]

14. Jönsson, P.; Eklundh, L. TIMESAT—A program for analyzing time-series of satellite sensor data. *Comput. Geosci.* **2004**, *30*, 833–845. [CrossRef]

15. Sobrino, J.A.; Julien, Y. Global trends in NDVI-derived parameters obtained from GIMMS data. *Int. J. Remote Sens.* **2011**, *32*, 4267–4279. [CrossRef]

16. Richardson, A.D.; Keenan, T.F.; Migliavacca, M.; Ryu, Y.; Sonnentag, O.; Toomey, M. Climate change, phenology, and phenological control of vegetation feedbacks to the climate system. *Agric. For. Meteorol.* **2013**, *169*, 156–173. [CrossRef]

17. Misra, G.; Asam, S.; Menzel, A. Ground and satellite phenology in alpine forests are becoming more heterogeneous across higher elevations with warming. *Agric. For. Meteorol.* **2021**, *303*, 108383. [CrossRef]

18. Broich, M.; Huete, A.; Paget, M.; Ma, X.; Tulbure, M.; Coupe, N.R.; Evans, B.; Beringer, J.; Devadas, R.; Davies, K.; et al. A spatially explicit land surface phenology data product for science, monitoring and natural resources management applications. *Environ. Model. Softw.* **2015**, *64*, 191–204. [CrossRef]

19. Julien, Y.; Sobrino, J.A. Global land surface phenology trends from GIMMS database. *Int. J. Remote Sens.* **2009**, *30*, 3495–3513. [CrossRef]

20. Zeng, L.; Wardlow, B.D.; Xiang, D.; Hu, S.; Li, D. A review of vegetation phenological metrics extraction using time-series, multispectral satellite data. *Remote Sens. Environ.* **2020**, *237*, 11511. [CrossRef]

21. Du, J.; Li, K.; He, Z.; Chen, L.; Lin, P.; Zhu, X. Daily minimum temperature and precipitation control on spring phenology in arid-mountain ecosystems in China. *Int. J. Climatol.* **2020**, *40*, 2568–2579. [CrossRef]

22. Zhang, X.; Zhai, P.; Huang, J.; Zhao, X.; Dong, K. Responses of ecosystem water use efficiency to spring snow and summer water addition with or without nitrogen addition in a temperate steppe. *PLoS ONE* **2018**, *13*, e0194198. [CrossRef]

23. Zhang, X.; Friedl, M.A.; Schaaf, C.B.; Strahler, A.H. Climate controls on vegetation phenological patterns in northern mid- and high latitudes inferred from MODIS data. *Glob. Chang. Biol.* **2004**, *10*, 1133–1145. [CrossRef]

24. Wang, Y.; Li, G.; Ding, J.; Guo, Z.; Tang, S.; Liu, R.; Chen, J. A combined GLAS and MODIS estimation of the global distribution of mean forest canopy height. *Remote Sens. Environ.* **2016**, *174*, 24–43. [CrossRef]

25. Araya, S.; Ostendorf, B.; Lyle, G.; Lewis, M. CropPhenology: An R package for extracting crop phenology from time series remotely sensed vegetation index imagery. *Ecol. Inform.* **2018**, *46*, 45–56. [CrossRef]

26. Hill, M.J.; Donald, G.E. Estimating spatio-temporal patterns of agricultural productivity in fragmented landscapes using AVHRR NDVI time series. *Remote Sens. Environ.* **2003**, *84*, 367–384. [CrossRef]

27. Lloyd, D. A phenological classification of terrestrial vegetation cover using shortwave vegetation index imagery. *Int. J. Remote Sens.* **1990**, *11*, 2269–2279. [CrossRef]

28. White, M.A.; Nemani, R.R. Real-time monitoring and short-term forecasting of land surface phenology. *Remote Sens. Environ.* **2006**, *104*, 43–49. [CrossRef]

29. Huang, X.; Liu, J.; Zhu, W.; Atzberger, C.; Liu, Q. The optimal threshold and vegetation index time series for retrieving crop phenology based on a modified dynamic threshold method. *Remote Sens.* **2019**, *11*, 2725. [CrossRef]

30. Friedl, M. MODIS Collection 5 global land cover: Algorithm refinements and characterization of new datasets. *Remote Sens. Environ.* **2010**, *114*, 168–182. [CrossRef]

31. Sharifi, E.; Steinacker, R.; Saghafian, B. Assessment of GPM-IMERG and other precipitation products against gauge data under different topographic and climatic conditions in Iran: Preliminary results. *Remote Sens.* **2016**, *8*, 135. [CrossRef]

32. Sun, S.; Song, Z.; Wu, X.; Wang, T.; Wu, Y.; Du, W.; Che, T.; Huang, C.; Zhang, X.; Ping, B.; et al. Spatio-temporal variations in water use efficiency and its drivers in China over the last three decades. *Ecol. Indic.* **2018**, *94*, 292–304. [CrossRef]

33. Huete, A.; Justice, C.; Liu, H. Development of vegetation and soil indices for MODIS-EOS. *Remote Sens. Environ.* **1994**, *49*, 224–234. [CrossRef]
34. Noumonvi, K.D.; Oblišar, G.; Žust, A.; Vilhar, U. Empirical approach for modelling tree phenology in mixed forests using remote sensing. *Remote Sens.* **2021**, *13*, 3015. [CrossRef]
35. Belda, S.; Pipia, L.; Morcillo-Pallarés, P.; Verrelst, J. Optimizing Gaussian Process Regression for Image Time Series Gap-Filling and Crop Monitoring. *Agronomy* **2020**, *10*, 618. [CrossRef]
36. Belda, S.; Pipia, L.; Morcillo-Pallarés, P.; Rivera-Caicedo, J.P.; Amin, E.; De Grave, C.; Verrelst, J. DATimeS: A machine learning time series GUI toolbox for gap-filling and vegetation phenology trends detection. *Environ. Model. Softw.* **2020**, *127*, 104666. [CrossRef]
37. Borges, E.F.; Sano, E.E.; Medrado, E. Radiometric quality and performance of TIMESAT for smoothing moderate resolution imaging spectroradiometer enhanced vegetation index time series from western Bahia State, Brazil. *J. Appl. Remote Sens.* **2014**, *8*, 083580. [CrossRef]
38. García-Haro, F.J.; Campos-Taberner, M.; Muñoz-Marí, J.; Laparra, V.; Camacho, F.; Sánchez-Zapero, J.; Camps-Valls, G. Derivation of global vegetation biophysical parameters from EUMETSAT Polar System. *ISPRS J. Photogramm. Remote Sens.* **2018**, *139*, 57–74. [CrossRef]
39. Yu, X.; Zhuang, D.; Chen, H.; Hou, X. Forest classification based on MODIS time series and vegetation phenology. *Int. Geosci. Remote Sens. Symp.* **2004**, *4*, 2369–2372. [CrossRef]
40. Shrestha, U.B.; Gautam, S.; Bawa, K.S. Widespread climate change in the Himalayas and associated changes in local ecosystems. *PLoS ONE* **2012**, *7*, e36741. [CrossRef]
41. Jönsson, P.; Eklundh, L. Seasonality extraction by function fitting to time-series of satellite sensor data. *IEEE Trans. Geosci. Remote Sens.* **2002**, *40*, 1824–1832. [CrossRef]
42. Myneni, R.B.; Keeling, C.D.; Tucker, C.J.; Asrar, G.; Nemani, R.R. Increased plant growth in the northern high latitudes from 1981 to 1991. *Nature* **1997**, *386*, 698–702. [CrossRef]
43. Li, H.; Jiang, J.; Chen, B.; Li, Y.; Xu, Y.; Shen, W. Pattern of NDVI-based vegetation greening along an altitudinal gradient in the eastern Himalayas and its response to global warming. *Environ. Monit. Assess.* **2016**, *188*, 186. [CrossRef]
44. Wang, X.; Piao, S.; Ciais, P.; Li, J.; Friedlingstein, P.; Koven, C.; Chen, A. Spring temperature change and its implication in the change of vegetation growth in North America from 1982 to 2006. *Proc. Natl. Acad. Sci. USA* **2011**, *108*, 1240–1245. [CrossRef]
45. Piao, S. Growing season extension and its impact on terrestrial carbon cycle in the Northern Hemisphere over the past 2 decades. *Glob. Biogeochem. Cycles* **2007**, *21*, 1–11. [CrossRef]
46. Zeng, Z. Regional air pollution brightening reverses the greenhouse gases induced warming-elevation relationship. *Geophys. Res. Lett.* **2015**, *42*, 4563–4572. [CrossRef]
47. Hart, R.; Salick, J.; Ranjitkar, S.; Xu, J. Herbarium specimens show contrasting phenological responses to Himalayan climate. *Proc. Natl. Acad. Sci. USA* **2014**, *111*, 10615–10619. [CrossRef]
48. Menzel, A.; Sparks, T.H.; Estrella, N.; Koch, E.; Aaasa, A.; Ahas, R.; Alm-Kübler, K.; Bissolli, P.; Braslavská, O.; Briede, A.; et al. European phenological response to climate change matches the warming pattern. *Glob. Chang. Biol.* **2006**, *12*, 1969–1976. [CrossRef]
49. Parmesan, C. Influences of species, latitudes and methodologies on estimates of phenological response to global warming. *Glob. Chang. Biol.* **2007**, *13*, 1860–1872. [CrossRef]
50. Fitter, A.H.; Fitter, R.S.R. Rapid changes in flowering time in British plants. *Science* **2002**, *296*, 1689–1691. [CrossRef]
51. Pellerin, M.; Delestrade, A.; Mathieu, G.; Rigault, O.; Yoccoz, N.G. Spring tree phenology in the Alps: Effects of air temperature, altitude and local topography. *Eur. J. For. Res.* **2012**, *131*, 1957–1965. [CrossRef]
52. Shen, M.; Zhang, G.; Cong, N.; Wang, S.; Kong, W.; Piao, S. Increasing altitudinal gradient of spring vegetation phenology during the last decade on the Qinghai-Tibetan Plateau. *Agric. For. Meteorol.* **2014**, *189*, 71–80. [CrossRef]
53. Du, J.; He, Z.; Piatek, K.B.; Chen, L.; Lin, P.; Zhu, X. Interacting effects of temperature and precipitation on climatic sensitivity of spring vegetation green-up in arid mountains of China. *Agric. For. Meteorol.* **2019**, *269*, 71–77. [CrossRef]
54. Peng, H.; Xia, H.; Chen, H.; Zhi, P.; Xu, Z. Spatial variation characteristics of vegetation phenology and its influencing factors in the subtropical monsoon climate region of southern China. *PLoS ONE* **2021**, *16*, e0250825. [CrossRef]
55. Suonan, J.; Classen, A.T.; Sanders, N.J.; He, J.-S. Plant phenological sensitivity to climate change on the Tibetan Plateau and relative to other areas of the world. *Ecosphere* **2019**, *10*, e02543. [CrossRef]
56. Linkosalo, T.; Lappalainen, H.K.; Hari, P. A comparison of phenological models of leaf bud burst and flowering of boreal trees using independent observations. *Tree Physiol.* **2008**, *28*, 1873–1882. [CrossRef]
57. Luedeling, E.; Gebauer, J.; Buerkert, A. Climate change effects on winter chill for tree crops with chilling requirements on the Arabian Peninsula. *Clim. Chang.* **2009**, *96*, 219–237. [CrossRef]
58. Green, P.B.; Cummins, W.R. Growth Rate and Turgor Pressure: Auxin Effect Studies with an Automated Apparatus for Single Coleoptiles 1. *Plant Physiol.* **1974**, *54*, 863. [CrossRef]
59. Bisht, V.K.; Kuniyal, C.P.; Bhandari, A.K.; Nautiyal, B.P.; Prasad, P. Phenology of plants in relation to ambient environment in a subalpine forest of Uttarakhand, western Himalaya. *Physiol. Mol. Biol. Plants* **2014**, *20*, 399–403. [CrossRef]
60. Chmielewski, F.M.; Rotzer, T. Response of tree phenology to climate change across Europe. *Agric. For. Meteorol.* **2001**, *108*, 101–112. [CrossRef]

61. Delbart, N.; Le Toan, T.; Kergoat, L.; Fedotova, V. Remote sensing of spring phenology in boreal regions: A free of snow-effect method using NOAA-AVHRR and SPOT-VGT data (1982-2004). *Remote Sens. Environ.* **2006**, *101*, 52–62. [CrossRef]

62. Yu, H.; Luedeling, E.; Xu, J. Winter and spring warming result in delayed spring phenology on the Tibetan Plateau. *Proc. Natl. Acad. Sci. USA* **2010**, *107*, 22151–22156. [CrossRef]

63. Henebry, G.M.; de Beurs, K.M. Remote Sensing of Land Surface Phenology: A Prospectus. In *Phenology: An Integrative Environmental Science*; Schwartz, M.D., Ed.; Springer Netherlands: Dordrecht, The Netherlands, 2013; pp. 385–411. ISBN 978-94-007-6925-0.

64. White, K.; Pontius, J.; Schaberg, P. Remote sensing of spring phenology in northeastern forests: A comparison of methods, field metrics and sources of uncertainty. *Remote Sens. Environ.* **2014**, *148*, 97–107. [CrossRef]

65. Liu, Q.; Fu, Y.H.; Zhu, Z.; Liu, Y.; Liu, Z.; Huang, M.; Janssens, I.A.; Piao, S. Delayed autumn phenology in the Northern Hemisphere is related to change in both climate and spring phenology. *Glob. Chang. Biol.* **2016**, *22*, 3702–3711. [CrossRef]

66. Zhao, J.; Wang, Y.; Zhang, Z.; Zhang, H.; Guo, X.; Yu, S.; Du, W.; Huang, F. The variations of land surface phenology in Northeast China and its responses to climate change from 1982 to 2013. *Remote Sens.* **2016**, *8*, 400. [CrossRef]

67. Tang, H.; Li, Z.; Zhu, Z.; Chen, B.; Zhang, B.; Xin, X. Variability and climate change trend in vegetation phenology of recent decades in the Greater Khingan Mountain area, Northeastern China. *Remote Sens.* **2015**, *7*, 11914–11932. [CrossRef]

68. Yang, Y.; Guan, H.; Shen, M.; Liang, W.; Jiang, L. Changes in autumn vegetation dormancy onset date and the climate controls across temperate ecosystems in China from 1982 to 2010. *Glob. Chang. Biol.* **2015**, *21*, 652–665. [CrossRef]

69. Stöckli, R.; Vidale, P.L. European plant phenology and climate as seen in a 20-year AVHRR land-surface parameter dataset. *Int. J. Remote Sens.* **2004**, *25*, 3303–3330. [CrossRef]

Article

Timescale Effects of Radial Growth Responses of Two Dominant Coniferous Trees on Climate Change in the Eastern Qilian Mountains

Changliang Qi [1,2], Liang Jiao [1,2,*], Ruhong Xue [1,2], Xuan Wu [1,2] and Dashi Du [1,2]

[1] College of Geography and Environmental Science, Northwest Normal University, No. 967, Anning East Road, Lanzhou 730070, China; sd2019212378@163.com (C.Q.); xrhnwnu@163.com (R.X.); 15002550410@163.com (X.W.); dashidu@163.com (D.D.)

[2] Key Laboratory of Resource Environment and Sustainable Development of Oasis, Gansu Northwest Normal University, Lanzhou 730070, China

* Correspondence: jiaoliang@nwnu.edu.cn; Tel.: +86-139-1935-0195

Abstract: To explore the difference in the response of the radial growth of *Pinus tabulaeformis* and *Picea crassifolia* on different timescales to climate factors in the eastern part of Qilian Mountains, we used dendrochronology to select four different timescales (day, pentad (5 days), dekad (10 days), and month) for exploration. The primary conclusions were as follows: (1) According to an investigation of the dynamic correlations between radial growth and climate conditions, drought during the growing season has been the dominant limiting factor for radial growth across both species in recent decades; (2) climate data at the dekad scale are best for examining the correlations between radial growth and climate variables; and (3) based on basal area increment, *P. tabuliformis* in the study area showed a trend of first an increase and then a decrease, while *P. crassifolia* showed a trend of continuous increase (BAI). As the climate continues to warm in the future, forest ecosystems in arid and semi-arid areas will be more susceptible to severe drought, which will lead to a decline in tree growth, death, and community deterioration. As a result, it is critical to implement appropriate management approaches for various species based on the peculiarities of their climate change responses.

Keywords: dendroecology; dominant coniferous tree; timescale; growth patterns; eastern Qilian Mountains

Citation: Qi, C.; Jiao, L.; Xue, R.; Wu, X.; Du, D. Timescale Effects of Radial Growth Responses of Two Dominant Coniferous Trees on Climate Change in the Eastern Qilian Mountains. *Forests* **2022**, *13*, 72. https://doi.org/10.3390/f13010072

Academic Editors: Any Mary Petritan and Mirela Beloiu

Received: 18 October 2021
Accepted: 31 December 2021
Published: 5 January 2022

Publisher's Note: MDPI stays neutral with regard to jurisdictional claims in published maps and institutional affiliations.

1. Introduction

AR6 points out that based on historical observations and model simulations in the 21st century, land warming at the global scale has increased the demand for atmospheric evaporation and the intensity of drought events. The increase in land temperature was higher than that in the ocean, which affected the atmospheric circulation and reduced the relative humidity near the surface, leading to the occurrence of regional drought events [1]. Forest ecosystems have produced a strong response to global climate change, which has caused changes in the growth of large areas of trees, forest composition, and carbon assimilation [2]. Most studies on the impact of climate change on biodiversity have focused on the direct effects of temperature and precipitation. Fewer studies have looked at indirect effects, such as how climate change affects the forest structure and composition, which has an impact on species presence and abundance [3].

The impact of climate warming on the radial growth of trees varies with environmental differences. In humid areas, higher temperatures can promote tree growth by accelerating the onset of xylem cells or accelerating the rate of photosynthesis [4]. Alternatively, a warming climate may trigger drought events, and in dry regions, trees have a reduced ability to absorb water due to reduced xylem carbohydrates, leading to reduced growth [5,6]. In addition, different forest species in different regions show different ecological adaptation mechanisms to climate change [7]. Some tree species have shown divergent responses to global climate change, indicating unstable responses to limiting environmental condition factors over time, like in spruce and Abies lasiocarpa in opaque gem parkland in Canada, black pine, and *P. tabuliformis* and *Sabina przewalskii* in northwest China [8–11]. However, European larch in the European Alps, *P. nigra* in Mediterranean hilly areas, and *P. crassifolia* in northwest China's mid-latitudes show fairly consistent responses to limiting climatic variables. Global warming is one of the major causes of reduced tree growth and death in arid and semi-arid regions [12–15]. There is a considerable number of studies that present a different perspective, and the results suggest that in semi-arid environments, elevated atmospheric CO_2 concentrations positively affect the radial growth rate of naturally growing trees and that the growth of trees is affected not only by climatic factors but also by other non-climatic factors [16–18]. Therefore, there is great significance in fully understanding the spatial and temporal responses patterns of forest ecosystems to climate transition, which can be tailored to provide a scientific basis for forest management schemes [19,20].

The responses of trees' radial growth to climate factors differ across timescales [21]. Traditional dendrochronology studies on growth–climate relationships usually focus on long monthly or annual timescales, excluding the short dekad (10 days), pentad (5 days), and daily timescales [10,22,23]. The trees' continuous development process is disturbed physically, and some climatic indicators of tree chronology are often lost [24]. Alternatively, a more accurate climate signal can be obtained from tree chronology if the tree response to climate is on an appropriate timescale that is consistent with tree growth patterns. The results of this study of growth patterns on short timescales provide interpretation and validation for studies of radial tree growth on long timescales. Daily climatic variables, particularly daily temperature in the pre- and early growth seasons, play an essential influence in shaping tree ring width [25]. Dendroclimatology has also used 5-day and 10-day time frames, in addition to daily timescales. For example, using the correlation between the tree ring width index and the mean temperature on a pentad scale, the important phenological interval of the tree-growing season in Eurasia was determined [26]. Tree growth and climate factors in early summer (17 June to 11 July) in Eastern Tamil and Putoran were reconstructed on a small timescale [27]. However, all the above studies were conducted within a single timescale. In addition, the results of tree radial growth response at different timescales remain uncertain. Therefore, it is necessary to analyze the relationship between tree growth and climate factors at different timescales.

Different tree species in the same habitat will lead to different responses to climate change [28]. For example, in northern Ontario, Canada, it was found that increases in temperature during the growing season promote the radial growth of the North American *Pinus banksiana* but cause a decline in the growth of *Picea mariana* [29]. Climate change in Europe has contributed to the growth of birch and beech, as the warming climate has caused the trees to start growing earlier and end later, lengthening the entire growing season [30]. In the Tianshan Mountains of northern China, *Picea schrenkiana* was more susceptible to drought than *Larix sibirica* because of variations in resource allocation trade-offs [11]. These findings suggest that there are significant variations in the response patterns of different tree species to climate change, as they relate to distinct ecological adaptation methods. As a result, additional tree species must be studied to determine the link between tree growth and climate change.

The Qilian Mountains (36–40° N, 94–103° E) are located on the northeastern edge of the Qinghai–Tibet Plateau in northwestern China, a transition zone between the arid west and the humid east. It is an important water source and runoff area for the oasis of the Hexi Corridor and the Yellow River Basin, as well as an important ecological security barrier in the western region of China. Forests of *P. crassifolia* and *P. tabuliformis* in the Qilian Mountains play important roles in water conservation and biodiversity protection. Therefore, it is important to study the responses of forest ecosystems in this region to climate change and global warming [31,32]. In recent decades, studies in the Qilian Mountains have focused on the evaluation of tree rings and reconstruction of the relationship between climate factors [33–36]. There are many studies on the middle and western parts of Qilian Mountains [37,38], and were few have been performed in the eastern Qilian Mountains [39–41]. The response patterns and mechanisms of radial growth under climate change are still not clear for different timescales and species. To determine the timescale effects of patterns in radial growth responses in two dominant coniferous trees under climate change in the eastern Qilian Mountains, we (1) analyzed the relationship between the radial growth of the two tree species at the daily, climatic, annual, and monthly timescales and the response to climate factors and (2) compared the differences in the interannual changes in the radial growth of the two tree species and clarified the growth of different tree species under climate change patterns and future development trends.

2. Material and Methods

2.1. Study Area and Climate

The study area was in Tulu Trench Forest Park (36°40′–36°44′ N, 102°36′–102°45′ E) in the eastern Qilian Mountains (Figure 1). It has a typical temperate continental monsoon climate, with rainy and hot periods (Figure 2). The Wushaoling meteorological station (37.2° N, 102.87° E; 3045.1 m), which is close to the sampling point, was selected for meteorological data. The mean annual precipitation was 414.8 mm from 1960 to 2018, and more than 60% of that occurred in June–September. The mean monthly temperature ranges from −11.8 °C in January to 11.6 °C in July. The annual total solar radiation is 469 kJ/cm^2, the frost-free period is only one-third of the year, and the annual evaporation is much greater than the annual precipitation. The study area is rich in vegetation resources and presents obvious vertical zoning, transitioning from low-altitude forests to high-altitude meadows; trees mainly include coniferous species, such as Qilian Qinghai spruce, *P tabulaeformis*, juniper, and *Picea wilsorii*, as well as *Populus davidiana* and *Betula albosinensis*. In addition, *Betula platyphylla* and other broad-leaved trees are present, and shrubs, such as Cotoneaster, Rhododendron and *Potentilla*, and herbaceous plants, such as *Kobresia pygmaea*, are also abundant [42].

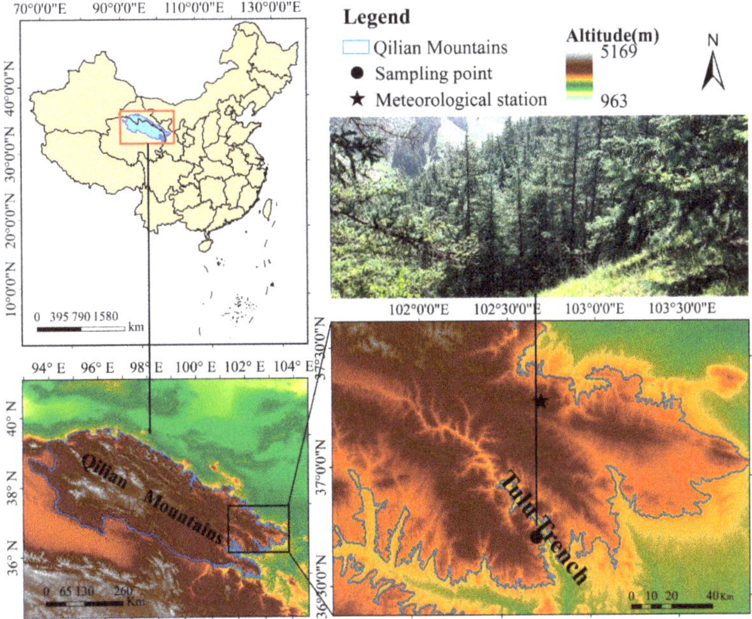

Figure 1. Locations of sampling sites and the nearest meteorological station (Wushaoling meteorological station).

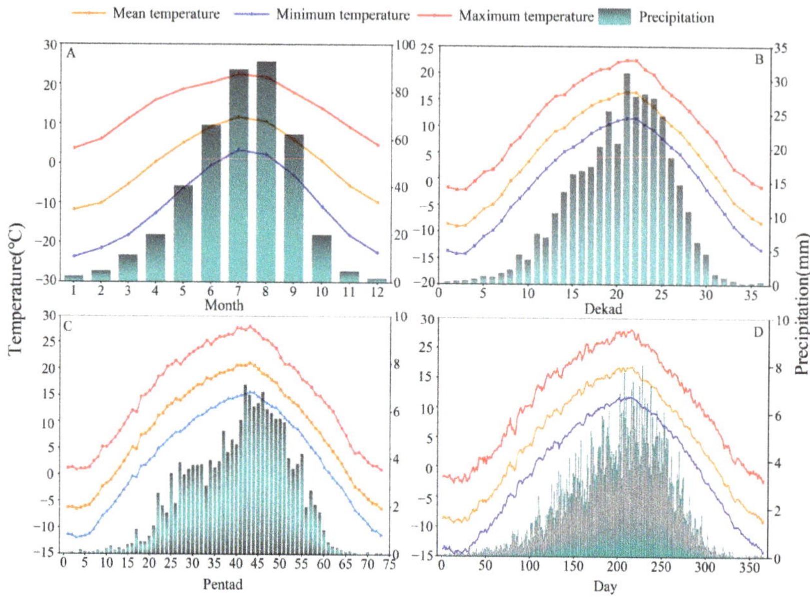

Figure 2. Variation trends in climate factors in the study area at daily, pentad, dekad, and monthly timescales during 1960–2018. (**A**) Trend in climate factors at the monthly scale, (**B**) change trend in climate factors at the dekad scale, (**C**) change trend in climate factors at the pentad scale, and (**D**) change trend in climate factors at the daily scale.

2.2. Field Sampling and Chronology Establishment

In August 2019, the collection of tree ring samples was carried out in the eastern part of Qilian Mountains. The sampling objects were the dominant species of *P. tabulaeformis* and *P. crassifolia* in this area, and their growth environment is similar. The specific sampling location is shown in Figure 1. To avoid the interference of non-climatic factors on the radial growth of trees, a total of 50 trees with good growth conditions were selected for the 2 tree species. Among them, there were 25 *P. tabulaeformis* and *P. crassifolia*, each with 2 cores. Sampling information is shown in Table 1. The collected sample cores were packaged in paper tubes, and after the relevant information about the sampling points was indicated, they were brought back to the laboratory for further processing.

Table 1. Information of tree ring sampling sites in this study.

Tree Species	*Pinus tabuliformis*	*Picea crassifolia*
Elevation (m)	2260 (m)	2903 (m)
Longitude (E)	102°44.25′ E	102°44.25′ E
Latitude (N)	36°41.18′ N	36°41.27′ N
Slope (°)	45°	36°
Aspect	North	North
Tree spacing (m)	3.0	4.5
Average diameter at breast height (cm)	32.4	33.8
Average tree height (m)	16.0	14.5
Canopy (m)	2.65	3.65

The tree ring samples were taken back to the laboratory for preliminary processing. First, the sample cores were fixed in a special wooden trough with latex. After the samples were dried naturally, they were polished with sandpaper until the rings were clearly visible, and then they were placed under a binocular microscope for visual dating. The tree ring width was measured with a LINTBA tree ring width meter with 0.001 mm accuracy (TM5, Rinntech, Heidelberg, Germany), followed by dating with the COFECHA program to ensure the accuracy of all sample dating and tree ring width measurements [43]. Finally, the ARSTAN program was used to establish the chronology [44], the spline function was used to fit, and influencing factors other than climate factors were removed. Finally, the tree ring width standard chronology (STD), difference chronology (RES), and autoregressive chronology (ARS) were established. The standard chronology was selected for analysis in this study.

2.3. Meteorological Data

For the meteorological data, the Wushaoling meteorological station (37°20′ N, 102°87′ E, 1960–2018, Figure 3), with an altitude of 3045 m, was nearest to the sampling point (about 57 km). Tree growth is not only affected by the climate of the current year but also by the climate of the previous year. Therefore, this study adopted September of the previous year to October of the current year as the analysis interval of meteorological indicators. The responses of tree growth to climate factors at different timescales were analyzed based on meteorological station data.

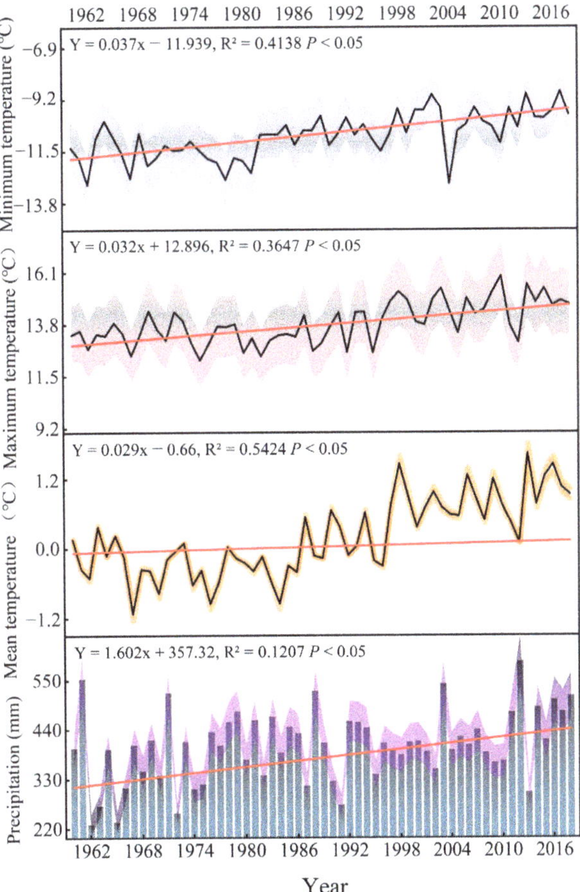

Figure 3. Interannual variations of temperature and precipitation in the study area from 1960 to 2018 (blue, red, and orange, and purple, respectively, indicate the inter-annual variation error range of temperature and precipitation).

2.4. Methods

SThe daily data were used to calculate the pentad, dekad, and monthly climatic values. We first generated the daily average of each climatic factor from 1960 to 2018 (Figure 2D). The total precipitation and mean temperature of 28 and 29 February in the leap year were used as precipitation and temperature of 28 February. There are 73 pentads in a year. For example, the first pentad is 1–5 January, and the 73rd pentad is 27–31 December [45]. The pentad precipitation is the total precipitation over a 5-day period, and the temperature is the mean temperature of those 5 days (Figure 2C). There are three dekads in a month and 36 dekads in a year. The first dekad of a month extends from the first to the tenth day of the month; the following 10 days are the second dekad; and the 21st to the last day of the month is the third dekad. Similar to the pentad, the dekad precipitation is the sum of the daily precipitation, and the temperature is the mean temperature of the 10 days (Figure 2B).

At the daily, pentad, dekad, and monthly timescales, the Pearson correlation analysis method was used to identify the growth response of the trees to climate and mainly control of climate factors. To assess the radial growth trend, we calculated the basal area increment (*BAI*) based on the original tree ring width chronology. *BAI* is a variable with biological significance that quantifies the radial growth rate and trend of trees without the influence of detrending that is generally used to produce the chronology [46]. *BAI* was calculated using the following formula:

$$BAI_t = BA_t - BA_{t-1} = \pi\left(\left(R_{t-1} + TRW_t\right)^2 - \left(R_{t-1}\right)^2\right)$$

where the *BA* variable represents consecutive cross-sectional basal areas, *R* is the core length measured for dated tree rings formed in year $t - 1$, and *TRW* is the measured raw width of the tree ring in year *t*.

3. Results

3.1. Chronological Parameter Analyses

Table 2 shows the chronological (STD) statistical parameters of both *P. crassifolia* and *P. tabuliformis* in the study area. The high SNR (21.334 and 15.410) and EPS (0.965 and 0.939, SSS > 0.850) values of the two tree species indicate that the chronologies established had high reliability and that the samples contained more climate information, making them suitable for the study of tree ring ecology. The high values of MS, PC1, SD, and SNR indicate that the annual fluctuations in the two chronologies were consistent and were sensitive to climate change. Among them, the SNR and EPS values of the chronology of *P. crassifolia* were higher than those of *P. tabuliformis*, indicating that the chronology of *P. crassifolia* contains more climate information.

Table 2. Statistical parameters of the standard chronology.

Parameters	*Pinus tabuliformis*	*Picea crassifolia*
Cores/trees	50/25	42/25
Chronology span	1898–2018	1836–2018
Standard deviation (SD)	0.239	0.771
Variance in the first principal comment (PC1)	0.407	0.318
Mean sensitivity (MS)	0.186	0.125
First-order serial autocorrelation (AC1)	0.613	0.696
Mean correlation for all series (R)	0.150	0.253
Mean correlation within trees (R1)	0.620	0.655
Mean correlation between trees (R2)	0.135	0.568
Signal-to-noise ratio (SNR)	15.410	21.334
Expressed population signal (EPS)	0.939	0.955
First year of SSS > 0.85 (number of trees)	1915(23)	1895(13)

3.2. Radial Growth Response of the Climate Factors of the Two Tree Species at Four Timescales

3.2.1. Growth–Climate Relationships at the Month Scale

The radial growth of *P. tabuliformis* was significantly negatively correlated with precipitation in February ($r = -0.289$, $p < 0.05$) and October ($r = -0.385$, $p < 0.01$) of the current year. Additionally, the radial growth of *P. tabuliformis* was significantly negatively correlated with temperature in the growing season, showing correlations with the mean temperature in September ($r = -0.431$, $p < 0.01$) of the previous year and June ($r = -0.357$, $p < 0.01$) and July ($r = -0.436$, $p < 0.01$) of the current year, the mean maximum temperature in May ($r = -0.341$, $p < 0.01$) and July ($r = -0.438$, $p < 0.01$) of the current year, and the mean minimum temperature in July ($r = -0.288$, $p < 0.05$) of the current year.

Radial growth of *P. crassifolia* was not significantly correlated with precipitation (Figure 4). The radial growth of Qinghai spruce was significantly negatively correlated with temperature at the end of the previous year's growing season and the current year's growing season (mean temperature P9: $r = -0.326$, $p < 0.01$, C1: $r = -0.326$, $p < 0.05$, C2: $r = -0.326$, $p < 0.01$, C3: $r = -0.326$, $p < 0.01$; maximum temperature C2: $r = -0.288$, $p < 0.05$; C3: $r = -0.379$, $p < 0.01$; C6: $r = -0.513$, $p < 0.01$; minimum temperature C6: $r = -610$, $p < 0.01$, C8: $r = -0.302$, $p < 0.05$).

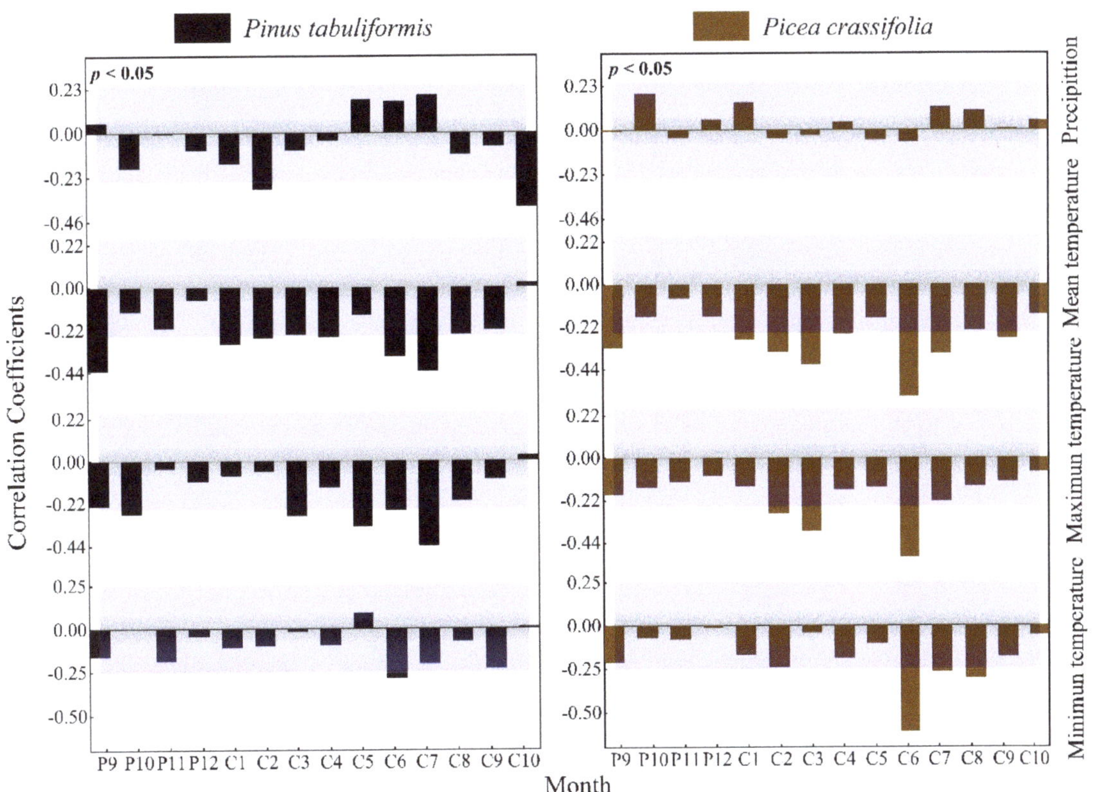

Figure 4. Response of the radial growth of trees to monthly values of climatic factors (Beyond the top of the purple shade, *p* values are positively correlated ($p < 0.05$), lower than the part in purple shadow indicates that *p* values are negatively correlated ($p < 0.05$).

3.2.2. Effect of Dekad Time Units on the Expression of Correlations

The radial growth of *P tabulaeformis* was significantly positively correlated with the precipitation in June and July of the current growing season, and it had a strong negative response to temperature mainly during the growing season; the radial growth of *P. crassifolia* was in the early third and fourth parts of the growing season. The monthly precipitation was negatively correlated. It had a strong negative response relationship with temperature mainly in the early and middle growing season (Figure 5).

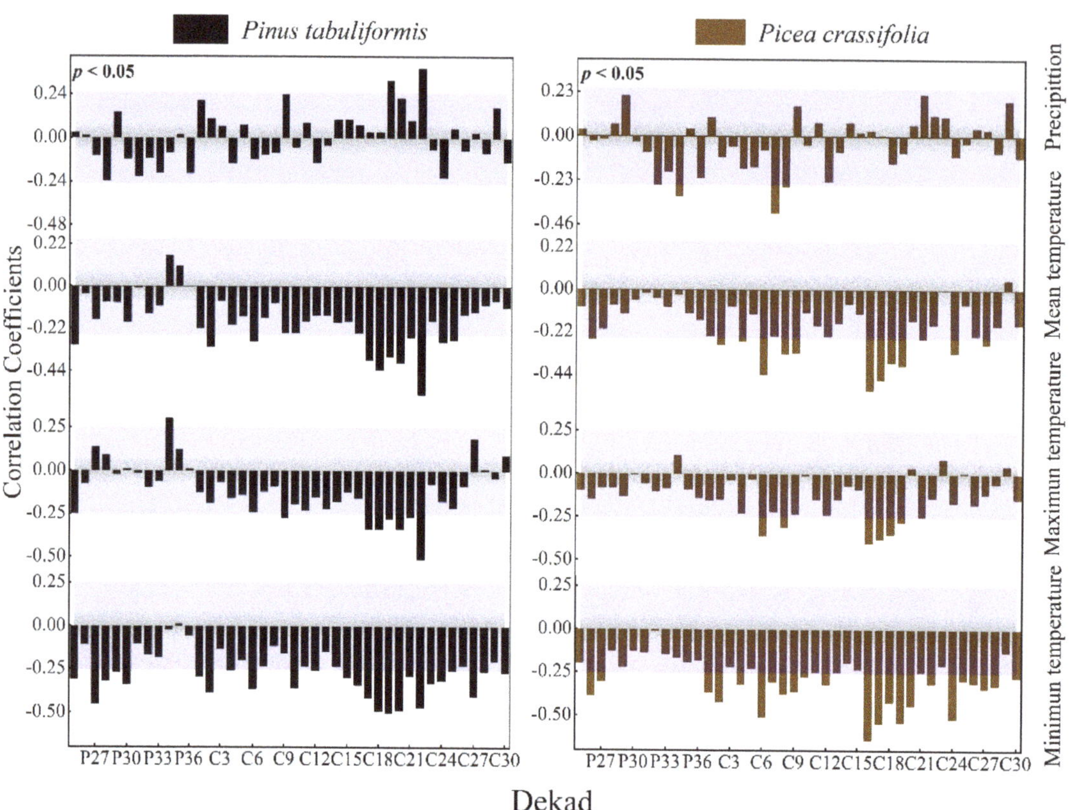

Figure 5. Response of the radial growth of trees to dekad values of climatic factors (Beyond the top of the purple shade, *p* values are positively correlated (*p* < 0.05), lower than the part in purple shadow indicates that *p* values are negatively correlated (*p* < 0.05).

3.2.3. The Effect of Waiting Time Units on the Expression of Correlation

The radial growth of *P tabulaeformis* was significantly positively correlated with the precipitation in the current growing season and negatively correlated with the temperature (average temperature, maximum temperature, minimum temperature) mainly in the current growing season and at the end of the previous growing season. *P. crassifolia* also had a similar response pattern (Figure 6).

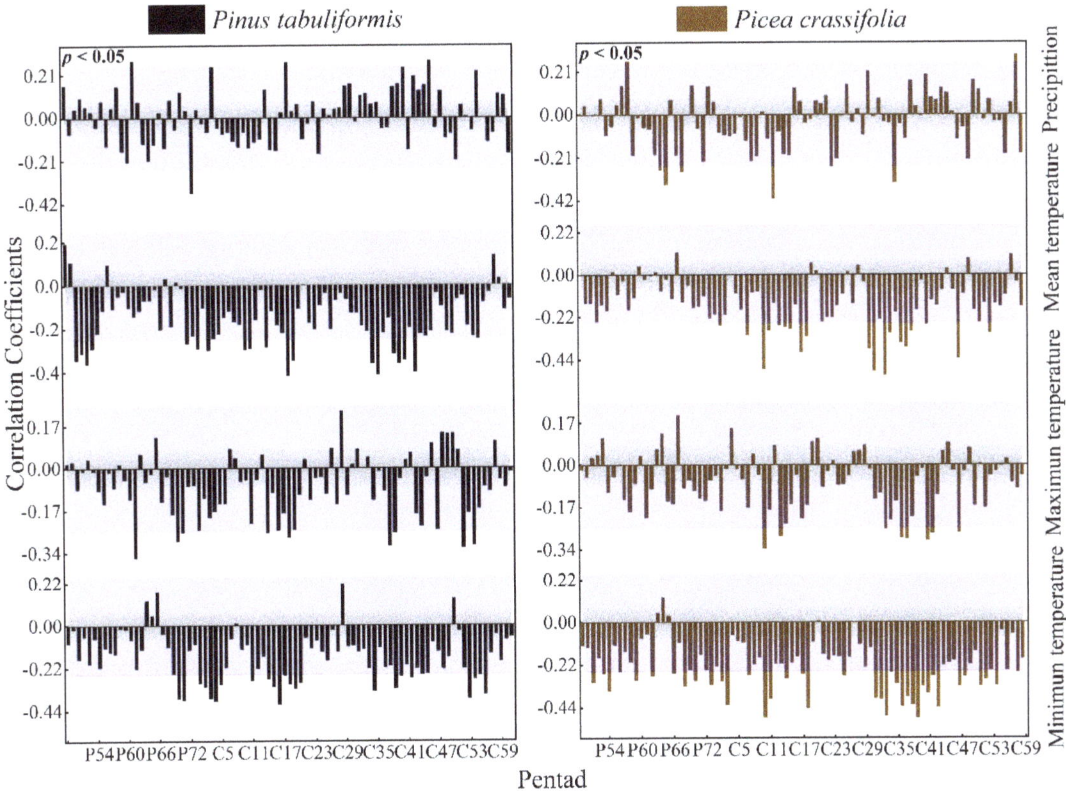

Figure 6. Response of the radial growth of trees to pentad values of climatic factors (Beyond the top of the purple shade, *p* values are positively correlated ($p < 0.05$), lower than the part in purple shadow indicates that *p* values are negatively correlated ($p < 0.05$).

3.2.4. Effect of Daily Time Units on Correlation Expression

The radial growth of the two dominant tree species in the study area showed similar correlations in daily climatic factors, dekad and pentad scales, mainly showing a positive correlation between precipitation in the growing season and a negative correlation between temperature in the first and middle of the growing season (Figure 7).

We calculated the correlation coefficients between the tree ring index and climate factors at various timescales to determine the key climate intervals that affect the growth of trees (Table 3). On all timescales, the tree ring index had the strongest correlation with the temperature index. At the same time, no matter how the timescale of precipitation or temperature changed, it showed that the absolute value of the correlation coefficient between the dekad-scale climate factor and the tree ring index was the largest.

3.2.5. Correlation between the Radial Growth of Different Trees and the Drought Index (SPEI)

The relationship between tree growth and the drought index (SPEI) is shown in Table 4. The results show that the radial growth of *P. tabulaeformis* was significantly positively correlated with the drought index at different timescales in the growing season, but there was no significant correlation between the radial growth of *P. crassifolia* and the drought index. This significant correlation indicated that the radial growth of *P. tabulaeformis* is more limited by drought stress in the sampling area.

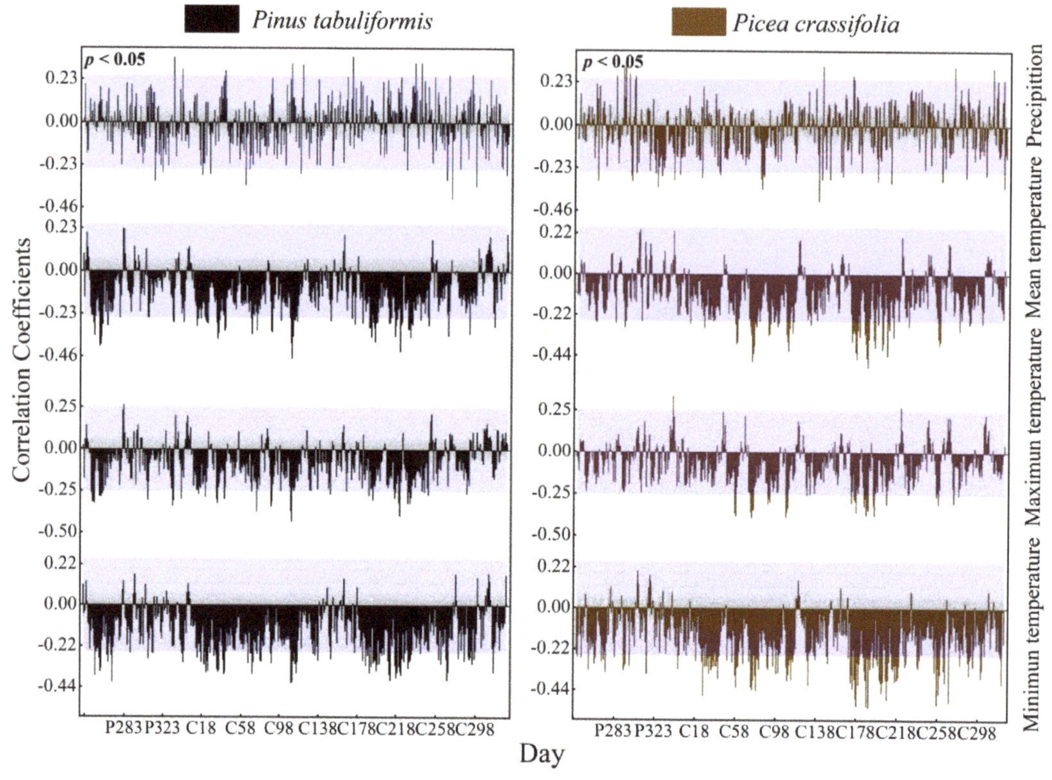

Figure 7. Response of the radial growth of trees to daily values of climatic factors (Beyond the top of the purple shade, p values are positively correlated ($p < 0.05$), lower than the part in purple shadow indicates that p values are negatively correlated ($p < 0.05$).

Table 3. The maximum correlation coefficient between the tree ring index and climate factors at different timescales.

Tree Species	Climate Element	Timescale	Correlation Coefficients
		Day	-0.431
		Pentad	-0.408
	Maximum temperature	Dekad	-0.499
		Month	-0.436
		Day	-0.503
		Pentad	-0.359
	Minimum temperature	Dekad	-0.511
		Month	-0.288
Pinus tabulaeformis		Day	-0.476
		Pentad	-0.418
	Mean temperature	Dekad	-0.563
		Month	-0.431
		Day	0.296
		Pentad	0.370
	Precipitation	Dekad	0.386
		Month	0.385

Table 3. *Cont.*

Tree Species	Climate Element	Timescale	Correlation Coefficients
Picea crassifolia	Maximum temperature	Day	−0.525
		Pentad	−0.493
		Dekad	−0.637
		Month	−0.379
	Minimum temperature	Day	−0.392
		Pentad	−0.303
		Dekad	−0.615
		Month	−0.610
	Mean temperature	Day	−0.500
		Pentad	−0.521
		Dekad	−0.532
		Month	−0.582
	Precipitation	Day	−0.377
		Pentad	−0.391
		Dekad	−0.402
		Month	0.190

Table 4. The correlation relationship between the radial growth of the two dominant coniferous species and the SPEI (1960–2018).

Month	3-Month Scale		6-Month Scale		12-Month Scale	
	PC	PT	PC	PT	PC	PT
Jan.	−0.151	0.017	−0.034	0.116	−0.154	0.061
Feb.	−0.097	0.032	−0.060	0.164	−0.159	0.026
Mar.	−0.016	0.119	−0.003	0.149	−0.104	0.035
Apr.	0.043	0.114	−0.003	0.149	−0.042	0.060
May	−0.095	0.294 *	−0.077	0.262 *	−0.048	0.160
Jun.	−0.215	0.284 *	−0.186	0.341 **	−0.052	0.312 *
Jul.	−0.164	0.394 **	−0.105	0.433 **	−0.095	0.405 **
Aug.	0.013	0.328 *	−0.028	0.410 **	−0.058	0.418 **
Sep.	0.080	0.281 *	−0.063	0.371 **	−0.039	0.430 **
Oct.	0.091	0.115	−0.052	0.337 **	−0.050	0.378 **

(* $p < 0.05$, ** $p < 0.01$; PC: *Picea crassifolia*, PT: *Pinus tabulaeformis*).

3.3. Radial Growth Characteristics of Trees

This study found that temperature is one in all the key factors limiting the radial growth of trees within the study area, which increased significantly at 0.428 °C/10a (Figure 3). However, different growth trends were observed in the two species (Figure 8). We saw an upward trend in the BAI of *P. crassifolia* from 1960 to 2018 (3.065 cm^2 yr^{-1}/10a/10a), while this factor increased first and then decreased for *P. tabuliformis*. Notably, the growth rate of *P. tabuliformis* has decreased significantly (−0.739 cm^2 yr^{-1}/10a) since the middle of the 20th century, given the significant increase in temperature.

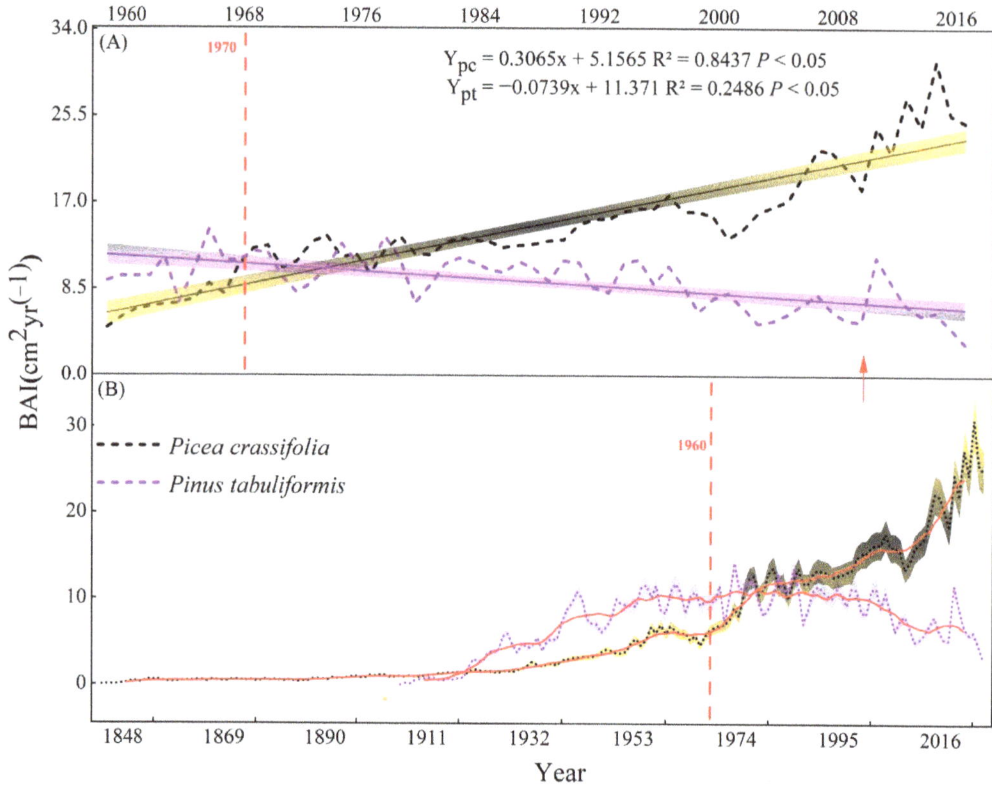

Figure 8. Interannual variation of the basal area increment (BAI) of two tree species. (**A**) Trend in growth since meteorological data have been recorded and (**B**) the entire time series of tree growth; the red line represents the 10-year moving average.

4. Discussion

4.1. Response Comparisons of the Radial Growth of Trees to Global Climate Change on Four Timescales

We found that there is a significant correlation between the radial growth of *P. tabuliformis* and precipitation in the study area, and the response of *P. crassifolia* to precipitation only appears on a smaller timescale (Figures 4–7). A significant negative correlation was found between precipitation and *P. tabuliformis*, while that with *P. crassifolia* was not obvious. However, different responses to climate factors may also be associated with differences between the growing environments of trees species. For example, *P. tabuliformis* grows mainly on sunny or semi-sunny slopes, while *P. crassifolia* grows mainly on shady slopes; this must be clear for the beginning. So the evaporation rates of the two species are different. Studies have shown that extremely high rainfall may affect environmental conditions, directly or indirectly limiting tree growth at specific times of the year [47]. The growth of the two dominant tree species in the study area was significantly and negatively correlated with temperature at four timescales. Although tree cambium activities stopped in the fall of the previous year, higher temperatures at this time may promote carbohydrate storage and the synthesis of organic substances in trees, which are beneficial to growth in the coming year [48,49]. In addition, the higher temperature in the early part of the growth period may cause the start date of the growing season to advance, making the trees grow wider rings [50]. Since the start and end dates of tree growth are basically determined, the correlation coefficients of trees growing on different timescales (day, pentad, dekad, month) and climate factors (temperature and precipitation) have certain similarities [39,46].

Many studies have confirmed the existence of transition zones between tree radial growth and climate on both temporal and spatial scales, and therefore, a single monthly scale climate change has not always been the best option in the past when studying the relationship between tree growth and climate change [51–53]. It has been reported that *Quecus ilex* growth in the Mediterranean is more correlated with precipitation at the small scale than on large timescales [54]. In Inner Mongolia, we also discovered more accurate results for the response connection of climate with tree growing at a time frame less than 1 month, where precipitation during 1 April–10 July was rebuilt [39]. For dendroclimatic models, the daily time frame was shown to be better. Tree growth models using daily climate data are better than monthly data in identifying special climate events that last only a few days but have a significant impact on the development of trees [55]. At the same time, the daily time unit also has some shortcomings, such as being too large a calculation. As shown in Table 3, the correlation coefficients between tree growth and climate factors at different timescales are similar. This indicates that the pentad and dekad timescales may be able to replace the daily scale. In addition, the pentad and dekad timescales have been widely used in studies of climate change and tree physiology [45,56–58]. However, according to our study, the response of tree growth to climate is more accurate on a dekad scale. To better comprehend the dekad scale's superiority, we discovered that the greatest correlation coefficient between the radial growth of trees and temperature in the 17–25th dekad period was larger than the monthly value for the same time. Meanwhile, the highest correlations between tree ring data and precipitation throughout the growing season increased from 0.296 to 0.386 at the dekad and monthly scales (Table 3). According to the principles of dendrophysiology and dendrophenology [47], the temperature from the 16th to the 25th dekad (June to August) is the main determinant of the radial growth of trees in the Qilian Mountains. Starting in early April every year, the average temperature is above 5 °C, and the trees will enter a period of rapid growth, the radial growth of trees into the rapid growth period [59]. In addition, the high temperature and more precipitation caused by climate change are conducive to the formation of cambium phenology, and the increase in and enlargement of cells are also conducive to the growth of trees and the formation of wide rings [34]. In conclusion, the assessment of growth responses in trees to climate factors at a smaller timescale avoids the problem of correlation relationships being concealed by an excessively large timescale. Therefore, climate factors at the dekad scale might be more appropriate than the monthly scale when assessing the correlation of *P. tabuliformis* with *P. crassifolia* in the Qilian Mountains.

4.2. Responses of Different Tree Species Growth Patterns to Climate Change

Our results showed that the growth patterns of the two tree species in the 20th century differed with climate change. The growth trend of *P. crassifolia* increased significantly, while that of *P. tabuliformis* decreased significantly after the 1980s (Figure 8). The two coniferous species' growth patterns and ecological adaption methods differed as a result of the sudden temperature increase. *P. crassifolia* responded to environmental warming by optimizing its water usage techniques and encouraged its own development by controlling CO_2 and nutrient absorption [60,61]. Under the circumstance of high atmospheric CO_2 concentration, the stomatal openings in the leaves of *P. crassifolia* reduced during photosynthesis [62], thus reducing the transpiration rate and improving the water use efficiency. Furthermore, after a sudden increase in temperature, the limiting impact of precipitation on *P. crassifolia* grew progressively, suggesting that increases in drought stress will similarly limit *P. crassifolia* development when future temperatures rise above the tree response critical value (Figure 8). Forests in arid and semi-arid regions reduce the adverse effects caused by climate change through their own re-adaptation to the environment, thereby maintaining their own growth and stability [35,63]. Drought stress, however, has become a major limiting factor in *P. tabuliformis* radial growth, and prolonged temperature rises will lead *P. tabuliformis* growth to slow in the future (Figure 8). A similar finding has been reached elsewhere, indicating that the reproduction rate of *P. tabuliformis* slowed with time and its range narrowed toward the end of the 20th century [64].

Different tree species in the same region respond differently to climate change, depending on their species and biological characteristics [65]. It was found that the radial growth of *P. crassifolia* and *P. tabuliformis* in the middle and east of the Qilian Mountains was significantly negatively correlated with the temperature during the growing season [41,66–70]. The available water could hardly meet the needs of plants due to the high temperatures and drought that occurred in the growing season in the northwest inland area of China, leading to severe drought stress on the radial growth of *P. tabuliformis* in the study area. This result is generally referred to as the drought-sensitive response of trees to climate warming [71,72]. Further comparison between the two species showed that the degree of correlation between the radial growth of *P. tabuliformis* and temperature is higher than that for *P. crassifolia*. *P. tabuliformis* is mainly distributed on sunny and semi-sunny slopes, where the environmental conditions are relatively hot and dry with strong illumination. High temperatures cause leaf stomatal closure and reduce the photosynthesis of trees. However, *P. crassifolia* is distributed on shady slopes and semi-shady slopes with weak illumination intensity, so it is less affected by high temperatures than *P. tabuliformis*. This study found that *P. tabuliformis* growing on the drier sunny and semi-sunny slopes of the Qilian Mountains is more prone to growth declines and deaths than *P. crassifolia* growing on shady slopes with increases in temperature and aridity. In the context of climate warming, the protection and management of *P. tabuliformis* forests should be emphasized.

In the past 50 years, the temperature in the northwest of China has continued to increase [73], and some models predict that with the increasing warming, resulting in severe and widespread drought in many land areas, drought stress caused by regional drought may be the main reason for the general increase in mortality of trees of different species, sizes, altitudes, and latitudes and longitudes, and the future increase in warming and drying may significantly affect the regional trees [74]. In addition, drought was also found to be the main factor limiting tree growth by calculating the drought index (SPEI) (Table 4). Thus, it can be seen that warming-induced drought stress has an extremely serious impact on normal tree growth and forest carbon uptake. At the same time, forests in arid and semi-arid environments will also suffer from severe drought impacts in the future, and this can lead to reduced tree growth and tree death [11]. The survival of different tree species will develop in the changing climate environment to form the best combination of tree species, which will lead to changes in forest composition, structure, and productivity.

5. Conclusions

The two dominant conifer species in the study area exhibited completely different growth patterns under the same climate background, and at the same time, they were under severe drought stress during the growing season. Meanwhile, *P. tabuliformis* radial growth exhibited a synchronous radial growth pattern, increasing initially and then reducing, but *P. crassifolia* radial development exhibited a synchronous radial growth pattern that considerably grew. Different tree species show different responses to climate change, which is conducive to providing more reasonable protection and management measures for the forest ecosystem and reducing the harm caused by environmental change. In addition, compared with the daily, pentad, and monthly timescales, the two tree species are more sensitive to climatic factors at the dekad timescale in the eastern Qilian Mountains. As a result, while researching the link between tree growth and climate, as well as historical climate reconstruction, we should pay greater attention to the effect of climatic parameters on short time periods, such as the dekad scale, over trees.

Author Contributions: Study conception and experimental design: L.J. and C.Q.; Acquisition and processing of samples: R.X., X.W., and D.D.; Analysis and interpretation of data: L.J. and C.Q.; Drafting of manuscript: L.J. and C.Q. All authors have read and agreed to the published version of the manuscript.

Funding: This research was supported by the National Natural Science Foundation of China (grant no. 41861006), the Natural Science Foundation of Gansu Province (no. 20JR10RA093), and the Research Ability Promotion Program for Young Teachers of Northwest Normal University (NWNU-LKQN2019-4).

Informed Consent Statement: Written informed consent has been obtained from the patient(s) to publish this paper.

Data Availability Statement: Not applicable.

Acknowledgments: We thank the anonymous referees for helpful comments on the manuscript.

Conflicts of Interest: The authors declare no conflict of interest.

References

1. Douville, H.; Raghavan, K.; Renwick, J.; Allan, R.P.; Arias, P.A.; Barlow, M.; Cerezo-Mota, R.; Cherchi, A.; Gan, T.Y.; Gergis, J.; et al. Water Cycle Changes [M/OL]//IPCC. In *Climate Change 2021: The Physical Science Basis*; Cambridge University Press: Cambridge, UK, 2021.
2. Engelbrecht, B.M.J. Plant ecology: Forests on the brink. *Nature* **2012**, *491*, 675–677. [CrossRef]
3. De Frenne, P.; Rodríguez-Sánchez, F.; Coomes, D.A.; Baeten, L.; Verstraeten, G.; Vellend, M.; Bernhardt-Römermann, M.; Brown, C.D.; Brunet, J.; Cornelis, J.; et al. Microclimate moderates plant responses to macroclimate warming. *Proc. Natl. Acad. Sci. USA* **2013**, *110*, 18561–18565. [CrossRef]
4. Sun, X.Y.; Wang, G.X.; Huang, M.; Hu, Z.Y.; Song, C.L. Effect of climate change on seasonal water use efficiency in subalpine Abies fabri. *J. Mt. Sci.* **2017**, *14*, 142–157. [CrossRef]
5. DesLauriers, A.; Beaulieu, M.; Balducci, L.; Giovannelli, A.; Gagnon, M.J.; Rossi, S. Impact of warming and drought on carbon balance related to wood formation in black spruce. *Ann. Bot.* **2014**, *114*, 335–345. [CrossRef] [PubMed]
6. Piao, S.L.; Zhang, X.P.; Chen, A.G.; Liu, Q.; Lian, X.; Wang, X.H.; Peng, S.S.; Wu, X.C. Effects of extreme climate events on carbon cycling in terrestrial ecosystems. *Sci. China Earth Sci.* **2019**, *49*, 1321–1334.
7. Liu, H.Y.; Yin, Y. Response of forest distribution to past climate change: An insight into future predictions. *Chin. Sci. Bull.* **2013**, *58*, 4426–4436. [CrossRef]
8. De Grandpre, L.; Tardif, J.C.; Hessl, A.; Pederson, N.; Conciatori, F.; Green, T.R.; Oyunsanaa, B.; Baatarbileg, N. Seasonal shift in the climate responses of Pinus sibirica, Pinus sylvestris, and *Larix sibirica* trees from semi-arid, north-central Mongolia. *Can. J. For. Res.* **2011**, *41*, 1242–1255. [CrossRef]
9. Lebourgeois, F.; Mérian, P.; Courdier, F.; Ladier, J.; Dreyfus, P. Instability of climate signal in tree-ring width in Mediterranean mountains: A multi-species analysis. *TreesC* **2012**, *26*, 715–729. [CrossRef]
10. Zhuang, L.; Axmacher, J.; Sang, W. Different radial growth responses to climate warming by two dominant tree species at their upper altitudinal limit on Changbai Mountain. *J. For. Res.* **2017**, *28*, 795–804. [CrossRef]
11. Jiao, L.; Jiang, Y.; Zhang, W.; Wang, M.; Wang, S.; Liu, X. Assessing the stability of radial growth responses to climate change by two dominant conifer trees species in the Tianshan Mountains, northwest China. *For. Ecol. Manag.* **2018**, *433*, 667–677. [CrossRef]
12. Büntgen, U.; Frank, D.; Wilson, R.; Carrer, M.; Urbinati, C.; Esper, J. Testing for tree-ring divergence in the European Alps. *Glob. Chang. Biol.* **2008**, *14*, 2443–2453. [CrossRef]
13. Hou, Y.; Niu, Z.; Zheng, F.; Wang, N.; Wang, J.; Li, Z.; Chen, H.; Zhang, X. Drought fluctuations based on dendrochronology since 1786 for the Lenglongling Mountains at the northwestern fringe of the East Asian summer monsoon region. *J. Arid. Land* **2016**, *8*, 492–505. [CrossRef]
14. Allen, C.D.; Breshears, D.D.; McDowell, N.G. On underestimation of global vulnerability to tree mortality and forest die-off from hotter drought in the Anthropocene. *Ecosphere* **2015**, *6*, art129. [CrossRef]
15. Chen, H.Y.H.; Yong, L. Net aboveground biomass declines of four major forest types with forest ageing and climate change in western Canada's boreal forests. *Glob. Chang. Biol.* **2015**, *21*, 3675–3684. [CrossRef]
16. Soulé, P.T.; Knapp, P.A. Radial growth rates of two co-occurring coniferous trees in the Northern Rockies during the past century. *J. Arid. Environ.* **2013**, *94*, 87–95. [CrossRef]
17. Graybill, D.A.; Idso, S.B. Detecting the aerial fertilization of atmospheric CO_2 enrichment in tree-ring chronologies. *Glob. Biogeochem. Cycles* **1993**, *7*, 81–95. [CrossRef]
18. Soulé, P.T.; Paul Knapp, A. Radial Growth and Increased Water-Use Efficiency for Ponderosa Pine Trees in Three Regions in the Western United States. *Prof. Geogr.* **2011**, *63*, 379–391. [CrossRef]
19. Anderegg, W.R.L.; Hicke, J.A.; Fisher, R.; Allen, C.D.; Aukema, J.; Bentz, B.; Hood, S.; Lichstein, J.W.; Macalady, A.K.; McDowell, N.; et al. Tree mortality from drought, insects, and their interactions in a changing climate. *New Phytol.* **2015**, *208*, 674–683. [CrossRef] [PubMed]
20. Kannenberg, S.A.; Schwalm, C.R.; Anderegg, W.R.L. Ghosts of the past: How drought legacy effects shape forest functioning and carbon cycling. *Ecol. Lett.* **2020**, *23*, 891–901. [CrossRef]
21. Turcotte, A.; Morin, H.; Krause, C.; Deslauriers, A.; Thibeault-Martel, M. The timing of spring rehydration and its relation with the onset of wood formation in black spruce. *Agric. For. Meteorol.* **2009**, *149*, 1403–1409. [CrossRef]

22. Zhang, Y.; Bergeron, Y.; Gao, L.; Zhao, X.; Wang, X.; Drobyshev, I. Tree growth and regeneration dynamics at a mountain ecotone on Changbai Mountain, northeastern China: Which factors control species distributions? *Écoscience* **2014**, *21*, 387–404. [CrossRef]
23. Zhang, L.; Jiang, Y.; Zhao, S.; Jiao, L.; Wen, Y. Relationships between tree age and climate sensitivity of radial growth in different drought conditions of Qilian Mountains, northwestern China. *Forests* **2018**, *9*, 135. [CrossRef]
24. Dong, M.-Y.; Jiang, Y.; Yang, H.-C.; Wang, M.-C.; Zhang, W.-T.; Guo, Y.-Y. Dynamics of stem radial growth of Picea meyeri during the growing season at the treeline of Luya Mountain, China. *Chin. J. Plant Ecol.* **2012**, *36*, 956–964. [CrossRef]
25. Grote, R.; Pretzsch, H.; Rötzer, T. The timing of bud burst and its effect on tree growth. *Int. J. Biometeorol.* **2004**, *48*, 109–118. [CrossRef]
26. Vaganov, E.A.; Hughes, M.K.; Kirdyanov, A.V.; Schweingruber, F.H.; Silkin, P.P. Influence of snowfall and melt timing on tree growth in subarctic Eurasia. *Nature* **1999**, *400*, 149–151. [CrossRef]
27. Naurzbaev, M.M.; Vaganov, E.A. Variation of early summer and annual temperature in east Taymir and Putoran (Siberia) over the last two millennia inferred from tree rings. *J. Geophys. Res. Space Phys.* **2000**, *105*, 7317–7326. [CrossRef]
28. Liu, M.; Mao, Z.-J.; Li, Y.; Li, X.-H.; Liu, R.-P.; Huang, W.; Sun, T.; Zhao, J. Climatic effects on radial growth of Korean pines with different bark forms in Liangshui Natural Reserve, Northeast China. *Yingyong Shengtai Xuebao* **2014**, *25*, 2511–2520. [PubMed]
29. Subedi, N.; Sharma, M. Climate-diameter growth relationships of black spruce and jack pine trees in boreal Ontario, Canada. *Glob. Chang. Biol.* **2012**, *19*, 505–516. [CrossRef]
30. Chmielewski, F.-M.; Rötzer, T. Response of tree phenology to climate change across Europe. *Agric. For. Meteorol.* **2001**, *108*, 101–112. [CrossRef]
31. Gao, L.; Gou, X.; Deng, Y.; Yang, M.; Zhang, F.; Li, J. Dendroclimatic reconstruction of temperature in the eastern Qilian Mountains, northwestern China. *Clim. Res.* **2015**, *62*, 241–250. [CrossRef]
32. Gao, L.; Gou, X.; Deng, Y.; Wang, Z.; Gu, F.; Wang, F. Increased growth of Qinghai spruce in northwestern China during the recent warming hiatus. *Agric. For. Meteorol.* **2018**, *260–261*, 9–16. [CrossRef]
33. Chen, F.; Yuan, Y.J.; Chen, F.H.; Wei, W.S.; Yu, S.L.; Chen, X.J.; Fan, Z.A.; Zhang, R.B.; Zhang, T.W.; Shang, H.M.; et al. A 426-year drought history for western Tianshan, Central Asia, inferred from tree rings and linkages to the North Atlantic and Indo-West Pacific Oceans. *Holocene* **2013**, *23*, 1095–1104. [CrossRef]
34. Chen, F.; Yuan, Y.J.; Yu, S.L.; Zhang, T.W.; Shang, H.M.; Zhang, R.B.; Qin, L.; Fan, Z.A. A 225-year long drought reconstruction for east Xinjiang based on Siberia larch (*Larix sibirica*) tree-ring widths: Reveals the recent dry trend of the eastern end of Tien Shan. *Quat. Int.* **2015**, *358*, 42–47. [CrossRef]
35. Xu, G.; Liu, X.; Trouet, V.; Treydte, K.; Wu, G.; Chen, T.; Sun, W.; An, W.; Wang, W.; Zeng, X.; et al. Regional drought shifts (1710–2010) in East Central Asia and linkages with atmospheric circulation recorded in tree-ring δ18O. *Clim. Dyn.* **2018**, *52*, 1–15. [CrossRef]
36. Zhang, T.; Zhang, R.; Jiang, S.; Bagila, M.; Ainur, U.; Yu, S. On the 'Divergence Problem' in the Alatau Mountains, Central Asia: A Study of the Responses of Schrenk Spruce Tree-Ring Width to Climate under the Recent Warming and Wetting Trend. *Atmosphere* **2019**, *10*, 473. [CrossRef]
37. Gou, X.; Chen, F.; Yang, M.; Li, J.; Peng, J.; Jin, L. Climatic response of thick leaf spruce (*Picea crassifolia*) tree-ring width at different elevations over Qilian Mountains, northwestern China. *J. Arid. Environ.* **2005**, *61*, 513–524. [CrossRef]
38. Liu, Y.; Sun, J.; Yang, Y.; Cai, Q.; Song, H.; Shi, J.; An, Z.; Li, X. Tree-ring-derived precipitation records from inner Mongolia, China, since A.D. 1627. *Tree-Ring Res.* **2007**, *63*, 3–14. [CrossRef]
39. Liang, E.; Eckstein, D.; Shao, X. Seasonal cambial activity of relict chinese pine at the northern limit of its natural distribution in north China—Exploratory results. *IAWA J.* **2009**, *30*, 371–378. [CrossRef]
40. Yang, B.; He, M.; Melvin, T.M.; Zhao, Y.; Briffa, K.R. Climate control on tree growth at the upper and lower treelines: A case study in the Qilian Mountains, Tibetan Plateau. *PLoS ONE* **2013**, *8*, e69065. [CrossRef]
41. Deng, Y.; Gou, X.; Gao, L.; Zhao, Z.; Cao, Z.; Yang, M. Aridity changes in the eastern Qilian Mountains since AD 1856 reconstructed from tree-rings. *Quat. Int.* **2013**, *283*, 78–84. [CrossRef]
42. Wang, Y.; Ma, Y.; Lu, R. Reconstruction of mean temperatures of January to August since AD 1895 based on tree-ring data in the eastern part of the Qilian Mountains. *Quat. Sci.* **2009**, *29*, 905–912.
43. Holmes, R.L. Computer-assisted quality control in tree-ring dating and measurement. *Tree-Ring Bull.* **1983**, *43*, 69–75.
44. Cook, E.R. A Time Series Approach to Tree-Ring Standardization. Ph.D. Thesis, The University of Arizona, Tucson, AZ, USA, 1985.
45. Zhao, P.; Zhang, R.H.; Liu, J.P.; Zhou, X.J.; He, J.H. Onset of southwesterly wind over eastern China and associated atmos-pheric circulation and rainfall. *Clim. Dyn.* **2007**, *28*, 797–811. [CrossRef]
46. Qi, Z.H.; Liu, H.Y.; Wu, X.C.; Hao, Q. Climate-driven speedup of alpine tree-line forest growth in the Tianshan Mountains, Northwestern China. *Glob. Chang. Biol.* **2015**, *21*, 816–826. [CrossRef] [PubMed]
47. Fritts, H.C. *Tree Rings and Climate*; Academic Press: London, UK, 1976; 567p.
48. Gou, X.; Chen, F.; Jacoby, G.; Cook, E.; Yang, M.; Peng, J.; Zhang, Y. Rapid tree growth with respect to the last 400 years in response to climate warming, northeastern Tibetan Plateau. *Int. J. Climatol.* **2007**, *27*, 1497–1503. [CrossRef]
49. Zhu, H.F.; Zheng, Y.H.; Shao, X.M.; Liu, X.H.; Xu, Y.; Liang, E.Y. Millennial temperature reconstruction based on tree-ring widths of Qilian juniper from Wulan, Qinghai Province, China. *Chin. Sci. Bull.* **2008**, *53*, 3914–3920. [CrossRef]

50. Duan, J.P.; Zhang, Q.B.; Lv, L.X.; Zhang, C. Regional-scale winter-spring temperature variability and chilling damage dynamics over the past two centuries in southeastern China. *Clim. Dyn.* **2012**, *39*, 919–928. [CrossRef]

51. Bai, X.; Zhang, X.; Li, J.; Duan, X.; Jin, Y.; Chen, Z. Altitudinal disparity in growth of Dahurian larch (*Larix gmelinii* Rupr.) in response to recent climate change in northeast China. *Sci. Total Environ.* **2019**, *670*, 466–477. [CrossRef]

52. Sun, J.; Liu, Y. Age-independent climate-growth response of Chinese pine (Pinus tabulaeformis Carrière) in North China. *Trees* **2014**, *29*, 397–406. [CrossRef]

53. Sidor, C.G.; Popa, I.; Vlad, R.; Cherubini, P. Different tree-ring responses of Norway spruce to air temperature across an altitudinal gradient in the Eastern Carpathians (Romania). *Trees* **2015**, *29*, 985–997. [CrossRef]

54. Gutiérrez, E.; Campelo, F.; Camarero, J.J.; Ribas, M.; Muntán, E.; Nabais, C.; Freitas, H. Climate controls act at different scales on the seasonal pattern of Quercus ilex L. stem radial increments in NE Spain. *Trees-Struct. Funct.* **2011**, *25*, 637–646. [CrossRef]

55. Duchesne, L.; Houle, D. Modelling day-to-day stem diameter variation and annual growth of balsam fir (*Abies balsamea* (L.) Mill.) from daily climate. *For. Ecol. Manag.* **2011**, *262*, 863–872. [CrossRef]

56. Chen, X.; Hu, B.; Yu, R. Spatial and temporal variation of phenological growing season and climate change impacts in temperate eastern China. *Glob. Chang. Biol.* **2005**, *11*, 1118–1130. [CrossRef]

57. Baxter, S.; Weaver, S.; Gottschalck, J.; Xue, Y. Pentad Evolution of Wintertime Impacts of the Madden–Julian Oscillation over the Contiguous United States. *J. Clim.* **2014**, *27*, 7356–7367. [CrossRef]

58. Tadross, M.A.; Hewitson, B.; Usman, M.T. The Interannual Variability of the Onset of the Maize Growing Season over South Africa and Zimbabwe. *J. Clim.* **2005**, *18*, 3356–3372. [CrossRef]

59. Yin, X.G.; Wu, X.D. Modelling analysis of Huanshan pine growth response to climate. *Q. J. Appl. Meteorol.* **1995**, *6*, 257–264.

60. Tian, Q.; He, Z.; Xiao, S.; Peng, X.; Ding, A.; Lin, P. Response of stem radial growth of Qinghai spruce (*Picea crassifolia*) to environmental factors in the Qilian Mountains of China. *Dendrochronologia* **2017**, *44*, 76–83. [CrossRef]

61. Han, H.; He, H.; Wu, Z.; Yu, C. Non-Structural Carbohydrate Storage Strategy Explains the Spatial Distribution of Tree-line Species. *Plants* **2020**, *9*, 384. [CrossRef]

62. Tognetti, R.; Cherubini, P.; Innes, J.L. Comparative stem-growth rates of Mediterranean trees under background and naturally enhanced ambient CO_2 concentrations. *New Phytol.* **2000**, *146*, 59–74. [CrossRef]

63. Gazol, A.; Camarero, J.J. Functional diversity enhances silver fir growth resilience to an extreme drought. *J. Ecol.* **2016**, *104*, 1063–1075. [CrossRef]

64. Cai, L.; Li, J.; Bai, X.; Jin, Y.; Chen, Z. Variations in the growth response of Pinus tabulaeformis to a warming climate at the northern limits of its natural range. *Trees* **2020**, *34*, 1–13. [CrossRef]

65. Wang, B.; Chen, T.; Xu, G.; Liu, X.; Wang, W.; Wu, G.; Zhang, Y. Alpine timberline population dynamics under climate change: A comparison between Qilian juniper and Qinghai spruce tree species in the middle Qilian Mountains of northeast Tibetan Plateau. *Boreas* **2016**, *45*, 411–422. [CrossRef]

66. Liang, E.; Shao, X.; Eckstein, D.; Huang, L.; Liu, X. Topography-and species-dependent growth responses of Sabina przewalskii and *Picea crassifolia* to climate on the northeast Tibetan Plateau. *For. Ecol. Manag.* **2006**, *236*, 268–277. [CrossRef]

67. Liang, E.; Shao, X.; Eckstein, D.; Liu, X. Speatial variability of tree growth along a latitudinal transect in the Qilian Mountains, northeastern Tibetan Plateau. *Can. J. For. Res.* **2010**, *40*, 200–211. [CrossRef]

68. Sun, J.; Liu, Y. Tree ring based precipitation reconstruction in the south slope of the middle Qilian Mountains, northeastern Tibetan Plateau, over the last millennium. *J. Geophys. Res.* **2012**, *117*. [CrossRef]

69. Gao, L.; Gou, X.; Deng, Y.; Yang, M.; Zhao, Z.; Cao, Z. Dendroclimatic response of *Picea crassifolia* along an altitudinal gradient in the eastern Qilian Mountains, Northwest China. *Arct. Antarct. Alp. Res.* **2013**, *45*, 491–499. [CrossRef]

70. Lu, J.; Xu, J.; Wu, Y.; Li, X.; Evans, R.; Downes, G.M. Climatic signals in wood property variables of *Picea crassifolia*. *Wood Fiber Sci. J. Soc. Wood Sci. Technol.* **2015**, *47*, 131–140.

71. Li, J.; Cook, E.; Chen, F.; Gou, X.; D'Arrigo, R.; Yuan, Y. An extreme drought event in the central Tien Shan area in the year 1945. *J. Arid. Environ.* **2010**, *74*, 1225–1231. [CrossRef]

72. Zhang, R.; Yuan, Y.; Gou, X.; Zhang, T.; Zou, C.; Ji, C.; Fan, Z.; Qin, L.; Shang, H.; Li, X. Intra-annual radial growth of Schrenk spruce (*Picea schrenkiana* Fisch. et Mey) and its response to climate on the northern slopes of the Tianshan Mountains. *Dendrochronologia* **2016**, *40*, 36–42. [CrossRef]

73. Huang, S.Y.; Zhang, M.J.; Wang, S.J.; Li, Y.J.; Pan, S.K. Spatial and temporal variability of summer 0 °C layer height and temperature in northwest China over the past 50 years. *J. Geogr.* **2011**, *66*, 1191–1199.

74. Wang, M.; Tao, D.L. Study on drought tolerance of main tree species in Changbai Mountain. *J. Appl. Ecol.* **1998**, *9*, 7–10.

Article

Dynamics of Vegetation Productivity in Relation to Surface Meteorological Factors in the Altay Mountains in Northwest China

Aishajiang Aili, Hailiang Xu *, Xinfeng Zhao, Peng Zhang and Ruiqiang Yang

State Key Laboratory of Desert and Oasis Ecology, Xinjiang Institute of Ecology and Geography, Chinese Academy of Sciences, Urumqi 830011, China
* Correspondence: xuhl@ms.xjb.ac.cn

Abstract: Vegetation productivity, as the basis of the material cycle and energy flow in an ecosystem, directly reflects the information of vegetation change. At the ecosystem level, the gross primary productivity (GPP) refers to the amount of organic carbon fixed by plant bodies. How to accurately estimate the spatiotemporal variation of vegetation productivity of the forest ecosystem in the Altay Mountains in northwest China has become a critical issue to be addressed. The Altay Mountains, with rich forest resources, are located in a semi-arid climate zone and are sensitive to global climate changes, which will inevitably have serious impacts on the function and structure of forest ecosystems in northwest China. In this paper, to reveal the variation trends of vegetation gross primary productivity (GPP) and its response to surface meteorological factors in the Altay Mountains in northwest China, daily temperature and precipitation data from the period of 2000–2017 were collected from seven meteorological stations in Altay prefecture and its surrounding areas; the data were analyzed by using the MODIS GPP model, moving average trend analysis, linear regression analysis and the climate tendency rate method. The results show that: (1) The spatial distribution pattern of GPP in the whole year was almost the same as that in the growing season of vegetation in the Altay Mountains. In the whole mountain range, the proportion of the area which had a GPP value of 400–600 g c/m^2 had the highest value; the proportion of the annual and growing season of this area was 41.10% and 40.88%, respectively, which was mainly distributed in the middle and west alpine areas of the Altay Mountains. (2) There was a big gap in the GPP value in the different stages of the vegetation growing season (April to September), which reached the highest value in July, the area with a GPP of 100–150 g c/m^2 was the highest, with 36.15%. (3) The GPP of the Altay Mountains showed an overall increasing trend, but the annual fluctuation was relatively large. In 2003, 2008, 2009 and 2014, the GPP showed lower values, which were 385.18 g c/m^2, 384.90 g c/m^2, 384.49 g c/m^2 and 393.10 g c/m^2, respectively. In 2007, 2011 and 2016, the GPP showed higher values, which were 428.49 g c/m^2, 428.18 g c/m^2 and 446.61 g c/m^2. (4) In 64.85% of the area of the Altay Mountains, the GPP was positively correlated with annual average temperature, and in 36.56% of the area, the correlation coefficient between temperature and GPP ranged from −0.2 to 0. In 71.61% of the area of the Altay Mountains, the GPP was positively correlated with annual accumulated precipitation, and in 28.39% of the area, the GPP was negatively correlated with annual accumulated precipitation. Under the scenario of global climate change, our study has quantitatively analyzed the long-term dynamics of vegetation GPP and its responses to meteorological factors in the Altay Mountains, which would be helpful for evaluating and estimating the variation trends of forest ecosystems in China, and has important guiding significance for policy formulation to protect forest resources and improve the local ecological environment.

Keywords: GPP; temporal and spatial change; climate tendency rate method; meteorological factors; Altay Mountains

Citation: Aili, A.; Xu, H.; Zhao, X.; Zhang, P.; Yang, R. Dynamics of Vegetation Productivity in Relation to Surface Meteorological Factors in the Altay Mountains in Northwest China. *Forests* **2022**, *13*, 1907. https://doi.org/10.3390/f13111907

Academic Editors: Any Mary Petritan and Mirela Beloiu

Received: 7 October 2022
Accepted: 8 November 2022
Published: 14 November 2022

Publisher's Note: MDPI stays neutral with regard to jurisdictional claims in published maps and institutional affiliations.

1. Introduction

Climate change is an important factor affecting forest growth. Therefore, approaching the impacts of climate change on forest ecosystems is of great significance to ameliorate degraded land and support forestry development. Gross primary productivity (GPP) refers to the amount of organic carbon fixed by organisms, mainly by green plants, through photosynthesis in a unit of time, also known as total primary productivity [1–4]. The GPP determines the initial amounts of material and energy entering the terrestrial ecosystem [5–8]. Climate is the main factor that determines the distribution of forests and other species. Temperature and precipitation are the two most significant climate factors that affect the characteristics and distribution of forest ecosystems [6–11]. Forest resources in the Altay Mountains are the key component of the Xinjiang terrestrial ecosystem and are a green reservoir for regulating and conserving water resources [12–14]. Therefore, revealing the internal relationship between climate change and vegetation productivity in the Altay Mountains has important practical significance for improving the quality and productivity of forest land in the Altay Mountains, improving the management level of forest land, and promoting the healthy development of forest land [15,16]. Due to the close relationship between forests and climate, climate change will inevitably have a certain impact on forests [17–19].

The accurate estimation of the gross primary production of vegetation is vital for understanding the global carbon cycle and predicting future climate change. Multiple GPP products are currently available based on different methods and research tools, but their performances vary substantially when validated against GPP estimates from eddy covariance data. Measuring or estimating vegetation productivity mainly includes direct harvesting, volume conversion, and biomass equation methods. The harvesting method was developed first, where the biomass of each component is directly obtained by cutting and weighing all trees or estimating the biomass by directly measuring the biomass of the standard trees [20]. Direct harvesting requires a lot of human and material resources while causing unavoidable damage to the forest and the environment. The volume conversion method can be conducted due to the significant correlation between the stand volume and biomass. Nevertheless, the output of this model cannot provide an accurate estimation of forest biomass as it is related to forest density, age and site conditions. Biomass growth models can describe changes in an individual tree or stand size over time. Although these methods are widely used to connect individual trees and stand levels [21,22], the errors in estimating the stand biomass using the individual tree biomass growth combined with the stand structure are almost not quantified.

Compared with other research tools, the MODIS data can effectively reflect the landscape information on a regional scale. The MODIS-GPP product (MOD-17) is a global GPP product using remote sensing data developed according to the light-use efficiency principle. The GPP product has been widely used in appraisal and application research on various vegetation productions [23].

Research on the impact of climate change on forest ecosystems began in 1990 in China. However, most of this research work focused mainly on the variation of the forest ecosystem and its response to the annual average change in climate indicators, and little or no consideration was given to the relationship between seasonal changes in meteorological factors and vegetation productivity [23–25]. Zhou et al. analyzed the vegetation vitality in northern Eurasia and northern North America by using satellite data from 1981 to 1999; the analysis indicated that the vegetation vitality increased significantly and the growth period was prolonged [26]. Wang Yingying analyzed the temporal and spatial changes in vegetation phenology and their impact on GPP in temperate regions of China by using MODIS data [27]. Yu Xiaozhou established the estimation equation of the daytime dark respiration of different tree species according to the characteristics of dark respiration and light inhibition intensity of a single leaf from each tree species, and revised the GPP estimation results of a broad-leaved Korean pine forest on Changbai Mountain in combination with the measurement results of the canopy leaf biomass [28]. The increase in temperature increases the NPP of the cold zone or subalpine forest ecosystem, and increases the decomposition rate and reduces the NEP of the forest ecosystem [29–31]. Cao et al. showed that the productivity of China's terrestrial ecosystem was highly sensitive to climate change from 1981 to 2000, and the interannual changes in NPP were significantly positively correlated with the temperature [32].

Due to the unique characteristics of the alpine climate, arid–semi-arid environment, and distinct vertical zonality, the vegetation productivity in forest ecosystems in the Altay Mountains are very sensitive to climate change and human disturbances. Therefore, an accurate assessment of the long-term dynamics of GPP in forest ecosystems in the context of global change and its mechanistic analysis of interannual variability would be helpful to estimate and predict the variation of Chinese forest ecosystems, and provide important guidance for policy formulation on the protection of forest resources. The variation of vegetation GPP in forest ecosystems is affected by complex interactions between climate, topography, forest structure, and soil fertilities. Among them, precipitation and temperature are determining factors affecting the vegetation GPP. However, detailed studies that take into account the decisive role of these factors are scarce. This study, therefore, was conducted to partly fill in this information gap. The MODIS-GPP product is a global GPP product using remote sensing data, developed according to the light-use efficiency principle. The MODIS data can effectively reflect the landscape information on a regional scale. To our knowledge, this study is the first research that analyzes the relationship between vegetation GPP and meteorological factors in the Altay Mountains area. In view of the significant decline in the ecological quality of the Altay Mountains in recent years and the unclear impact of climate change on the ecosystem, this study uses the MODIS GPP model, moving average trend analysis, linear regression analysis, and climate tendency rate method to reveal the spatial–temporal variation of vegetation productivity in the Altay Mountains and its response to meteorological factors, such as temperature and precipitation, by the means of field monitoring, computer simulation, and remote sensing technology. By combining the study with meteorological data and field observation data, we are able to accurately evaluate the variation trends of vegetation GPP in relation to meteorological factors by using the MODIS-GPP model. The methods used in this study to evaluate the variation trends of vegetation GPP can be used to assess ecosystem degradation, design the key restoration areas, as well as mitigate the impact of various meteorological disasters on mountain forest ecosystems. Due to the unavailability of systematic data on the impact of natural disasters and human disturbance, we are not able to carry out systematic research on the reasons of degradation of the forest ecosystem in the Altay Mountains. Further research can be conducted by combining the study with field investigations and plant physiology experiments on the basis of a longer period of remote sensing data.

2. Material and Methods

2.1. Description of Study Area

The Altay Mountains spans China, Kazakhstan, Russia and Mongolia, with a total length of approximately 2000 km. It runs from northwest to southeast. In this study, the Altay Mountains in China is selected as the study area. This area belongs to the south slope of the middle section of the Altay Mountains, with a length of approximately 450 km from east to west and a width of approximately 80–150 km from north to the south. The mountain gradually becomes narrow from northwest to southeast, showing the topographic characteristics of high and wide in the northwest and low and narrow in the southeast (Figure 1), with a total area of 2.6×10^4 km^2 [33].

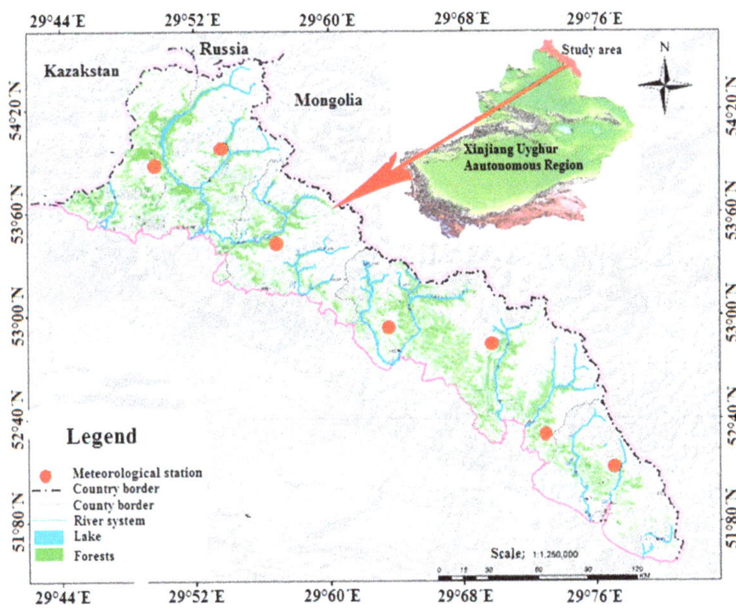

Figure 1. Location of Study Area (Driving number of map: GS (2019) 1823).

The Altay Mountains has a vast area of forests, grasslands and abundant wetlands, which are important components to maintain the ecological functions of the mountain area. Among them, the forest area is 9802 km^2, accounting for 37.7%, and the total area of the wetland is approximately 800 km^2, accounting for 3% of the total area of the Altay Mountains. The forest is mainly composed of Siberian larch (*Larix sibirica*), and the main arbors include Siberian spruce (*Picea obovata*) and Betula pendula, which, respectively, constitute the larch forest, spruce forest, larch–spruce mixed forest and Betula pendula mixed forest. There is a large number of peat swamps. It is the largest and most complete peat swamp and the key "carbon pool" for water conservation in Xinjiang.

The Altay Mountains, with continental climate characteristics, are located in the hinterland of Eurasia and far from the sea. The spring temperature rises quickly and is windy, the summer is cool and short, the autumn temperature drops quickly and is sunny, and the winter is cold and long. The annual average temperature is approximately -2 °C, and the extreme maximum temperature is 33.3 °C. The mountain area with an altitude of 3100~3300 m is covered with snow all year. The annual average temperature in the middle- and high-mountain area, with an altitude of 1400~2600 m, is below -9 °C, and is the hottest in July, which is only approximately 15 °C. The annual temperature difference is approximately 30 °C, and the daily temperature difference is approximately 12 °C. The annual average temperature in the low-mountain and hilly area is below 4 °C. The altitude increases by 100 m, and the annual precipitation increases by 30/80 mm. The annual precipitation is 200~300 mm in the low-mountain belt, 300~500 mm in the middle-mountain belt, and 600~800 mm in the high-mountain belt, and decreases from north to south and from west to east. The Altay Mountains are an important natural forest area in Xinjiang, known as the ecological protective screen in the north of Xinjiang. The total area of the forest is 69×10^4 hm^2, and the standing forest volume is 9.197×10^4 m^3, accounting for 47% of the natural forest in Xinjiang [14,34,35].

2.2. Data Sources

The daily meteorological data of seven meteorological stations around the Altay Mountains (Altay Meteorological Bureau, Burqin County Meteorological Bureau, Habahe County Meteorological Bureau, Jimunai County Meteorological Bureau, Fuhai County Meteorological Bureau, Fuyun County Meteorological Bureau, Qinghe County Meteorological Bureau) from 2000 to 2017 were obtained from China Meteorological Data Sharing Service Network (http://data.cma.gov.cn/, 12 October 2018), and the meteorological data were strictly controlled to maintain the accuracy of the research results. To ensure the accuracy of meteorological data, the inverse distance weighted (IDW) method was used to interpolate the site data of the meteorological data. The selected meteorological stations were evenly distributed in the whole study area. The daily average temperature and precipitation data were used to reveal the relationship between meteorological factors and vegetation GPP. The NDVI (Normalized Differentiated Vegetation Index) data used in this paper were the MODIS MOD13Q1 products, with a spatial resolution of 250 m and a time series from January 2000 to December 2017. First, the obtained remote sensing data were preprocessed by data format conversion, mosaic, projection conversion and study area extraction. Further, in order to reduce the impact of noise information on the data, Savitzky Golay filtering [36] and MVC synthesis processing [37] were also performed on the NDVI data to obtain annual NDVI data representing the best condition of vegetation growth. Savitzky Golay filtering was used in this study because of its analytical and computational simplicity, and good smoothing capabilities. The MVC (Maximum Value Composite) is similar to that used in the AVHRR-NDVI product, whereby the pixel observation with the highest NDVI value is selected to represent the entire period. The MOD13Q1 data used in this study were provided every 16 days at a 250 m spatial resolution as a gridded level 3 product. Furthermore, the vegetation indices were used for the global monitoring of vegetation conditions and were used for displaying the land cover changes. These data can be used for characterizing the biophysical properties of land surface, including primary production and land cover conversion [36,37].

2.3. Statistical Methods

2.3.1. Climate Tendency Method

Using the time series of meteorological factors, with time as the independent variable and meteorological factors as the dependent variable, set Y as the meteorological variable and t as the time, and establish a linear regression equation between Y and t. The Pearson correlation coefficient was calculated to obtain the linear relationship between vegetation GPP and meteorological factors. The climate tendency rate of meteorological factors is expressed by the following formula:

$$Y_i = a_0 + a_1 t_i \tag{1}$$

where Y_i is meteorological factor, t_i is time, α_1 is linear trend term, $\alpha_1 \times 10$ is the climate tendency rate of meteorological elements every 10 years, with the unit of 10 years. When it is less than 0, it means that the meteorological factor sequence decreases with time, otherwise it increases. The larger the absolute value of α_1, the more significant the trend [38–41].

2.3.2. Correlation Analysis

In this study, the dimensionless climate trend coefficient r_{xt} is obtained by using the following equation:

$$r_{xt} = \frac{\sum_{t=1}^{n} (x_t - x)\left(t - \frac{n+1}{2}\right)}{\sqrt{\sum_{t=1}^{n} (x_t - x)^2 \left(t - \frac{n+1}{2}\right)^2}} \tag{2}$$

In the equation, r_{xt} is the trend coefficient, and its significance can be tested by t distribution statistics. The relationship between tendency rate b and trend coefficient r_{xt} is as follows:

$$b = r_{xt}(\sigma_x/\sigma_t) \tag{3}$$

where σ_x and σ_t are the mean square deviations of factor sequence and natural sequence, respectively.

The Pearson correlation coefficient (R) is employed to present the relationship of vegetation productivity with precipitation and temperature. A high R-value indicates a strong correlation, and a low R-value represents a weaker correlation. The correlation analysis for each grid is performed through Matlab software by writing code programs.

2.3.3. GPP Inversion of Total Primary Productivity

In this study, the GPP inversion of total primary productivity is obtained by using the MODIS GPP model. The precise quantification of GPP at the landscape level is however challenging as there is no direct measurement technique beyond the leaf level. Forest productivity is primarily estimated using the data gathered through temporal forest inventories and established empirical allometric equations in terms of biomass. The MODIS GPP model is established based on the linear relationship between GPP and the photosynthetically active radiation absorbed by vegetation. The calculation process is as follows:

$$GPP = APAR \times \varepsilon_{max} \times f(a\min) \times f(VPD) \tag{4}$$

$$APAR = SWRad \times 0.45 \times (1 - e^{k \times LAI}) \tag{5}$$

where GPP is the total primary productivity, the unit is g $c \cdot m^{-2} \cdot s^{-1}$; ε_{Max} is the maximum light energy utilization rate, the unit is kg C/MJ. The light energy utilization rate (E_{max}) is the ratio of the energy stored to the energy of light absorbed. The amount of energy stored can only be estimated because many products are formed, and they vary with the plant species and environmental conditions. APAR (Absorbed Photosynthetic Active Radiation) is the photosynthetically active radiation absorbed by vegetation, and the unit is $MJm^{-2} \cdot s^{-1}$, thus, representing the product with a 45% incident short-wave radiation (SWRad) and a photosynthetically active radiation ratio which is absorbed by the vegetation canopy. The SWRad is radiation at wavelengths shorter than 4 microns. Usually, radiation is in the visible and near-infrared wavelengths. This radiation ratio is calculated using LAI (Leaf Area Index) through simple Beer's law; this study used the GlobMap LAI data from 2000 to 2017. GlobMapLAI is a global long-term series LAI product generated based on AVHRR/MODIS data. k is the canopy extinction coefficient—generally, its value is 0.5. f (VPD) and f ($a\min$) are the correction factors of the vapor pressure difference and air temperature at a 2 m height—they are calculated using the following equation:

$$f(VPD) = \frac{VPD_{max} - VPD}{VPD_{max} - VPD_{min}} \tag{6}$$

$$f(T_{min}) = \frac{T_{min} - T_{min\_min}}{T_{min\_max} - T_{min\_min}} \tag{7}$$

where VPD_{max} and T_{min_max} are the maximum daily vapor pressure difference (Pa) and maximum daily air temperature (°C) at the time of maximum photosynthetic efficiency, respectively; VPD_{min} and T_{min_min} are the minimum vapor pressure difference (Pa) and daily minimum temperature (°C) when photosynthesis is 0, respectively. These parameters are default parameters in the BPLUT table [42–44]. In this study, the 500 m MODIS reflectance data (MOD09GA) were used for spectral unmixing. The MOD09GA product was provided along with daily 500 m spatial resolution surface reflectance data, which were generated from Terra-MODIS bands 1 to 7 (620 nm–2155 nm). The MODIS data collection and resolution methods used in this study are available and well explained in the website (https://modis-images.gsfc.nasa.gov/_docs/CMUSERSGUIDE.pdf, 12 October 2018), which is proposed by Kathleen in 2012 [45].

3. Results and Discussions

3.1. General Characteristics of Vegetation Productivity in the Altay Mountains

In this study, both the annual GPP and growing season GPP of vegetation in different parts of the Altay Mountains were compared (Figure 2).

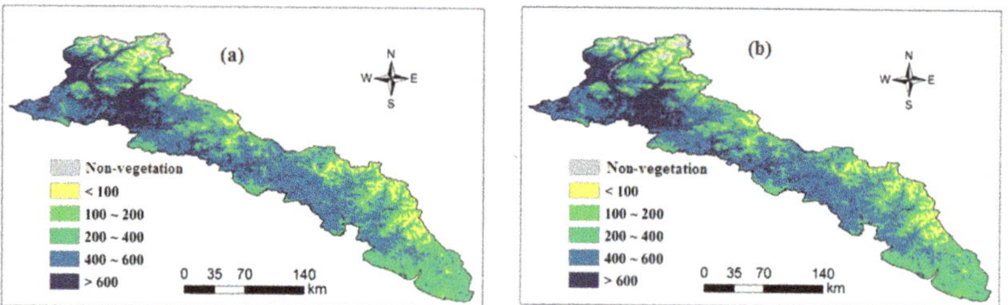

Figure 2. Overall characteristics of annual GPP (**a**) and growing season GPP (**b**) in the Altay Mountains.

It can be seen from Figure 2 that the spatial distribution pattern of annual GPP and growth season GPP of vegetation in the Altay Mountains is almost the same, and the proportion of each grade is also very similar. In the whole mountain area, the area with a GPP of 400~600 g c/m^2 has the highest proportion, with annual and growing season proportions of 41.10% and 40.88%, respectively (Table 1). These areas are mainly distributed in the middle- and high-mountain areas in the middle and west parts of the Altay Mountains. The area with a GPP of 200~400 g c/m^2 ranks the second, with annual and growing season GPP proportions of 29.21% and 29.96%, respectively, which are mainly distributed in the areas with a higher altitude in the east and west parts of the Altay Mountains. The area with a GPP of <100 g c/m^2 is the smallest, with annual and growing season GPP proportions of 6.17% and 6.28%, respectively, which are mainly distributed in high-altitude areas in the west and east parts of the Altay Mountains.

Table 1. Proportion of the area with a different GPP in annual and growing season.

Period	GPP (g c·m^{-2})				
	<100	100~200	200~400	400~600	>600
Annual (%)	6.17	9.18	29.21	41.10	14.34
Growing season (%)	6.28	9.26	29.96	40.88	13.62

The GPP of different months in the growing season is significantly different. In this study, the proportion of the area with a different GPP from April to September was compared, and the distributions of the higher GPP area and lower GPP area were analyzed (Figure 3).

It can be seen from Figure 3 that, in April, there is still a large area of snow cover in the high-mountain area. In the areas with vegetation coverage, the proportion of the area with a GPP of 1~5 g c/m^2 is the highest, which is 36.73%, and is mainly distributed in the areas below the snow line in the middle part of the mountain. The GPP with 83.73% of the area is less than 15 g c/m^2.

In May, the vegetation coverage area expanded and the vegetation GPP increased; the proportion of area with a GPP of 80~120 g c/m^2 is the largest, accounting for 39.04%, which was mainly distributed at the west part of middle- and lower-altitude area. However, there are still lower vegetation coverage areas in this month; the areas with a GPP < 10 g c/m^2 are mainly distributed under the snow line at a high altitude.

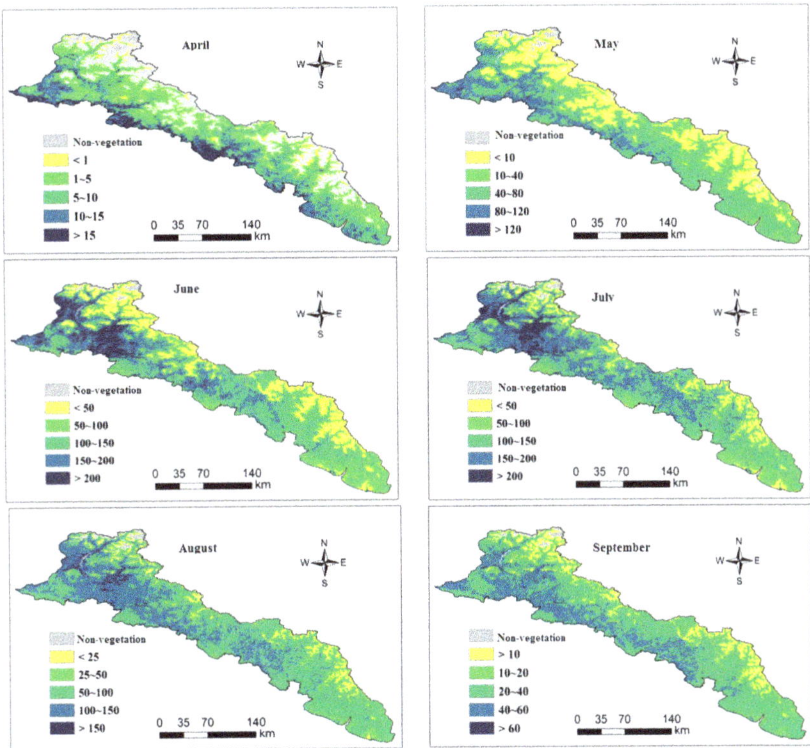

Figure 3. Distributions of GPP in the growing season in the Altay Mountains.

In June, the vegetation grew vigorously, and the vegetation GPP gradually increased. The proportion of area with a GPP of 100~150 g c/m^2 is the highest, which is 32.58%, and it is widely distributed in the east and middle parts of the mountain. The proportion of the highest vegetation coverage areas with a GPP > 200 g c/m^2 is 8.18%, which is mainly distributed in the western part of the mountain with a medium altitude.

In July, the vegetation GPP of the Altay Mountains was further improved, and the vegetation GPP of most parts of the mountain has a higher value and is relatively consistent. The proportion of area with a GPP of 100~150 g c/m^2 is the highest, which is 36.15%, and is mainly distributed at the east and middle parts of the mountain. In this month, there are still some lower vegetation coverage areas with a GPP < 50 g c/m^2, accounting for 8.75%, and are mainly distributed at the high-altitude areas in the east and west sections of the Altay Mountains.

In August, the GPP of vegetation in whole mountains decreased gradually, and the proportion of area with a GPP of 50~100 g c/m^2 is the highest, accounting for 49.48%, and is mainly distributed at the east and the middle parts of the mountain. The GPP of most parts of the mountain ranges from 50 g c/m^2 to 150 g c/m^2. However, there are still some lower-vegetation coverage areas; the proportion of area with a GPP < 25 g c/m^2 accounts for 5.51% and is mainly distributed at the higher-altitude area in the west and east parts of the mountains.

In September, the GPP of whole mountain rapidly decreased; the area with a GPP of 20~40 g c/m^2 accounts for 51.38%. There were only small areas with a higher GPP; the proportion of area with a GPP > 60 g c/m^2 accounts for 0.65% (Table 2).

Table 2. The proportion of the areas with different GPP in the growing season in the Altay Mountains.

April		May		June		July		August		September	
GPP/g c·m^{-2}	Proportion/%	GPP/g c·m^{-2}	Proportion/%	GPP/g c·m^{-2}	Proportion/%	GPP/g c·m^{-2}	Proportion/%	GPP/g c·m^{-2}	Proportion/%	GPP/g c·m^{-2}	Proportion/%
<1	4.2	<10	17.27	<50	18.23	<50	8.75	<25	5.51	<10	9.46
1~5	36.73	10~40	10.95	50~100	20.72	50~100	18.98	25~50	10.83	10~20	14.04
5~10	18.37	40~80	31.19	100~150	32.58	100~150	36.15	50~100	49.48	20~40	51.38
10~15	24.42	80~120	39.04	150~200	20.29	150~200	28.88	100~150	31.61	40~60	24.47
>15	16.27	>120	1.55	>200	8.18	>200	7.25	>150	2.57	>60	0.65

3.2. Annual Variation of Vegetation Productivity in the Altay Mountains

From 2000 to 2017, the change in vegetation GPP in the Altay Mountains fluctuated greatly, showing an increasing trend (Figure 4).

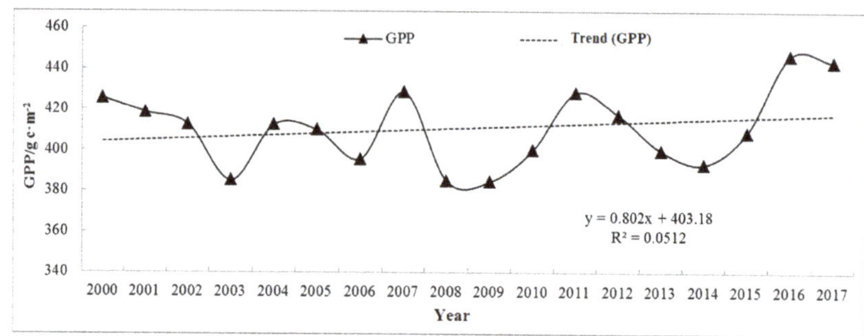

Figure 4. Annual variation of GPP in the Altay Mountains (2000–2017).

It can be seen from Figure 4 that the vegetation GPP of the Altay Mountains presents an overall increasing trend, but the interannual fluctuation is still large. The annual average value peaked in 2007, 2011 and 2016, and their values were 428.49 g c/m^2, 428.18 g c/m^2 and 446.61 g c/m^2. In 2003, 2008, 2009 and 2014, the values were relatively low, with 385.18 g c/m^2, 384.90 g c/m^2, 384.49 g c/m^2 and 393.10 g c/m^2, respectively.

In this study, the significance of variation trend of the vegetation GPP in the period of 2000–2017 was also analyzed, and the results are shown in Figure 5.

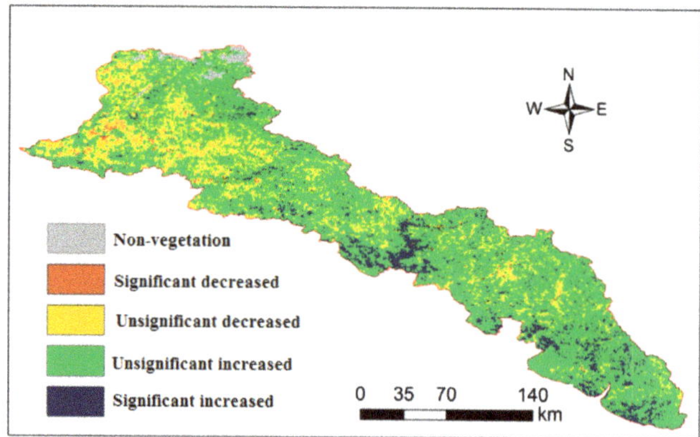

Figure 5. Significance of variation of GPP in the Altay Mountains.

It can be seen from Figure 5 that in 73.16% of the study area, the vegetation GPP has an increasing trend; the significantly increased area accounts for 7.72%, which is mainly distributed in the middle and eastern parts of the mountain. However, the vegetation GPP shows a decreasing trend in 26.84% of the study area, and is mainly distributed at the western part of the mountain; the significantly decreased area accounts for 0.98%.

The results of the annual variation rate of the vegetation GPP during the study period are presented in Figure 6.

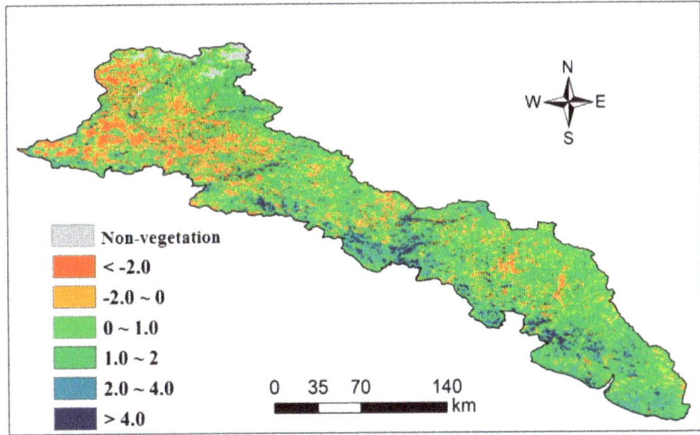

Figure 6. Variation rate of GPP in the Altay Mountains.

It can be seen from Figure 6 that the variation rates of the vegetation GPP in the study area are significantly different. The area with a GPP variation rate of $-2.0{\sim}0$ g·c·m^{-2}·a^{-1}, $0{\sim}1.0$ g c·m^{-2}·a^{-1} and $1.0{\sim}2.0$ g c·m^{-2}·a^{-1} accounts for 21.23%, 27.46% and 23.43%. The areas with a GPP variation rate of $-2.0{\sim}0$ g c·m^{-2}·a^{-1} are mainly distributed in the middle-height area of the western part of the Altay Mountains; the areas with a GPP variation rate of $0{\sim}1.0$ g c·m^{-2}·a^{-1} are mainly distributed in the western and eastern parts of the mountain; and the areas with a GPP variation rate of $1.0{\sim}2.0$ g c·m^{-2}·a^{-1} are distributed discretely in the whole part of the mountain. The areas with a GPP variation rate of $2.0{\sim}4.0$ g c·m^{-2}·a^{-1} account for 18.76%, and are mainly distributed at the lower-mountain area of the middle and east parts of the Altay Mountains. The areas with a GPP variation rate of <-2.0 g c·m^{-2}·a^{-1} account for 6.03%, and are mainly distributed in the middle-latitude area of the western section. The area with a GPP variation rate of >4.0 g c·m^{-2}·a^{-1} accounts only for 3.09%, which is scattered in the low-latitude areas in the middle and east sections.

3.3. Impact of Meteorological Factors on Vegetation GPP

The variation of the vegetation GPP of the forest is the outcome of the coupling influence of various factors including climate change, natural disasters and human disturbance. The changes in surface meteorological factors will inevitably affect the forest to a certain extent; the influence of temperature and precipitation on vegetation GPP is especially more significant. In this study, the impact of temperature and precipitation on vegetation GPP in different parts of the mountain were analyzed by comparing the correlation coefficient, which indicates the strength of relevance between them. Since the annual GPP and growing season GPP showed a similar distribution pattern, as mentioned above, the relationship between the annual average GPP in different parts of the mountain and meteorological factors were examined by using correlation analysis. However, the correlation between two independent variables, temperature and precipitation, was not considered because this was not relevant to our study. Temperature is the determining factor for the growth and development of vegetation. The stronger the impact of temperature on the vegetation GPP, the greater the correlation coefficient between them (Figure 7a).

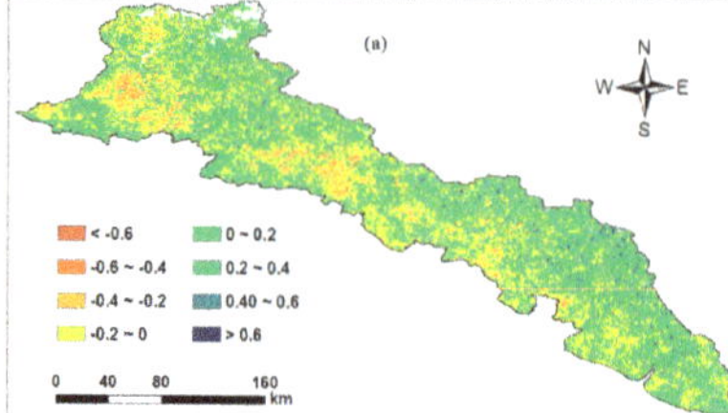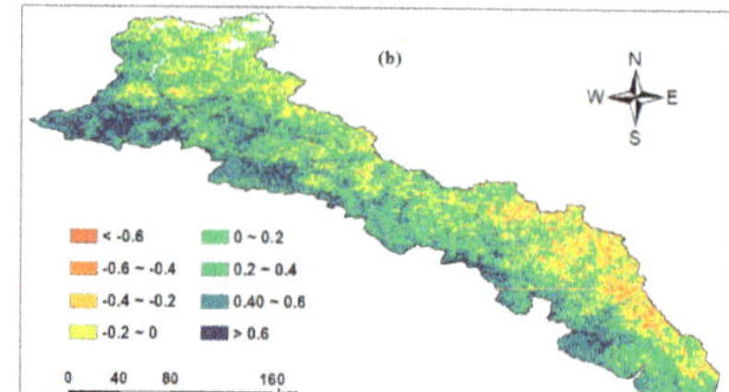

Figure 7. Correlation of vegetation GPP with temperature (**a**) and precipitation (**b**) in the Altay Mountains.

It can be seen from Figure 7a that vegetation GPP is positively correlated with the annual average temperature, with a mean correlation coefficient of 0.19. The area with positive correlations accounted for 64.85% of the whole, with a mean value of 0.29—this showed a wide distribution but was more intensive in the middle-altitude area of the southern slope of the Altay Mountains. The area with a correlation coefficient of 0–0.2 accounts for 30.23%, the area with a correlation coefficient of 0.2–0.4 accounts for 10.48%. However, in 35.15% of the study area, the correlation coefficient between temperature and the vegetation GPP has a negative value. The area with a correlation coefficient of −0.2–0 accounts for 13.2%. The areas with a negative correlation coefficient were mainly distributed in the north and southeast parts of the mountain with a lower altitude.

Similarly, precipitation also plays an essential role in vegetation GPP. It can be seen from Figure 7b that a GPP of 71.61% of the Altay Mountain area is positively correlated with the annual average precipitation, with a mean correlation coefficient of 0.39. Regarding the distribution of the area, it has a similar distribution pattern to temperature but a relatively wider distribution of the positively correlated areas. The areas with a correlation coefficient of 0–0.2, 0.2–0.4 and 0.4–0.6 account for 16.49%, 16.07% and 13.39%. However, a vegetation GPP of 28.39% of the study area is negatively correlated with the annual average precipitation, of which most were located in the north part of the mountain with a higher altitude.

Overall, the result of the correlation analysis indicates a strong linkage between temperature and precipitation with vegetation GPP in the study area. An increase in these variables will likely increase the vegetation productivity. This study's outcome aligns with some previous studies that established a connection between vegetation productivity and meteorological factors. Du et al. analyzed the influence of temperature and precipitation on forest ecosystems in the vegetation growing season in the Qilian Mountains of northwestern China, and found a significant interaction between temperature and precipitation, contributing up to 30% of total variability in the predicted ecosystem level in the vegetation growing season [46]. Meteorological factors determine the growth and development of vegetation in the Altay Mountains. Conversely, because the forest ecosystem is a huge carbon pool, it acts as a source or sink of CO_2 in the atmosphere, thus, further strengthening or offsetting future climate change. Benjamin et al. evaluates the sensitivity of forest productivity to precipitation and air temperature in inner Asian forests by using satellite remote sensing, dendrochronology and dynamic global vegetation model (DGVM) simulations, and indicated that in cool arid environments, precipitation is not the only limitation to forest productivity. Interactions between changes in precipitation and air temperature may enhance soil moisture stress while simultaneously extending the growing season length, with unclear consequences for net carbon uptake [47]. Wang et al. studied the impacts of climate change on forest growth in saline-alkali land of the Yellow River Deltain North China by using standard site methods and tree ring sampling, and indicated that precipitation is the main meteorological factor affecting tree growth, while temperature and air pressure are also significantly correlated with tree growth [48]. The Altay Mountains are located in the hinterland of Eurasia, far away from the sea, and have a few water vapor sources. In the situation of global atmospheric circulation, they are located in the westerly zone. The westerly flow enters this area along the Ertish river valley, raising the condensation cloud to cause rain. Therefore, precipitation is relatively large, and has become one of the wetter regions in the Xinjiang. The variation trends of vegetation coverage in different parts of the mountain were quite different. In general, the areas where the vegetation coverage has decreased are mainly distributed in the west and middle parts of the Altay Mountains, the Piedmont of the mountains, as well as the plain areas in the southwest, central and northeast parts of the mountain.

4. Conclusions

The MODIS-GPP product is a global GPP product using remote sensing data, and is developed according to the light-use efficiency principle. The MODIS data can effectively reflect the landscape information on a regional scale. Combined with meteorological data and field observation data, the variation trends of vegetation GPP in relation to meteorological factors can be accurately evaluated by using the MODIS-GPP model.

The spatial distribution pattern of annual vegetation GPP and growing season GPP of the Altay Mountains was almost consistent. In the whole study area, the proportion of the areas with a vegetation GPP of 400~600 g c/m^2 was the highest, and was mainly distributed in the middle- and high-altitude areas of the middle and west parts of the Altay Mountains with the best plant growth conditions.

Due to the differences in meteorological factors such as temperature and precipitation in the Altay Mountains, the GPP in different stages of the vegetation growth season (April to September) had a great difference, reaching the highest value in July; the area with a GPP of 100~150 g c/m^2 had the highest proportion, with 36.15%. In April, there was still a large area of snow cover in the high-altitude area of the Altay Mountains, and the vegetation growth conditions were poor. Among the areas with vegetation coverage, the proportion of area with a GPP of 1~5 g c/m^2 was the highest, with 36.73%, and was mainly distributed in the areas below the snow line in the middle-altitude area.

The vegetation GPP of the Altay Mountains had an overall increasing trend in the study period, but there was a large gap in GPP values in different years—the peak value appeared in 2007, 2011 and 2016, and the lower values appeared in 2003, 2008 and 2009, respectively. In 73.16% of the study area, the vegetation GPP showed an increasing trend, and 7.72% of the area was in the middle and east parts of the mountain, this increasing trend was more significant. However, the decreasing trend appeared in 26.84% of the study area, and was mainly distributed in the west part of the mountain.

Temperature and precipitation were the most important meteorological factors affecting vegetation GPP in the Altay Mountains. In 71.61% of the area of the Altay Mountains, the vegetation GPP was positively correlated with precipitation, while the vegetation GPP in 64.85% of the study area was positively correlated with the annual average temperature.

Our study has quantitatively analyzed the long-term dynamics of vegetation GPP and its responses to meteorological factors in the Altay Mountains. The results of our study would be helpful in carrying out further research on the assessment of the degradation of the forest ecosystem, evaluation of the ecosystem function, identification of the key restoration areas, as well as mitigating the impact of various meteorological disasters on mountain forest ecosystems.

Author Contributions: Conceptualization, methodology, validation, formal analysis, investigation, A.A. and H.X.; Software visualization, A.A., P.Z., X.Z. and R.Y. All authors have read and agreed to the published version of the manuscript.

Funding: This study was jointly supported by the Natural Science Foundation of Xinjiang Uyghur Autonomous Region «Study on the Cooperative Configuration of Oasis Wind break and Sand fixation System in Extremely Arid Areas» and «Key technologies for natural forest protection and restoration in 2022 (ZX-2022040)».

Data Availability Statement: Data could be provided on reasonable request from the first author.

Acknowledgments: The authors are grateful to the staff of the Department of Forestry of Altay Prefecture for providing the necessary facilities during the study. We are also thankful to anonymous reviewers for their valuable suggestions and comments on the overall improvement of the manuscript.

References

1. Evans, J.; Geerken, R. Discrimination between climate and human-induced dryland degradation. *J. Arid Environ.* **2004**, *57*, 535–554. [CrossRef]
2. Supannika, P.; Yoshifumi, Y. Application of the 3-PG Model for Gross Primary Productivity Estimation in Deciduous Broadleaf Forests: A Study Area in Japan. *Forests* **2011**, *2*, 590–609.
3. Wan, J.Z.; Yum, J.H.; Yin, G.J.; Song, Z.M.; Wei, D.X.; Wang, C.J. Effects of soil properties on the spatial distribution of forest vegetation across China. *Glob. Ecol. Conserv.* **2019**, *1*, 635. [CrossRef]
4. He, Z.; Zhao, W.; Zhang, L.; Liu, H. Response of tree recruitment to climatic variability in the Alpine treeline ecotone of the Qilian Mountains, northwestern China. *Forest Sci.* **2013**, *59*, 118–126. [CrossRef]
5. Urbanov, A.M.; Snajdr, J.; Baldrian, P. Composition of fungal and bacterial communities in forest litter and soil is largely determined by dominant trees. *Soil Biol. Biochem.* **2015**, *84*, 53–64. [CrossRef]
6. Wang, Y.; Yan, X.D. The response of the forest ecosystem in China to global climate change. *Chin. J. Atmos. Sci.* **2006**, *30*, 1009–1018. (In Chinese)
7. Molina, A.J.; Llorens, P.; Garcia-Estringana, P.; Heras, M.M.; Cayuela, C.; Gallart, F.; Latron, J. Contributions of throughfall, forest and soil characteristics to near-surface soil water-content variability at the plot scale in a mountainous Mediterranean area. *Sci. Total Environ.* **2019**, *647*, 1421–1432. [CrossRef]
8. Melissa, M.; Kreye, D.; Adams, C.; Francisco, J.E. The Value of Forest Conservation for Water Quality Protection. *Forests* **2014**, *5*, 862–884.
9. Lin, W.C.; Lin, Y.P.; Lien, W.Y.; Wang, Y.C.; Lin, C.T.; Chiou, C.R.; Johnathen, A.; Neville, D.C. Expansion of Protected Areas under Climate Change: An Example of Mountainous Tree Species in Taiwan. *Forests* **2014**, *5*, 2882–2904. [CrossRef]
10. Yu, S. A review on the impact of climate change on forest tree mortality. *Protect. Forest Sci. Technol.* **2014**, *12*, 67–69.
11. Trindade, W.C.F.; Santos, M.H.; Artoni, R.F. Climate change shifts the distribution of vegetation types in South Brazilian hotspots. *Reg. Environ. Chang.* **2020**, *20*, 90. [CrossRef]
12. Palombo, C.; Chirici, G.; Marchetti, M.; Tognetti, R. Island abandonment affecting forest dynamics at high elevation in Mediterranean mountains more than climate change? *Plant Biosyst.* **2013**, *147*, 1–11. [CrossRef]
13. Aishajiang, A.; Xu, H.L.; Yuan, K.Y. Assessment of the effects of artificial restoration measures in abandoned gold mining area in Altay Mountains using PCA and monitoring data. *Arid Land Geog.* **2019**, *42*, 288–294. (In Chinese)
14. Xu, F.J.; Xu, H.L.; Aleta, I.T. *Study on the Integration and Model of Mining Area Restoration Technology in Altay Area*; China Forestry Press: Beijing, China, 2019. (In Chinese)
15. Jian, N. Forest productivity of the Altay and Tianshan Mountains in the dryland, northwestern China. *Forest Ecol. Manag.* **2004**, *202*, 13–22.
16. Cheng, F.; Yuan, Y.J.; Wei, W.S.; Zhang, T.W.; Shang, H.M.; Zhang, R.B. Precipitation reconstruction for the southern Altay Mountains (China) from tree rings of Siberian spruce, reveals recent wetting trend. *Dendrochronologia* **2014**, *32*, 266–272. [CrossRef]
17. Sohngen, B.; Mendelsohn, R.; Sedjo, R. A global model of climate change impacts on timber markets. *J. Agric. Res. Econ.* **2001**, *26*, 326–343.
18. Boisvenue, C.; Running, S.W. Impacts of climate change on natural forest productivity-evidence since the middle of the 20th century. *Glob. Chang. Biol.* **2006**, *12*, 862–882. [CrossRef]
19. Jiang, F.Q.; Yu, Z.Y.; Zeng, D.H. Impact of climate change on the Three-north Shelter Forest Program and corresponding. *Chin. J. Ecol.* **2009**, *28*, 1702–1705. (In Chinese)
20. Castedo-Dorado, F.; Gómez-García, E.; Diéguez-Aranda, U.; Barrio-Anta, M.; Crecente-Campo, F. Aboveground stand-level biomass estimation: A comparison of two methods for major forest species in northwest Spain. *Ann. Forest Sci.* **2012**, *69*, 735–746. [CrossRef]
21. Harrison, W.C.; Daniels, R.F. A New Biomathematical Model for Growth and Yield of Loblolly Pine Plantations. In *USDA Forest Service General Technical Report NC-North Central Forest Experiment Station (USA)*; USDA Forest Service: Washington, DC, USA, 1988.
22. Qin, J.; Cao, Q.V. Using disaggregation to link individual-tree and whole-stand growth models. *Can. J. Forest Res.* **2006**, *36*, 953–960. [CrossRef]
23. Cheng, X.X.; Yan, X.D. Effects of climate change on typical forest in the Northeast of China. *Acta Ecol. Sin.* **2008**, *28*, 534–543. (In Chinese)
24. Li, J.Q.; Li, Z.Y.; Yi, H.R. Interaction relation between forest and global climate change J. *Northwest Forest Univ.* **2010**, *25*, 23–28. (In Chinese)
25. Zhao, F.Y.; Wang, M.Y.; Shu, L.F. Progress of studies on influences of climate change on forest fire regime. *Adv. Clim. Chang. Res.* **2009**, *5*, 50–55. (In Chinese)

26. Zhou, L.M.; Tucker, C.J.; Kaufmann, R.K. Variations in northern vegetation activity inferred from satellite data of vegetation index during 1981 to 1999. *J. Geophys. Res.-Atmos.* **2001**, *106*, 20069–20083. (In Chinese) [CrossRef]

27. Wang, Y.Y. Temporal and Spatial Variation of Vegetation Phenology in Temperate China and Its Influence on GPP Based on MODIS Data. Master's Dissertation, Hebei Normal University, Shijiazhuang, China, 2011. (In Chinese).

28. Yu, X.Z. Dark Respiration Characteristics of Forest and Its Influence on the Estimation of GPP in Ecosystem—A Case Study of Broad-Leaved Korean Pine Forest in Changbai Mountain. Master's Dissertation, University of Chinese Academy of Sciences, Beijing, China, 2011. (In Chinese).

29. Barber, V.A.; Juday, G.P.; Finney, B.P. Reduced growth of Alaskan white spruce in the twentieth century from temperature-induced drought stress. *Nature* **2000**, *405*, 668–673. [CrossRef]

30. Giardina, C.P.; Ryan, M.G. Evidence that decomposition rates of organic carbon in mineral soil do not vary with temperature. *Nature* **2000**, *404*, 858–861. [CrossRef] [PubMed]

31. Clark, D.A.; Piper, S.C.; Keeling, C.D. Tropical rain forest tree growth and atmospheric carbon dynamics linked to interannual temperature variation during 1984-2000. *Proc. Natl. Acad. Sci. USA* **2003**, *100*, 5852–5857. [CrossRef]

32. Cao, M.K.; Prince, S.D.; Small, J. Remotely sensed interannual variations and trends in terrestrial net primary productivity 1981 to 2000. *Ecosystems* **2004**, *7*, 233–242. [CrossRef]

33. Yang, H.F.; Gang, C.C.; Mu, S.J. Analysis of the spatio-temporal in net primary productivity of grassland during the past 10 years in Xinjiang. *Acta Prataculturae Sin.* **2014**, *23*, 39–50. (In Chinese)

34. Tan, B.W.; Luo, Z.L. Natural forest protection and forest ecological benefits in Xinjiang. *Henan Agric.* **2017**, *28*, 37–41. (In Chinese)

35. Zheng, S.L.; Xu, W.Q.; Yang, L. Carbon density and storage of forest ecosystem in Altay Mountain, Xinjiang. *J. Nat. Res* **2016**, *31*, 1553–1663. (In Chinese)

36. Ricardo, J.; Morgado, -D. Savitzky–Golay filtering as image noise reduction with sharp color reset. *Microprocess. Microsyst.* **2020**, *74*, 103006.

37. Alberto, F.; Pilar, L.; Jose Luis, C. Automatic mapping of surfaces affected by forest fires in Spain using AVHRR NDVI composite image data. *Remote Sens. Environ.* **1997**, *60*, 153–162.

38. Intergovernmental Panel on Climate Change. Climate Change 2007: Synthesis Report, Text, 2008; Geneva, Switzerland. Available online: https://digital.library.unt.edu/ark:/67531/metadc29351/ (accessed on 11 May 2022).

39. Parmesan, C.; Yohe, G. A globally coherent fingerprint of climate change impacts across natural systems. *Nature* **2003**, *421*, 37–42. [CrossRef] [PubMed]

40. Hirota, M.; Nobre, C.; Oyama, M.D. The climatic sensitivity of the forest, savanna and forest-savanna transition in tropical South America. *New Phytol.* **2010**, *187*, 707–719. [CrossRef]

41. Liu, S.R.; Guo, Q.S.; Wang, B. Prediction of net primary productivity of forests in China in response to climate change. *Acta Ecol. Sin.* **1998**, *16*, 32–37. (In Chinese)

42. Du, J.S.; Yu, D.Y. Impacts of climate change and human activities on net primary productivity of grassland in agro-pastoral transitional zone in northern China. *J. Beijing Norm. Univ. (Nat. Sci.)* **2018**, *54*, 365–372. (In Chinese)

43. Wang, Y.H.; Zhou, G.S. Responses of temporal dynamics of aboveground net primary productivity of Leymus chinensis community to precipitation fluctuation in Inner Mongolia. *Acta Ecol. Sin.* **2004**, *24*, 1140–1145. (In Chinese)

44. Yang, F.Q.; Geng, X.X.; Wang, R.; Zhang, Z.X.; Guo, X.J.A. synthesis of mineralization styles and geodynamic settings of the Paleozoic and Mesozoic metallic ore deposits in the Altay Mountains, NW China. *J. Asian Earth Sci.* **2018**, *159*, 233–258. [CrossRef]

45. Kathleen, S. MODIS Cloud Mask User's Guide. 2011. Available online: https://modis-images.gsfc.nasa.gov/_docs/CMUSERSGUIDE.pdf (accessed on 23 January 2022).

46. Du, J.; He, Z.B.; Piatek, K.B.; Chen, L.F.; Lin, P.F.; Zhu, X. Interacting effects of temperature and precipitation on climatic sensitivity of spring vegetation green-up in arid mountains of China. *Agric. Forest Meteorol.* **2019**, *269*, 71–77. [CrossRef]

47. Benjamin, P.; Neil, P.; Liu, H.Y.; Zhu, Z.C.; Rosanne, D.A.; Philippe, C.; Nicole, D.; David, F.; Caroline, L.; Ranga, M.; et al. Recent trends in Inner Asian forest dynamics to temperature and precipitation indicate high sensitivity to climate change. *Agric. Forest Meteorol.* **2013**, *178*, 31–45.

48. Wang, R.J.; Zhang, J.F.; Zhang, D.S.; Dong, L.S.; Qin, G.H.; Wang, S.F. Impacts of climate change on forest growth in saline-alkali land of Yellow River Delta, North China. *Dendrochronologia* **2022**, *74*, 125975. [CrossRef]

Severe Drought Still Affects Reproductive Traits Two Years Later in a Common Garden Experiment of *Frangula alnus*

Kristine Vander Mijnsbrugge *, Marc Schouppe, Stefaan Moreels, Yorrick Aguas Guerreiro, Laura Decorte and Marie Stessens

Department of Forest Ecology and Management, Research Institute for Nature and Forest, 9500 Geraardsbergen, Belgium
* Correspondence: kristine.vandermijnsbrugge@inbo.be

Abstract: Longer periods of intensified droughts in Western Europe are predicted due to ongoing climate change. Studying the responses of woody species during intense drought events can help toward understanding the consequences for forest ecosystems. We studied the effects of an intense summer water limitation on several reproductive traits, two years after the treatment, in *Frangula alnus* Mill. shrubs grown in a common garden. Drought-treated shrubs produced more berries one and two years after the drought event, while the height increment of the second post-treatment year was still significantly retarded. The mean weight of stones from berries picked two years after the drought treatment and their germination percentage, which was corrected for mean stone weight, were higher for the treated shrubs. These results indicate a resource re-allocation toward reproduction, rather than toward growth, which was still in action two years after the water limitation. The higher germination success, which is a transgenerational effect, and which has already been suggested to be an adaptation to survival in more stressful growth conditions, is also still detectable two years after the severe drought. *F. alnus* produces mature berries continuously during the whole summer. From the middle of July till the end of August, the counts of mature berries, the mean stone weight and the germination percentage, corrected for mean stone weight, decreased, whereas the timing of seedling emergence, also corrected for stone weight, advanced slightly. The timing of seedling emergence correlated weak but significantly with the timing of bud burst in the mother shrubs, with a variance analysis indicating a stronger genetic control for bud burst in comparison to seedling emergence. Several results corroborated previous findings. Population differentiation in the common garden was observed for mature berry counts and for several phenological traits. In conclusion, longer-term effects of drought on reproductive traits in woody species may add more complexity to the consequences of climate change on tree species distributions and survival of forest ecosystems.

Keywords: glossy buckthorn; provenance trial; berry count; seedling emergence; germination; water limitation; transgenerational effect; stone weight; variance analysis; bud burst

Citation: Vander Mijnsbrugge, K.; Schouppe, M.; Moreels, S.; Aguas Guerreiro, Y.; Decorte, L.; Stessens, M. Severe Drought Still Affects Reproductive Traits Two Years Later in a Common Garden Experiment of *Frangula alnus*. *Forests* **2023**, *14*, 857. https://doi.org/10.3390/f14040857

Academic Editor: Any Mary Petritan

Received: 23 February 2023
Revised: 18 April 2023
Accepted: 19 April 2023
Published: 21 April 2023

1. Introduction

Forests worldwide face increasing challenges, not least driven by climate change [1,2]. The frequency and the duration of drought and heat stress are rising and will continue to do so [3]. Climate change is already responsible for increasing tree mortality, which can lead to an altered composition and structure of forests [3,4]. To predict how forest ecosystems will evolve in the future, knowledge is needed on the impact of climate change on tree reproduction patterns [5]. It is clear that the reproduction of trees is influenced by climate change [6]. Still, our understanding of the mechanisms that govern tree fecundity is restricted [7]. Observational studies indicate that reproduction in trees can both increase and decrease when studied over a longer time period [8–10], but uncertainty remains in terms of how far the observed clines are responses to an altering climate [11]. Therefore, experiments that stress plants with environmental conditions that are predicted by climate change are justified to assess the impact on reproduction of forest trees and shrubs. Still, one has to keep in mind that the impact of, e.g., warming, fertilization or drought, should not be extrapolated without caution, as on average, these effects may be dampened over time [12].

Natural selection and phenotypic plasticity are two well-known mechanisms for plants, as sedentary organisms, to respond to an altering growth environment. Less investigated is the phenomenon of transgenerational plasticity as a way for plants to cope with changing growth conditions. Transgenerational effects in plants can be described as the impact that the parental environment can have on the offspring performance, excluding any influence of the genes that are transmitted from the parent to the descendants [13]. The inherited adjustments in the offspring can be an adaptation to the parental environmental conditions that caused these effects. The offspring that experiences a comparable environment than the parental one can therefore display a better fitness [14]. Diverse mechanisms can lead to adaptive transgenerational adjustments including altered seed provisioning and epigenetic modifications [13]. Perennial species have longer life cycles than annuals, and it can be suggested that transgenerational effects may play a more important role in their responses to altering growth environments [13]. The question remains as to how far they will help woody vegetations in keeping pace with the rapid climate change.

Because of their long life span, studies on the impact of an altering climate on tree fecundity and recruitment tend to remain observational in nature. It often needs many years before a tree starts to flower and fructify. Shrubs have the advantage of flowering and fructifying more quickly, permitting shrubs grown in a container to reproduce and thus allowing experimental studies specifically focusing on reproduction. Our study species was *Frangula alnus* Mill. (glossy buckthorn), which is an insect-pollinated and bird-dispersed shrub to small tree, characterized by an extended natural distribution range in Europe [15]. Similarly to most other shrub species, *F. alnus* has no economic significance, although the ecological importance in its natural range is beyond doubt, and is therefore not much studied. *F. alnus* flowers and produces berries on the year's shoot, which grows indeterminately, implying that mature berries are formed from early summer to early autumn.

Two drought experiments in a common garden of *F. alnus* have been described before. The experimental set-up consisted of a common garden with three provenances, including a local Belgian, a more northernly located Swedish and a more southernly located Italian one. In the first experiment, a summer drought was imposed on potted *F. alnus* plants, by withholding any watering for 27 days [16]. Due to the severity of the drought stress, 41% of the plants died off. Here, the Italian provenance suffered earlier and more from the imposed stress in comparison to the Belgian and Swedish provenances. This provenance displayed a retarded leaf senescence in the subsequent autumn. One year later, both bud burst and leaf senescence were advanced for all provenances that experienced the drought treatment. No effect was detectable anymore in the leaf phenological traits two years later. In the year of the drought stress, evidently, height growth was retarded, an effect still visible one year later. In the second drought experiment, a milder early summer drought stress was imposed on potted *F. alnus* plants, with plants displaying the first symptoms of leaf wilting, but without any plants dying off [17]. The drought treatment significantly reduced the amount of mature berry production in the same year of the treatment, and it did not affect stone weight. The germination percentage of the berries collected in the same year of the drought treatment, was higher among the drought-treated mother plants [18]. In addition, the timing of seedling emergence was advanced for the berries that were collected at the time of maximum berry production (middle of July). The timing of seedling emergence among the three provenances followed the same order as the timing of bud burst, possibly suggesting a common genetic basis.

The question remains whether drought stress may still display legacies on reproductive traits in the years after the water limitation. The main objective of the study here presented was to measure the effects of the severe water withholding treatment in the first experiment on *F. alnus* [16], as described above, specifically on the reproductive traits two years after the drought treatment. We observed mainly reproductive traits up to two years after the drought treatment. The following research questions were put forward. (i) Did the severe drought treatment influence reproductive traits up to two years after the treatment? (ii) Were transgenerational effects visible two years after the drought treatment? (iii) Were the after-effects provenance dependent?

2. Materials and Methods

2.1. Common Garden

This study builds on a common garden experiment of the shrub *F. alnus* that was established as described before [16]. In short, mature berries were harvested in three natural populations located in Italy (lat. 43.12181, lon. 11.17654), Belgium (lat. 51.08424, lon. 4.793124) and Sweden (lat. 62.44210, lon. 17.23451) (Figure S1). The sites are characterized by an annual mean temperature and an annual precipitation of 13.7 °C and 706 mm for the Italian provenance, 10.1 °C and 785 mm for the Belgian provenance and 2.9 °C and 682 mm for the Swedish provenance, respectively (data from WorldClim [19] and already reported in [16]). Plants were grown in the nursery of the Research Institute of Nature and Forest (Geraardsbergen, Belgium), following standard nursery techniques. In 2016, 8 cuttings were taken from every genotype (8 genotypes for the Italian, 17 genotypes for the Belgium and 14 genotypes for the Swedish provenance) and further raised as container plants using standard potting soil (organic matter 20%, pH 5.0–6.5, Electrical Conductivity 450 μS cm^{-1}, dry matter 25%, 1.5 kg m^{-3} powdered compound fertilizer NPK 12 + 14 + 24). In this way, the common garden was established with 309 plants, and a drought experiment in a greenhouse was performed in 2018 as described before [16]. In summary, half of the plants (with four clones for each genotype) did not receive any water for 28 days in the summer of 2018, after which they were again well watered. In this process, 64 plants died off in the drought-treated group of plants, leaving 245 plants in total in the remaining common garden. After the drought experiment, plants were individually intermingled and further raised as container plants on an outdoor container field. In the spring of 2020, immediately after bud burst, all plants were pruned at 10 cm above soil level.

2.2. Measurements

Height of the plants was measured at the end of 2019 and 2020. As the plants were pruned in the beginning of 2020, the height of the plants at the end of 2020 was a proxy for the height increment of this year.

F. alnus produces berries from an early age onwards, so reproductive traits can be studied on young container plants. As plants flower on the current year of growth, berries ripen continuously in summer and early autumn. Mature berries of *F. alnus* are purple-black colored. All mature berries were counted on each plant separately on 12 August 2019, one year after the drought treatment, in both the control and the water-limited group.

In 2020, mature berries were counted on each plant separately throughout the summer, starting when the first black berries appeared and lasting till no new black berries were produced (counting days on 3, 9, 16 and 24 July and 7 and 21 August 2020). On every counting day, all black berries were also picked, avoiding double counting. Except for the first two counting days, the picked berries were kept (from 16 July till 21 August) and were pooled for the ramets belonging to the same genotype and having received the same treatment (drought or control) two years before. Stones were extracted from each batch of berries by hand, washed and air dried for a few days. Then, the batches of stones were weighted. About two weeks later, the batches were stratified using sand and nursery potting soil in equal volumes. Tray cells were filled with the sand-soil mixture up to one cm from the top. The batches of stones were scattered on the mixture in every cell and were covered with the same mixture. The filled trays were placed in a refrigerator at 4 °C for a cold stratification, until the end of January 2021. During the cold stratification, the soil mixture was kept moist. Together, 218 different batches of cleaned stones were stratified, containing all together 7406 stones (Table 1).

Table 1. Number of stratified batches ($n_b\,^\circ$) and sum of stones in the batches ($n_s\,^\circ$), for each provenance of the mother plants, for each treatment of the mother plants in 2020 (control and drought) and for each berry collection day in 2020.

Berry Collection Day	Belgian				Italian				Swedish				Total	
	Control		Drought		Control		Drought		Control		Drought		Control	Drought
	$n_b\,^\circ$	$n_s\,^\circ$	$n_b\,^\circ$	$n_s\,^\circ$	$n_b\,^\circ$	$n_s\,^\circ$	$n_b\,^\circ$	$n_s\,^\circ$	$n_b\,^\circ$	$n_s\,^\circ$	$n_b\,^\circ$	$n_s\,^\circ$	$n_s\,^\circ$	$n_s\,^\circ$
16 July	17	1214	16	954	7	442	5	118	14	648	13	564	2304	1636
24 July	17	724	14	729	8	234	4	47	13	445	8	119	1403	895
7 August	17	462	16	332	6	101	3	30	7	115	3	12	678	374
21 August	9	52	12	42	2	4	1	2	1	3	5	13	59	57

In February 2021, the trays were transferred from the refrigerator to a non-heated but frost-free greenhouse. Germination advanced in the beginning of March and emerging seedlings were counted on a regular basis until all seedlings had appeared. Counting was performed on 8, 12, 15, 19, 22, 26 and 29 March, and on 2, 6, 9 and 16 April 2021.

Germination percentage was calculated by dividing the count of emerged seedlings on the last counting day in a given batch by the number of stratified stones in the batch. Germination percentage is a proxy for the viability of the stones. The sequence of counts over time allowed us to study the timing of emergence. For this, percentages of emergence were calculated by dividing each count of each batch on every counting day by the count of this batch on the last counting day.

In the autumn of 2020, leaf senescence was evaluated on the mother shrubs in the common garden using a 5-level scoring protocol as follows: 1: green leaves; 2: light green leaves; 3: yellowing leaves; 4: leaves turning brown; 5: leaves falling off. The process of leaf senescence was observed on 18 September and on 5 and 26 October 2020.

In the spring of 2021, bud burst was evaluated on the mother shrubs in the common garden using a 5-level scoring protocol as follows: 1: buds in winter rest; 2: buds opening and first green tissue visible; 3: green tissue sliding out of the bud but not yet unfolding; 4: first leaves opening but not yet fully opened; 5: leaves fully opened. Bud burst scoring was performed on a regular basis on 9, 16, 23 and 30 April, and on 7, 14 and 21 May 2021.

2.3. Statistical Analysis

The obtained data were processed in R (version 4.2.1, Vienna, Austria [20]). The first focus of the analysis was on the putative influence of the treatment of the mother shrubs in 2018 on the different response variables. Additionally, we focused on a putative population differentiation in the common garden.

Linear and generalized linear mixed models were applied using the package nlme [21]. The variable T represented the drought treatment of the mother shrubs in 2018 with "control" and "drought" as levels. The variable P denoted the provenance of the mother shrubs with "Be" for Belgian, "It" for Italian and "Sw" for Swedish. The variables H19 and H20 represented the height of the mother plants at the end of 2019 and 2020, respectively. The variable C indicated the days of berry counts on the mother plants in 2020. The variable D represented the days that emerged seedlings were counted or that bud burst was scored on the mother shrubs, both in the spring of 2021. The variable S denoted the mean stone weight.

In the random part of all the models, a unique identifier for every genotype was present. A unique identifier for every mother plant was added in the random part of the model for the response variable berry count in 2020, and also for the response variable timing of bud burst in the spring of 2021, as repeated counts or observations were performed on the same plants. Another unique identifier for pooled ramets within a genotype (pooled for the drought treatment and pooled for the control) was added to the random part of the models for the response variables mean weight of a stone, germination percentage and timing of seedling emergence, to account for the repeated collections of berries on the same mother shrubs. Finally, a unique identifier for every batch of stones was added to the model of the timing of seedling emergence to account for the different counting events on the same batches of stones.

A linear mixed model was fit to the height increment of the mother plants in 2020 (H20):

$$H20 = \alpha_{H20} + \beta_{PH20}P + \beta_{TH20}T \tag{1}$$

A generalized mixed model with Poisson distribution was fit to the counts of mature berries on every shrub that produced mature berries on 12 August 2019 (B19):

$$B19 = \alpha_{B19} + \beta_{PB19}P + \beta_{TB19}T + \beta_{H19B19}H19 \tag{2}$$

A generalized mixed model with Poisson distribution was fit to the counts of mature berries on every shrub in 2020 (B20):

$$\begin{aligned} B20 = {} & \alpha_{B20} + \beta_{PB20}P + \beta_{TB20}T + \beta_{H20B20}H20 + \beta_{CB20}C + \beta_{C2B20}C^2 + \beta_{C3B20}C^3 \\ & + \beta_{PCB20}PC + \beta_{PC2B20}PC^2 + \beta_{PC3B20}PC^3 \end{aligned} \tag{3}$$

The genotypes for which only ramets in the control group of mother shrubs survived the treatment in 2018, were excluded from the dataset. A polynomial to the third degree for day of berry collection (C) was added in the model as the raw data indicated a non-linear curve of mature berry counts over time. An interaction term between day of berry collection and provenance allowed the berry counts to vary over time in a different way for the different provenances.

Mean stone weight (S) was modeled using a linear mixed model:

$$S = \alpha_S + \beta_{PS}P + \beta_{TS}T + \beta_{CS}C + \beta_{C2S}C^2 \tag{4}$$

A polynomial to the second degree for day of berry collection was added in the model as the raw data indicated a non-linear curve of mean stone weight over time.

Germination percentage (Gp) was modelled using a linear mixed model:

$$Gp = \alpha_{Gp} + \beta_{PGp}P + \beta_{TGp}T + \beta_{CGp}C + \beta_{SGp}S \tag{5}$$

The chance (p_1) was modeled for a seedling to have already emerged on a given day. A binomial generalized linear mixed model was fit to the seedling count data over time, with the count of germinated seedlings on the last count day as the "weight" argument in the model. The model formula is as follows:

$$\log(p_1/(1 - p_1)) = \alpha_{p1} + \beta_{Pp1}P + \beta_{Tp1}T + \beta_{Dp1}D + \beta_{Cp1}C + \beta_{Sp1}S \tag{6}$$

To allow a correlation analysis between the timing of seedling emergence and the timing of bud burst of the mother plants in the spring of 2021, two models were run. As treatment in 2018 was not significant in either of the models, it was omitted from the fixed part of the models. The model for bud burst was based on cumulative logistic regression, as the response variable bud burst was of an ordinal type. The package 'ordinal' was used [22]. The chance (p_2) was modeled for a mother plant to have reached at minimum a given bud burst score (i.e., to have reached a bud burst score or a score higher than this score on a specific day). Model formula is as follows:

$$\log(p_2/(1 - p_2)) = \alpha_{p2} - \beta_{Pp2}P - \beta_{Dp2}D - \beta_{H20p2}H20 - \text{random effects} \tag{7}$$

The random part consisted of a unique identifier for every genotype and a unique identifier for every mother plant.

Timing of germination was modeled for the berry collection on 16 July 2020 using a binomial generalized linear mixed model with the formula:

$$\log(p_3/(1 - p_3)) = \alpha_{p3} + \beta_{Pp3}P + \beta_{Dp3}D + \beta_{Sp3}S + \text{random effects} \tag{8}$$

The random part consisted of a unique identifier for every genotype and a unique identifier for every batch of collected stones on 16 July 2020.

Based on Formulas (7) and (8), the modelled day was calculated when the chance was 50% to have reached a bud burst score of 3 or higher (bud burst scorings on the mother shrubs, D_{50bb}), and to have emerged (germinating seedlings, D_{50ge}), respectively. The random effects were added in this calculation so that a D_{50} was calculated for every mother shrub (bud burst) and for every batch of stratified stones that was collected on the 16th of July 2020 (seedling emergence), respectively. For the calculations, the mean height of a mother shrub in 2020 (mH20) and the mean weight of a seed (mS) were used.

$$D_{50bb} = (\alpha_{p2} - \beta_{Pp2} - \beta_{H20p2}mH20 - \text{random effects})/\beta_{Dp2} \tag{9}$$

$$D_{50ge} = (\alpha_{p3} + \beta_{Pp3} + \beta_{Sp3}mS + \text{random effects})/\beta_{Dp3} \tag{10}$$

A Pearson correlation coefficient was calculated between the D_{50} values for bud burst of the mother plants and the D_{50} values for seedling emergence.

To allow a partitioning of variance analysis, both models (7) and (8) were run as linear mixed models in which the variance of the random part was partitioned in variance between genotypes, in variance between ramets (bud burst) or pooled ramets (seedling emergence) within genotypes, and in residual variance. As the plants shared the same growth environment, the variance between genotypes had a genetic base, which was not the case for the variance between (pooled) ramets and the residual variance.

Leaf senescence in 2020 was modelled using cumulative logistic regression. The chance (p_4) was modeled for a mother plant to have reached at maximum a given leaf senescence score on a given day:

$$\log(p_4/(1 - p_4)) = \alpha_{p4} - \beta_{Pp4}P - \beta_{Tp4}T - \beta_{Dp4}D - \beta_{H20p4}H20 \tag{11}$$

The random part consisted of a unique identifier for every genotype and a unique identifier for every mother plant.

3. Results

3.1. Height Increment, Two Years after the Summer Drought Treatment

Height increment growth in 2020, two years after the summer water withholding treatment, was still clearly negatively affected by the treatment (treatment *p*-value < 0.001 in Table 2 and Figure 1).

Table 2. Statistical analysis of height increment in 2020. In the model, the Belgian provenance is the standard to which the Italian (It) and Swedish (Sw) provenances are compared, and for the treatment in 2018 (T), the control is the standard to which the water limitation is compared.

Variable	Estimate	Std. Error	df	*t*-Value	*p*-Value
(Intercept)	64.948	2.404	196	27.021	<0.001 ***
It	4.528	4.625	33	0.979	0.335
Sw	−6.117	3.480	33	−1.758	0.088
T	−9.095	1.830	196	−10.434	<0.001 ***

*** $p < 0.001$.

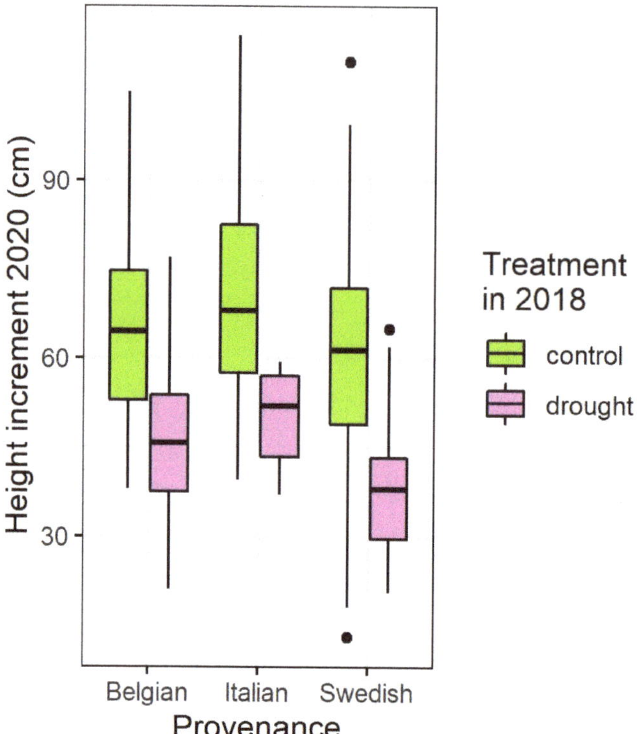

Figure 1. Boxplot representing the height increment of the shrubs in the common garden in 2020, according to the provenances and treatment in 2018.

3.2. Mature Berry Counts, One and Two Years after the Drought Treatment

One year after the drought treatment (2019), counts of berries were significantly larger on the drought-treated mother shrubs compared to the controls (p-value < 0.001 for treatment in Table 3 and Figure 2a,b).

In 2020, berries were counted on a regular basis in the growing season, facilitating the study of the berry production over time. Mature berry production displayed a clear maximum in the middle of the growing season and was still slightly increased by the drought treatment in 2018 (treatment p-value = 0.040 in Table 3 and Figure 3). The Italian and the Swedish provenances both produced less mature berries compared to the local Belgian provenance (p-values = 0.002 and < 0.001 for the Italian and Swedish provenance, respectively, in Table 3 and Figure 3a). In addition, the Swedish provenance displayed an earlier maximum of berry production in time (p-values < 0.001 for the interaction between day of berry collection in 2020 and the Swedish provenance in Table 3 and Figure 3a). Higher plants produced slightly more mature berries (p-value = 0.042 in Table 3 and Figure 3b).

3.3. Mean Stone Weight

The mean weight of a stone for the pooled ramets within a genotype with the same treatment in 2018, was slightly higher for the drought-treated mother shrubs (p-value = 0.031 for treatment in Table 4 and in Figure 4a and Figure S3). The day when the berries were collected from which the stones were extracted, significantly influenced the mean stone weight, showing a maximum mean weight in the middle of the growing season (p-values < 0.001

for C and C^2 in Table 4 and Figure 4a). No population differentiation was present between the different provenances.

Table 3. Statistical analysis of the mature berry counts in 2019 and 2020. In the model, the Belgian provenance is the standard to which the Italian (It) and Swedish (Sw) provenances are compared, and for the treatment in 2018 (T), the control is the standard to which the water limitation is compared.

Year	Variable	Estimate	Std. Error	z-Value	*p*-Value
2019	(Intercept)	0.840	0.377	2.23	0.026 *
	It	−0.153	0.391	−0.392	0.695
	Sw	−0.592	0.370	−1.600	0.110
	T	0.754	0.123	6.119	<0.001 ***
	H19	0.002	0.004	0.540	0.589
2020	(Intercept)	0.45	0.26	1.76	0.079
	It	−0.97	0.31	−3.13	0.002 **
	Sw	−0.83	0.22	−3.74	<0.001 ***
	T	0.22	0.11	2.05	0.040 *
	H20	0.01	0.00	2.03	0.042 *
	C	−18.67	1.45	−12.89	<0.001 ***
	C^2	−28.35	1.09	−26.08	<0.001 ***
	C^3	9.08	0.95	9.58	<0.001 ***
	C:It	−7.62	4.95	−1.54	0.123
	C^2:It	−6.55	3.47	−1.89	0.059
	C^3:It	5.77	2.94	1.97	0.049 *
	C:Sw	−29.60	3.98	−7.44	<0.001 ***
	C^2:Sw	5.50	2.78	1.98	0.048 *
	C^3:Sw	1.56	1.95	0.80	0.423

*** $p < 0.001$; ** $p < 0.01$; * $p < 0.05$. H19 and H20: height measurements in 2019 and 2020, respectively; C: day of berry counting.

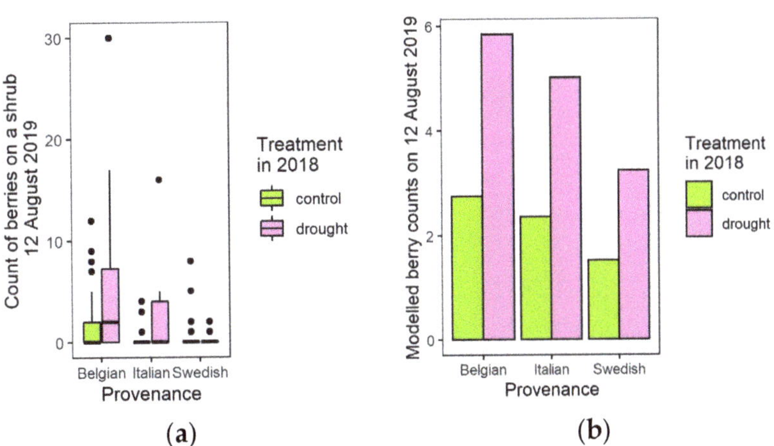

Figure 2. Actual (**a**) and modelled count of berries (**b**) on a shrub according to the treatment in 2018 and the provenance.

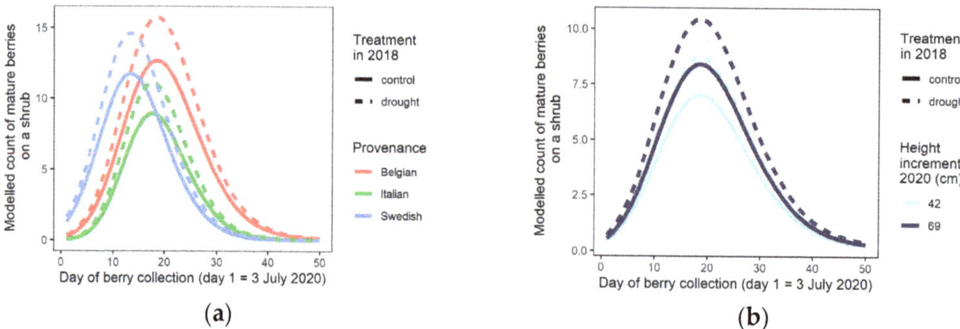

Figure 3. Modelled mature berry production on a shrub over time in 2020 for drought-treated and control plants, according to the provenance (**a**) and to the height of the plants (**b**).

Table 4. Statistical analysis of the mean weight of a stone in 2020. In the model, the Belgian provenance is the standard to which the Italian (It) and Swedish (Sw) provenances are compared, and for the treatment in 2018 (T), the control is the standard to which the water limitation is compared.

Variable	Estimate	Std. Error	df	*t*-Value	*p*-Value
(Intercept)	0.0166	0.0006	131	28.15	<0.001 ***
It	0.0018	0.0012	31	1.56	0.128
Sw	0.0008	0.0009	31	0.96	0.343
T	0.0008	0.0004	34	2.25	0.031 *
C	−0.0205	0.0022	131	−9.27	<0.001 ***
C^2	−0.0172	0.0022	131	−7.84	<0.001 ***

*** $p < 0.001$; * $p < 0.05$. C: day of berry collection.

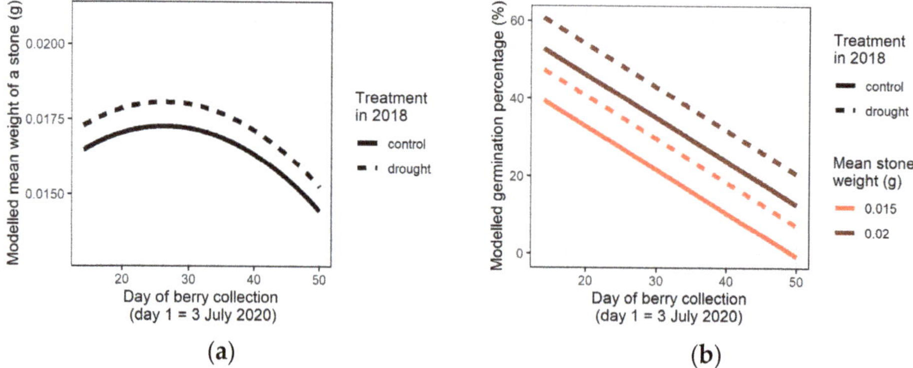

Figure 4. Modelled mean weight of a stone, according to the treatment in 2018 and to the day of berry collection in 2020 (**a**) and modelled germination percentage according to the treatment in 2018, to the day of berry collection in 2020 and to the mean stone weight (**b**).

3.4. Germination of the Stones in the Spring of 2021

Germination in the different batches of stones was observed carefully in the spring of 2021. The germination percentage, as the ratio between the number of germinated stones and the total number of stones in a batch, was significantly higher for the batches of stones derived from the mother shrubs that experienced the drought treatment in 2018 (*p*-value = 0.003 for T in Table 5 and in Figure 4b and Figure S4). Stones originating from the berry collections in the middle of July till the end of August 2020 displayed a decrease in germination percentage over time (p-value < 0.001 for C in Table 5 and Figure 4b). Heavier stones showed also a higher germination (p-value < 0.001 for S in Table 5).

Table 5. Statistical analysis of the germination percentage and the timing of seedling emergence in the spring of 2021. In the model, the Belgian provenance is the standard to which the Italian (It) and Swedish (Sw) provenances are compared, and for the treatment in 2018 (T), the control is the standard to which the water limitation is compared.

	Germination Percentage					Timing of Seedling Emergence			
Variable	Estimate	Std. Error	df	*t*-Value	*p*-Value	Estimate	Std. Error	*z*-Value	*p*-Value
(Intercept)	15.22	9.39	128	1.62	0.107	−4.04	0.46	−8.80	<0.001 ***
It	−8.02	5.20	31	−1.54	0.133	1.01	0.28	3.60	<0.001 ***
Sw	−1.11	3.85	31	−0.29	0.775	0.75	0.20	3.73	<0.001 ***
T	7.93	2.44	33	3.25	0.003 **	−0.24	0.17	−1.44	0.151
S	2661.75	434.73	128	6.12	<0.001 ***	57.89	21.57	2.68	0.007 **
C	−1.12	0.11	128	−9.86	<0.001 ***	0.02	0.01	2.24	0.025 *
D						0.21	0.00	90.95	<0.001 ***

*** $p < 0.001$; ** $p < 0.01$; * $p < 0.05$. S: mean weight of a stone; C: day of berry collection in 2020; D: counting day of emerged seedlings in the spring of 2021.

The treatment of the mother shrubs in 2018 did not influence the timing of germination (no significant p-value in Table 5). The first stone harvest originating from the berries collected in the middle of July displayed the latest germination, while the later the berries were collected, the earlier the seedlings emerged (p-value = 0.025 for C in Table 5 and Figure 5a). Heavier stones also displayed an advancement of the germination (p-value = 0.007 for S in Table 5 and Figure 5b). Additionally, contrary to the germination percentage, the Italian and Swedish provenances germinated earlier than the local Belgian provenance (p-values < 0.001 for both provenances in Table 5 and Figure 5b).

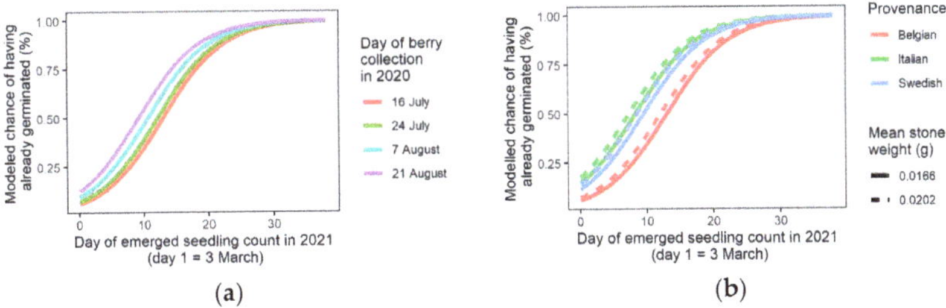

Figure 5. Modelled timing of seedling emergence in the spring of 2021, according to the day of berry collection in 2020 (**a**) and the provenance and mean stone weight (**b**).

3.5. Timing of Bud Burst and Seedling Emergence

Two models were run to allow a comparison between the timing of bud burst on the mother shrubs in the common garden and the timing of seedling emergence, two phenological traits that were observed in the spring of 2021. As the treatment of the mother shrubs in 2018 had no effect anymore on these two phenological traits, this variable was omitted from the models. Bud burst in the Italian provenance occurred earlier than the Belgian provenance (p-value = 0.002 for It in Table 6 and Figure 6a, whereas the Swedish provenance only showed a tendency for earlier bud burst (p-value = 0.070 for Sw in Table 6 and Figure 6a). Seedlings from the Italian and Swedish provenances emerged earlier than the Belgian seedlings (p-values = 0.016 and 0.010, respectively, in Table 6).

Table 6. Statistical analysis of the timing of bud burst on the mother shrubs and the timing of seedling emergence, both in the spring of 2021. In the model, the Belgian provenance is the standard to which the Italian (It) and Swedish (Sw) provenances are compared.

Variable	Timing of Bud Burst on Mother Shrubs				Timing of Seedling Emergence			
	Estimate	Std. Error	z-Value	*p*-Value	Estimate	Std. Error	z-Value	*p*-Value
It	−1.68	0.55	−3.03	0.002 **	0.87	0.36	2.42	0.016 *
Sw	−0.83	0.46	−1.81	0.070	0.71	0.28	2.58	0.010 *
H20/S	0.04	0.01	6.41	<0.001 ***	100.33	60.47	1.66	0.097
D	−0.16	0.01	−29.36	<0.001 ***	0.19	0.00	70.59	<0.001 ***

*** $p < 0.001$; ** $p < 0.01$; * $p < 0.05$. H20: height of the mother shrubs in 2020 (bud burst); S: the mean weight of a stone (seedling emergence); D: day of bud burst observation or the counting day of emerging seedlings.

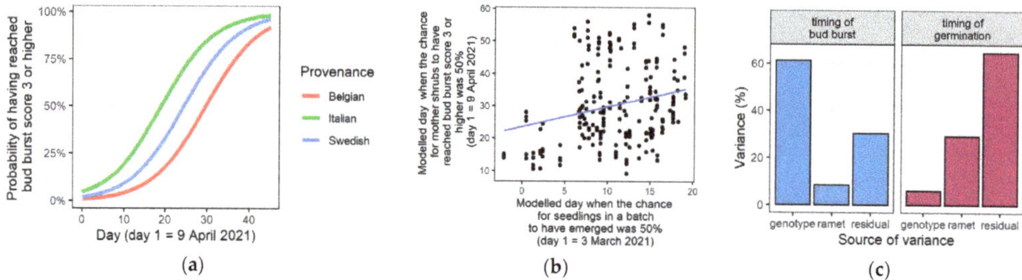

Figure 6. Modelled timing of bud burst on the mother shrubs in the spring of 2021, according to the provenance (**a**). Scatter plot comparing the modelled days when the chance for seedlings to have emerged was 50% with the modelled days when the chance for the mother shrubs to have reached at least bud burst score 3 was 50%. Linear regression line is shown (**b**). Partitioning of variance for bud burst and seedling emergence in the spring of 2021 (**c**).

For every mother shrub in the common garden, the modelled day was calculated when the chance to have reached at least bud burst score 3 was 50% (D_{50bb}). Similarly, for every batch of stones collected on 16 July 2020, the modelled day was calculated when the chance to have already emerged was 50% (D_{50ge}). The correlation coefficient between the D_{50bb} and the D_{50ge} values was 0.25 (p-value < 0.001) (Figure 6b).

The nested structure of the data allowed us to conduct a partitioning of variance analysis. For this, the two models used for the correlation analysis were run once more, but as linear models. This approximation for the ordinal and binomial data was applied as only linear models allow a partitioning of variance analysis. Because of the approximation, results should be taken with care. Still, there was a clear difference between the variance structure of the two models. For timing of bud burst, the relative variance between genotypes was larger than the relative variance between the ramets within a genotype together with the residual variance (Figure 6c). For the timing of seedling emergence, the relative variance between genotypes was very small in comparison to the relative variance which had no genetic base (variance between ramets within a genotype and residual variance).

3.6. Leaf Senescence

The timing of leaf senescence in the autumn of 2020 was not influenced anymore by the treatment in 2018 (no significant p-value in Table 7). The Swedish provenance displayed an advanced leaf senescence when compared to the other two provenances (p-value < 0.001 for Sw in Table 7 and Figure 7).

Table 7. Statistical analysis of the timing of leaf senescence on the mother shrubs in the autumn of 2020. In the model, the Belgian provenance is the standard to which the Italian (It) and Swedish (Sw) provenances are compared, and for the treatment in 2018 (T), the control is the standard to which the water limitation is compared.

Variable	Estimate	Std. Error	z-Value	p-Value
It	0.009	0.455	0.019	0.984
Sw	3.244	0.401	8.088	<0.001 ***
T	−0.049	0.210	−0.236	0.813
H20	0.023	0.006	3.747	<0.001 ***
D	0.205	0.010	20.287	<0.001 ***

*** $p < 0.001$. H20: height of the mother shrubs in 2020; D: day of leaf senescence observation.

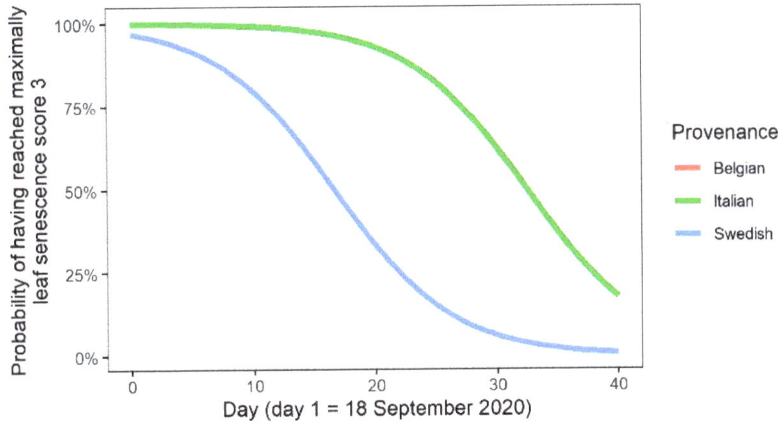

Figure 7. Modelled timing of leaf senescence on the mother shrubs in the autumn of 2020, according to the provenance. Belgian and Italian lines overlap.

4. Discussion

4.1. Influence of Former Water Limitation on Growth and Reproduction

A carbon isotope study showed that fruit production in several temperate tree species relies on carbohydrates that were synthesized through photosynthesis during the months preceding the fruit production [23], allowing a plastic response of fructification to variable growing conditions in the foregoing period. In our experiment, up to two years after the water withholding, the drought-treated shrubs in the common garden produced more berries than the non-treated shrubs, suggesting that a longer-lasting signal diverted newly synthesized carbohydrates toward berry production. In the second post-drought year, the shrubs still displayed a clear retarded height growth, pointing to a re-allocation of resources to reproduction rather than to growth. It should be noted that berry production in *F. alnus* in the year of a drought period, was found to be decreased and not yet increased in the treated group of plants [16], suggesting that a re-allocation of resources toward berry production is a longer-term effect. A similar observation is described for an 18-year-lasting rainfall-reduction experiment in a Mediterranean setting, where *Quercus ilex* and *Arbutus unedo* produced more fruits in comparison to controls [24]. On the other hand, an observational study of monospecies stands of *Fagus sylvatica*, *Q. petraea* and *Q. robur*, during a time period of 8 years, displayed no evidence for a trade-off between vegetative growth and reproduction, with seasonal temperatures positively correlating mainly with crown cover and fruit production [25]. This phenomenon was also described in other studies: e.g., in recent decades, a clear concurring increase in vegetative and reproductive growth was observed in *Q. petraea* and *Q. robur* stands in central Europe [26]. It is likely that favorable meteorological conditions may promote the buildup of resources, thus facilitating an allocation to both growth and reproduce together, whereas the maintenance of reproduction at the expense of growth, as observed in our common garden, may happen in more stressful growth conditions or as a response to it. In a tree ring analysis of *F. sylvatica*, the largest reductions in radial growth were observed in years with massive seed production coinciding with summer drought [27]. By shifting the allocation of resources to reproduction rather than to growth, woody plants may raise the probability of successful offspring at the expense of a higher probability of mortality, as the resources not allocated to growth may decrease stress defenses [28]. Stress-induced reproduction has mainly been studied in herbaceous plant species [29]. Seed production in temperate forest tree species often rely on mast seeding, the synchronous intermittent production of large seed crops, a mechanism that reduces the amount of seeds that are consumed by predators [30]. The responses of reproduction to stress events in mast seeding tree species is rather complex [31]. Still, some studies have found a possible link between drought events and mast seeding. Drought in early summer is a good predictor for masting in the following year for *F. sylvatica* and *F. grandifolia* [32]. In *Abies alba* Mill, a dry spring that was followed by a humid spring the next year could be correlated with high cone production two years later [33].

Two years after the drought treatment, stones extracted from the berries picked from drought-treated mother shrubs were slightly heavier than the controls. Additionally, these stones displayed a slightly higher germination percentage than controls. It should be noted that stone weight was included in the germination percentage model as a co-factor, indicating that the germination percentage was higher for stones derived from the drought-treated mother plants when considering equal weights of stones. A milder drought experiment in *F. alnus* (no plants died from the experienced stress) already showed that the germination percentage of stones derived from berries picked from mother shrubs in the same year that drought stress was imposed on these mother shrubs was higher in comparison to the controls [18], whereas the stone weight was not affected [17]. In the experiment here described, the imposed drought stress on the mother shrubs was more severe, as 41% of the plants had died off during the drought period, which may be a reason why stones derived from the drought-treated mother plants were heavier. The drought stress experienced by the mother shrubs, which affected the germination percentage of the derived stones, suggests a transgenerational effect of the drought, both in the same year of the drought treatment [18] and two years after the treatment (this study). It is likely that this transgenerational effect raised the chances of seedling survival in case the growth environment would remain stressful. This result may be related to the observation that the offspring of drought-treated grassland species produced higher amounts of seed when compared to offspring from control mother plants [34,35]. The higher amount of seeds would be needed to keep proper chances of survival in a continuing stressful growth environment. Similarly, in several annual plant species, dry-grown parents were shown to produce seeds with higher germination percentages [35,36]. Heavier stones derived from mother shrubs that experienced drought stress may be related to the finding that drought tolerance and heavy seeds are correlated functional traits, likely reflecting a general plant strategy [37]. A higher seed mass facilitates a quicker root development, in turn allowing germinating seedlings to rapidly escape a dry surface of the soil, thus enhancing the chances of survival [38].

4.2. Population Differentiation in the Common Garden

Berry counts and some phenological traits that we studied in the common garden displayed population differentiation, corroborating earlier findings. The local Belgian provenance produced more mature berries than the other two provenances, both in an earlier study [17] and in the experiment here described. As hypothesized before, this may suggest that in *F. alnus*, locally adapted provenances can invest more in reproduction. A higher reproductive output as an expression of local adaptation has been detected in reciprocal transplant experiments of several herbaceous plants [39,40].

The peak of berry production in 2020 was advanced in the Swedish provenance, while the Belgian and Italian provenances peaked congruently, as was already observed in a comparable experiment [17]. Based on the berry and leaf phenologies, it can be suggested that the phenology of berry maturation may be correlated with the timing of leaf senescence, as for the latter, the Swedish provenance was also advanced when compared to the other two provenances that displayed a similar timing. In monocarpic plants, the timing of seed maturation is closely associated with the timing of leaf senescence, as nutrients from the leaves are re-allocated to maturing seeds [41]. An indirect indication of a correlation between the phenologies of leaf senescence and seed maturation in a woody species was found in alternate-bearing pistachio (*Pistacia vera* L.), where 61% of the N storage originated from soil uptake in a non-reproductive year, whereas all the N was drawn from senescing leaves in a reproductive year [42].

Bud burst in the mother shrubs of the Italian and Swedish provenances started earlier (Italian) or tended to start earlier (Swedish) than the Belgian provenance. Interestingly, seedlings from the Italian and Swedish provenances likewise emerged earlier than the Belgian seedlings. These phenomena were already observed in a similar common garden experiment in *F. alnus* [18]. As was postulated, these findings may corroborate the hypothesis that bud and seed phenology in general may share a common genetic basis [43]. Some evidence has already been published. In apple fruit tree (*Malus x domestica* Borkh.) research, timing of bud burst and timing of seed germination were found to be related, and it was suggested that buds and seeds needed a similar chilling requirement [44]. Likewise, the chilling requirements for bud burst and seed germination were found to be correlated in walnut (*Juglans regia* L.) [45]. Here, we studied the putative relationship between the two phenological traits in more detail, and we found only a weak but significant correlation between the two phenophases. From a variance partitioning analysis, it could be deduced that the timing of bud burst was more strongly genetically defined than the timing of seedling emergence. The timing of bud burst is adapted to the local growth environment, leading to a higher relative genetic control [46] and high heritability values for this recurrent phenophase [47,48]. Seedling emergence, a process that only happens once in the lifetime of a tree, needs strong fine tuning with the very local micro-climate, and therefore likely benefits more from plasticity in its timing than bud burst. Likewise, it was already shown in a variance partitioning analysis that the timing of seedling emergence was less genetically determined in comparison to germination percentage [18]. It can be argued that once established, the rare occurrence of late spring frost damage on bursting buds and unfolding leaves may be less detrimental for the survival of the shrub or tree in comparison to an emerging seedling, justifying the need for a stronger plasticity in seedling emergence, which likely has its cost. This can be related to the finding in a reciprocal transplant experiment of *F. sylvatica*, including seven populations from south to north in its European distribution range, that germination was highly plastic in response to environmental conditions [49]. Additionally, environmental cues were found to tightly control the timing of germination in three *Nothofagus* species [50].

4.3. Phenology of Berry Maturation

As *F. alnus* flowers and fructifies on the current years' shoots, mature berries are produced continuously during the whole summer. From the middle of July, the peak moment of berry production, till the end of August, the counts of mature berries, the mean stone weight and the germination percentage decreased in our study, whereas the timing of seedling emergence advanced slightly. A higher berry count co-occurring with a higher germination percentage may be related to silver birch (*Betula pendula* Roth), where seeds originating from trees that produced higher amounts of seeds also displayed a higher germination success [51]. It remains an intriguing question why *F. alnus* invests in berry production after the peak moment in the middle of July, for which survival chances of the offspring are lowered over time. As many tree species invest in inter-annual irregular seed production, i.e., the mast seeding, as a strategy to avoid seed losses [30], it could be hypothesized that berry production over a longer time intra-annually may also serve as a type of safeguard against seed loss at a certain time within the seasonal berry maturation period.

The advanced seedling emergence of the berries that were picked later in the growing season in our experiment, an effect that was corrected for stone weight (mean stone weight was present as a co-factor in the model), could be related to an earlier germination for seeds originating from genotypes that displayed later fruit maturation, as already found in apple (*M. x domestica*) [52].

5. Conclusions

We found that a severe water limitation can show legacy effects at least up to two years after the stress event. Height increment was still reduced while berry count, stone weight and germination percentage were raised in the treated plants, indicating a post-drought re-allocation of resources. The higher stone weight and the higher germination percentage two years after the drought treatment of the mother plants can be considered as transgenerational effects, which may help plants to survive in deteriorating environments. The observation that the local provenance produced more berries in comparison to a more northernly and a more southernly originating provenance suggests that translocation of provenances as an anticipation to the warming climate (assisted migration) should be considered with care.

Supplementary Materials: The following supporting information can be downloaded at: https://www.mdpi.com/article/10.3390/f14040857/s1. Figure S1: Distribution map of *Frangula alnus* in Europe, freely available at euforgen.org (accessed on 12 October 2022), with indication of the sampling sites (asterisk); Figure S2: Boxplots of mature berry counts in 2020, according to the treatment in 2018, to the provenance and to the berry collection days in 2020; Figure S3: Boxplots of the mean weight of a stone, according to the treatment in 2018, the provenance and the day of berry collection in 2020; Figure S4: Boxplots of the germination percentage, according to the day of berry collection in 2020, to the provenance and to the treatment in 2018.

Author Contributions: The described experiment was conceptualized by S.M. and K.V.M. The methodology was also demarcated by S.M. and K.V.M. Plants were grown by S.M. and M.S. (Marc Schouppe). Data collection and formal analysis were executed by M.S. (Marc Schouppe), Y.A.G., S.M., M.S. (Marie Stessens) and L.D. Data processing was completed by K.V.M., M.S. (Marie Stessens) and L.D. Manuscript preparation was performed by K.V.M. The manuscript was reviewed and edited by K.V.M., L.D. and M.S. (Marie Stessens). All authors have read and agreed to the published version of the manuscript.

Funding: This research received no external funding.

Data Availability Statement: Data are available at https://doi.org/10.5281/zenodo.7651300.

Acknowledgments: We greatly appreciate Hanne De Kort for the original seed collection. Contribution to the data acquisition: Lisa Carnal, Amy Lauwers, Cédric Van Dun and Matthieu Gallin. Nico De Regge took care of the plants.

Conflicts of Interest: The authors declare no conflict of interest.

References

1. Seidl, R.; Thom, D.; Kautz, M.; Martin-Benito, D.; Peltoniemi, M.; Vacchiano, G.; Wild, J.; Ascoli, D.; Petr, M.; Honkaniemi, J.; et al. Forest disturbances under climate change. *Nat. Clim. Chang.* **2017**, *7*, 395–402. [CrossRef] [PubMed]
2. Patacca, M.; Lindner, M.; Lucas-Borja, M.E.; Cordonnier, T.; Fidej, G.; Gardiner, B.; Hauf, Y.; Jasinevičius, G.; Labonne, S.; Linkevičius, E.; et al. Significant increase in natural disturbance impacts on European forests since 1950. *Glob. Chang. Biol.* **2022**, *29*, 1359–1376. [CrossRef] [PubMed]
3. Allen, C.D.; Macalady, A.K.; Chenchouni, H.; Bachelet, D.; McDowell, N.; Vennetier, M.; Kitzberger, T.; Rigling, A.; Breshears, D.D.; Hogg, E.H.; et al. A global overview of drought and heat-induced tree mortality reveals emerging climate change risks for forests. *For. Ecol. Manag.* **2010**, *259*, 660–684. [CrossRef]
4. Allen, C.D.; Breshears, D.D.; McDowell, N.G. On underestimation of global vulnerability to tree mortality and forest die-off from hotter drought in the Anthropocene. *Ecosphere* **2015**, *6*, 1–55. [CrossRef]

5. Bogdziewicz, M.; Kelly, D.; Thomas, P.A.; Lageard, J.G.; Hacket-Pain, A. Climate warming disrupts mast seeding and its fitness benefits in European beech. *Nat. Plants* **2020**, *6*, 88–94. [CrossRef]

6. Monks, A.; Monks, J.M.; Tanentzap, A.J. Resource limitation underlying multiple masting models makes mast seeding sensitive to future climate change. *New Phytol.* **2016**, *210*, 419–430. [CrossRef]

7. Bogdziewicz, M.; Ascoli, D.; Hacket-Pain, A.; Koenig, W.D.; Pearse, I.; Pesendorfer, M.; Satake, A.; Thomas, P.; Vacchiano, G.; Wohlgemuth, T.; et al. From theory to experiments for testing the proximate mechanisms of mast seeding: An agenda for an experimental ecology. *Ecol. Lett.* **2020**, *23*, 210–220. [CrossRef]

8. Richardson, S.J.; Allen, R.B.; Whitehead, D.; Carswell, F.E.; Ruscoe, W.A.; Platt, K.H. Climate and Net Carbon Availability Determine Temporal Patterns of Seed Production by Nothofagus. *Ecology* **2005**, *86*, 972–981. [CrossRef]

9. Redmond, M.D.; Forcella, F.; Barger, N.N. Declines in pinyon pine cone production associated with regional warming. *Ecosphere* **2012**, *3*, 1–14. [CrossRef]

10. Allen, R.B.; Hurst, J.M.; Portier, J.; Richardson, S.J. Elevation-dependent responses of tree mast seeding to climate change over 45 years. *Ecol. Evol.* **2014**, *4*, 3525–3537. [CrossRef]

11. Pesendorfer, M.B.; Bogdziewicz, M.; Szymkowiak, J.; Borowski, Z.; Kantorowicz, W.; Espelta, J.M.; Fernández-Martínez, M. Investigating the relationship between climate, stand age, and temporal trends in masting behavior of European forest trees. *Glob. Chang. Biol.* **2020**, *26*, 1654–1667. [CrossRef]

12. Leuzinger, S.; Luo, Y.; Beier, C.; Dieleman, W.; Vicca, S.; Körner, C. Do global change experiments overestimate impacts on terrestrial ecosystems? *Trends Ecol. Evol.* **2011**, *26*, 236–241. [CrossRef]

13. Herman, J.; Sultan, S. Adaptive Transgenerational Plasticity in Plants: Case Studies, Mechanisms, and Implications for Natural Populations. *Front. Plant Sci.* **2011**, *2*, 102. [CrossRef]

14. Galloway, L.F.; Etterson, J.R. Transgenerational Plasticity Is Adaptive in the Wild. *Science* **2007**, *318*, 1134–1136. [CrossRef]

15. Zecchin, B.; Caudullo, G.; de Rigo, D. Frangula alnus in Europe: Distribution, habitat, usage and threats. In *European Atlas of Forest Tree Species*; Publication Office of the European Union: Luxembourg, 2016.

16. Vander Mijnsbrugge, K.; De Clerck, L.; Van der Schueren, N.; Moreels, S.; Lauwers, A.; Steppe, K.; De Ligne, L.; Campioli, M.; Van den Bulcke, J. Counter-Intuitive Response to Water Limitation in a Southern European Provenance of Frangula alnus Mill. in a Common Garden Experiment. *Forests* **2020**, *11*, 1186. [CrossRef]

17. Vander Mijnsbrugge, K.; Schouppe, M.; Moreels, S.; Aguas Guerreiro, Y.; Decorte, L.; Stessens, M. Influence of Water Limitation and Provenance on Reproductive Traits in a Common Garden of Frangula alnus Mill. *Forests* **2022**, *13*, 1744. [CrossRef]

18. Vander Mijnsbrugge, K.; Schouppe, M.; Moreels, S.; De Leenheer, S. Transgenerational Effects of Water Limitation on Reproductive Mother Plants in a Common Garden of the Shrub Frangula alnus. *Forests* **2023**, *14*, 348. [CrossRef]

19. Fick, S.E.; Hijmans, R.J. WorldClim 2: New 1-km spatial resolution climate surfaces for global land areas. *Int. J. Climatol.* **2017**, *37*, 4302–4315. [CrossRef]

20. Team, R.C. *R: A Language and Environment for Statistical Computing*; R Foundation for Statistical Computing: Vienna, Austria, 2019; Available online: http://www.R-project.org/ (accessed on 1 February 2020).

21. Bates, D.; Machler, M.; Bolker, B.M.; Walker, S.C. Fitting Linear Mixed-Effects Models Using lme4. *J. Stat. Softw.* **2015**, *67*, 1–48. [CrossRef]

22. Christensen, R.H.B. *Ordinal: Regression Models for Ordinal Data*, R Package Version 2015.6-28; R Foundation for Statistical Computing: Vienna, Austria, 2015. Available online: http://www.cran.r-project.org/package=ordinal/ (accessed on 1 February 2020).

23. Hoch, G.; Siegwolf, R.T.W.; Keel, S.G.; Körner, C.; Han, Q. Fruit production in three masting tree species does not rely on stored carbon reserves. *Oecologia* **2013**, *171*, 653–662. [CrossRef]

24. Bogdziewicz, M.; Fernández-Martínez, M.; Espelta, J.M.; Ogaya, R.; Penuelas, J. Is forest fecundity resistant to drought? Results from an 18-yr rainfall-reduction experiment. *New Phytol.* **2020**, *227*, 1073–1080. [CrossRef] [PubMed]

25. Vergotti, M.J.; Fernández-Martínez, M.; Kefauver, S.C.; Janssens, I.A.; Peñuelas, J. Weather and trade-offs between growth and reproduction regulate fruit production in European forests. *Agric. For. Meteorol.* **2019**, *279*, 107711. [CrossRef]

26. Caignard, T.; Kremer, A.; Firmat, C.; Nicolas, M.; Venner, S.; Delzon, S. Increasing spring temperatures favor oak seed production in temperate areas. *Sci. Rep.* **2017**, *7*, 8555. [CrossRef] [PubMed]

27. Hacket-Pain, A.J.; Lageard, J.G.A.; Thomas, P.A. Drought and reproductive effort interact to control growth of a temperate broadleaved tree species (*Fagus sylvatica*). *Tree Physiol.* **2017**, *37*, 744–754. [CrossRef] [PubMed]

28. Lauder, J.D.; Moran, E.V.; Hart, S.C. Fight or flight? Potential tradeoffs between drought defense and reproduction in conifers. *Tree Physiol.* **2019**, *39*, 1071–1085. [CrossRef]

29. Takeno, K. Stress-induced flowering: The third category of flowering response. *J. Exp. Bot.* **2016**, *67*, 4925–4934. [CrossRef]

30. Zwolak, R.; Celebias, P.; Bogdziewicz, M. Global patterns in the predator satiation effect of masting: A meta-analysis. *Proc. Natl. Acad. Sci. USA* **2022**, *119*, e2105655119. [CrossRef]

31. Bogdziewicz, M. How will global change affect plant reproduction? A framework for mast seeding trends. *New Phytol.* **2022**, *234*, 14–20. [CrossRef]

32. Piovesan, G.; Adams, J.M. Masting behaviour in beech: Linking reproduction and climatic variation. *Can. J. Bot.* **2001**, *79*, 1039–1047. [CrossRef]

33. Davi, H.; Cailleret, M.; Restoux, G.; Amm, A.; Pichot, C.; Fady, B. Disentangling the factors driving tree reproduction. *Ecosphere* **2016**, *7*, e01389. [CrossRef]

34. Mojzes, A.; Kalapos, T.; Kröel-Dulay, G. Drought in maternal environment boosts offspring performance in a subordinate annual grass. *Environ. Exp. Bot.* **2021**, *187*, 104472. [CrossRef]

35. Germain, R.M.; Caruso, C.M.; Maherali, H. Mechanisms and Consequences of Water Stress–Induced Parental Effects in an Invasive Annual Grass. *Int. J. Plant Sci.* **2013**, *174*, 886–895. [CrossRef]

36. Tielbörger, K.; Valleriani, A. Can seeds predict their future? Germination strategies of density-regulated desert annuals. *Oikos* **2005**, *111*, 235–244. [CrossRef]

37. Stahl, U.; Kattge, J.; Reu, B.; Voigt, W.; Ogle, K.; Dickie, J.; Wirth, C. Whole-plant trait spectra of North American woody plant species reflect fundamental ecological strategies. *Ecosphere* **2013**, *4*, art128. [CrossRef]

38. Leishman, M.R.; Westoby, M. The Role of Seed Size in Seedling Establishment in Dry Soil Conditions—Experimental Evidence from Semi-Arid Species. *J. Ecol.* **1994**, *82*, 249–258. [CrossRef]

39. Rice, K.J.; Knapp, E.E. Effects of Competition and Life History Stage on the Expression of Local Adaptation in Two Native Bunchgrasses. *Restor. Ecol.* **2008**, *16*, 12–23. [CrossRef]

40. Hamann, E.; Kesselring, H.; Armbruster, G.F.J.; Scheepens, J.F.; Stöcklin, J. Evidence of local adaptation to fine- and coarse-grained environmental variability in Poa alpina in the Swiss Alps. *J. Ecol.* **2016**, *104*, 1627–1637. [CrossRef]

41. Schippers, J.H.M.; Schmidt, R.; Wagstaff, C.; Jing, H.-C. Living to Die and Dying to Live: The Survival Strategy behind Leaf Senescence. *Plant Physiol.* **2015**, *169*, 914–930. [CrossRef]

42. Rosecrance, R.C.; Weinbaum, S.A.; Brown, P.H. Alternate Bearing Affects Nitrogen, Phosphorus, Potassium and Starch Storage Pools in Mature Pistachio Trees. *Ann. Bot.* **1998**, *82*, 463–470. [CrossRef]

43. Rohde, A.; Bhalerao, R. Plant dormancy in the perennial context. *Trends Plant Sci.* **2007**, *12*, 217–223. [CrossRef]

44. Mehlenbacher, S.A.; Voordeckers, A.M. Relationship of flowering time, rate of seed germination, and time of leaf budbreak and usefulness in selecting for late-flowering apples. *J. Am. Soc. Hortic. Sci.* **1991**, *116*, 565–568. [CrossRef]

45. Vahdati, K.; Aslani Aslamarz, A.; Rahemi, M.; Hassani, D.; Leslie, C. Mechanism of seed dormancy and its relationship to bud dormancy in Persian walnut. *Environ. Exp. Bot.* **2012**, *75*, 74–82. [CrossRef]

46. Vander Mijnsbrugge, K.; Moreels, S. Varying Levels of Genetic Control and Phenotypic Plasticity in Timing of Bud Burst, Flower Opening, Leaf Senescence and Leaf Fall in Two Common Gardens of *Prunus padus* L. *Forests* **2020**, *11*, 1070. [CrossRef]

47. Soularue, J.-P.; Kremer, A. Assortative mating and gene flow generate clinal phenological variation in trees. *BMC Evol. Biol.* **2012**, *12*, 79. [CrossRef]

48. Savolainen, O.; Pyhäjärvi, T.; Knürr, T. Gene Flow and Local Adaptation in Trees. *Annu. Rev. Ecol. Evol. Syst.* **2007**, *38*, 595–619. [CrossRef]

49. Muffler, L.; Schmeddes, J.; Weigel, R.; Barbeta, A.; Beil, I.; Bolte, A.; Buhk, C.; Holm, S.; Klein, G.; Klisz, M.; et al. High plasticity in germination and establishment success in the dominant forest tree *Fagus sylvatica* across Europe. *Glob. Ecol. Biogeogr.* **2021**, *30*, 1583–1596. [CrossRef]

50. Arana, M.V.; Gonzalez-Polo, M.; Martinez-Meier, A.; Gallo, L.A.; Benech-Arnold, R.L.; Sánchez, R.A.; Batlla, D. Seed dormancy responses to temperature relate to Nothofagus species distribution and determine temporal patterns of germination across altitudes in Patagonia. *New Phytol.* **2016**, *209*, 507–520. [CrossRef]

51. Rousi, M.; Heinonen, J.; Neuvonen, S. Intrapopulation variation in flowering phenology and fecundity of silver birch, implications for adaptability to changing climate. *For. Ecol. Manag.* **2011**, *262*, 2378–2385. [CrossRef]

52. Pauwels, E.; Eyssen, R.; Keulemans, J. Seed dormancy and bud break of apple seedlings. *Acta Hortic.* **1998**, *484*, 119–122. [CrossRef]

MDPI

Article

Seed Harvesting and Climate Change Interact to Affect the Natural Regeneration of *Pinus koraiensis*

Kai Liu [1], Hang Sun [1], Hong S. He [2,*] and Xin Guan [3]

[1] Key Laboratory of Geographical Processes and Ecological Security in Changbai Mountains, Ministry of Education, School of Geographical Sciences, Northeast Normal University, Changchun 130024, China
[2] School of Natural Resources, University of Missouri, 203 ABNR Bldg, Columbia, MO 65211, USA
[3] Jilin Provincial Experimental School, Changchun 130022, China
* Correspondence: heh@missouri.edu

Abstract: The poor natural regeneration of *Pinus koraiensis* is a key limitation for restoring the primary mixed *Pinus koraiensis* forests. Seed harvesting and climate change are the important factors that influence the natural regeneration of *Pinus koraiensis*; however, it is hard to illustrate how, in synergy, they affect its regeneration at the landscape scale. In this study, we coupled an ecosystem process model, LINKAGES, with a forest landscape model, LANDIS PRO, to evaluate how seed harvesting and climate change influenced the natural regeneration of *Pinus koraiensis* over large temporal and spatial scales. Our results showed that seed harvesting decreased the abundance of *Pinus koraiensis* juveniles by 1, 14, and 18 stems/ha under the historical climate, and reduced by 1, 17, and 24 stems/ha under the future climate in the short- (years 0–50), medium- (years 60–100), and long-term (years 110–150), respectively. This indicated that seed harvesting intensified the poor regeneration of *Pinus koraiensis*, irrespective of climate change. Our results suggested that seed harvesting diminished the generation capacity of *Pinus koraiensis* over the simulation period. Seed harvesting reduced the abundance of *Pinus koraiensis* at the leading edge and slowed down its shift into high-latitude regions to adapt to climate change. Our results showed that the effect magnitudes of seed harvesting, climate change, their interaction and combination at the short-, medium- and long-term were −61.1%, −78.4%, and −85.7%; 16.5%, 20.9%, and 38.2%; −10.1%, −16.2% and −32.0%; and −54.7%, −73.8%, and −79.5%, respectively. Seed harvesting was a predominant factor throughout the simulation; climate change failed to offset the negative effect of seed harvesting, but the interactive effect between seed harvesting and climate change almost overrode the positive effect of climate change. Seed harvesting, climate change, and their interaction jointly reduced the natural regeneration of *Pinus koraiensis*. We suggest reducing the intensity of seed harvesting and increasing silvicultural treatments, such as thinning and artificial plantation, to protect and restore the primary mixed *Pinus koraiensis* forests.

Keywords: seed harvesting; climate change; interactive effect; regeneration; *Pinus koraiensis*

Citation: Liu, K.; Sun, H.; He, H.S.; Guan, X. Seed Harvesting and Climate Change Interact to Affect the Natural Regeneration of *Pinus koraiensis*. *Forests* **2023**, *14*, 829. https://doi.org/10.3390/f14040829

Academic Editors: Any Mary Petritan and Mirela Beloiu

Received: 3 March 2023
Revised: 13 April 2023
Accepted: 16 April 2023
Published: 18 April 2023

1. Introduction

Pinus koraiensis is a climax species in the temperate forests of Northeast China [1,2]. It has high ecological and economic values, including carbon sequestration, timber, and seed products [3]. Due to excessive timber harvesting in the past fifty years, the dominance of *Pinus koraiensis* has evidently decreased, and the primary mixed *Pinus koraiensis* forests have degraded into secondary forests [4,5]. The Chinese government implemented the Natural Forest Conservation Project in 1998 and the Commercial Harvesting Exclusion Policy in 2014 to restore the primary forests and increase forest stocks [6,7]; however, these projects and policies remarkedly reduced local incomes from timber. *Pinus koraiensis* seeds are large and edible, with a high content of pinolenic acid, and thus they have various health effects; this determines their high economic value [8]. Given the economic value of

Pinus koraiensis seeds, predatory seed harvesting gradually became an important income source for the local population [9]; however, this harvesting might further aggravate the poor natural regeneration of *Pinus koraiensis*. Previous studies showed that predatory seed harvesting significantly decreased seed provenance and would reduce the dominance of *Pinus koraiensis* in the long term [10]; this would hinder the restoration of the primary mixed *Pinus koraiensis* forests under future climate change scenarios.

The temperate forests in Northeast China are being affected by climate change. Predictions based on both the General Circulation Models (GCMs) and the regional Climate-Weather Research and Forecasting model (CWRF) indicated that Northeast China would have a warm and wet climate in the future, especially in summer and that precipitation would significantly increase [11,12]. The warming climate has caused tree species to migrate towards regions with a higher latitude and elevation in order to adapt to climate change [13]. A recent dendrochronological study showed that the growth of *Pinus koraiensis* at the trailing edge and in the young age cohort was limited by the warm and dry climate conditions [14], which might have adverse effects on its migration. Another study showed that *Pinus koraiensis* would shift to higher latitude regions as a result of a warm and wet climate [11]. However, seed harvesting reduces the provenance of *Pinus koraiensis* and the probability of its regeneration, which may be unfavorable for its migration into suitable regions. Given the uncertainties of climate and the negative effect of seed harvesting, it is essential to clarify how seed harvesting and climate change impact on its migration for the successful adaptation of *Pinus koraiensis* to the future climate.

Climate change generally interacts with disturbances to impact forest landscapes [15]. A previous study found that climate change and tree harvesting had interactive effects on species distribution and that the interaction between these factors accelerated the contraction or expansion of distributions induced by climate change [16]. *Pinus koraiensis* and its associated species are both symbiotic and competitive; climate change and seed harvesting can alter these interspecific relationships. Climate change altered forest composition by promoting species that were adaptive to future climate while suppressing maladaptive species [11]. Seed harvesting directly reduced the quantity of potential germination seed, thereby potentially decreasing the regeneration of *Pinus koraiensis* and indirectly promoting the recruitment of species with a similar ecological niche. [10]. Whether there is an interaction between climate change and seed harvesting is unknown.

It is challenging to assess the effects of seed harvesting and climate change on the natural regeneration of *Pinus koraiensis* at the landscape scale. Seed harvesting, a landscape process, synergizes with competition at the stand scale, seed dispersal at the landscape scale, and climate change at the regional scale to affect *Pinus koraiensis* regeneration. Most previous studies investigated and monitored the adverse effects of seed harvesting on *Pinus koraiensis* seed or seedling banks at the plot or stand scale [4,9,17,18]. Still, it is hard to interpret how seed harvesting functioned over larger temporal and spatial scales, in particular under climate change scenarios. A forest landscape model can simulate the synergistic effects of population dynamics, seed dispersal, and harvesting on forests at large temporal and spatial scales. In addition, it can be coupled with an ecosystem process model to incorporate the effect of climate change [19,20]. Thus, coupling a forest landscape model with an ecosystem process model provides a feasible tool for evaluating the effects of seed harvesting and climate change on the natural regeneration of *Pinus koraiensis* at the landscape scale.

In this study, we evaluated how seed harvesting and climate change impact the natural regeneration of *Pinus koraiensis* at the landscape scale using a forest landscape model, LANDIS PRO, in the Xiaoxing'an Mountain, Northeast China. Specifically, we aimed to address (1) how the natural regeneration of *Pinus koraiensis* varies under alternative scenarios of seed harvesting and change climate combination; (2) whether seed harvesting reduces the regeneration capacity of *Pinus koraiensis* at the landscape scale and slows down its migration into high-latitude regions regardless of climate scenarios; and (3) whether there is an interactive effect between seed harvesting and climate change on the natural regeneration of *Pinus koraiensis*. Additionally, if it exists, we investigated the magnitudes of the relative contribution of seed harvesting, climate change, and their interaction and combination in the short- (years 0–50), medium- (years 60–100), and long-term (years 100–150), respectively.

2. Materials and Methods

2.1. Study Area

We chose the Xiaoxing'an Mountains as the study area, which cover 1.5 million hectares, with longitude and latitude ranges of 127°50′ E to 130°10′ E and 47°05′ N to 49°10′ N, and with elevations ranging from 139 to 1429 m (Figure 1). The study area belongs to a temperate continental monsoon climate with a mean annual temperature from 1.0 to −1.0 °C and precipitation of 700 to 550 mm from the south to the north, respectively. The dominant soil is *Haplic Luvisols*, and distributes widely throughout the region. The study area is located in the transitional zone between a temperate forest and boreal forest, and the leading edges of some species are in this region, such as *Pinus koraiensis*. The typical primary mixed *Pinus koraiensis* forests include *Quercus mongolica* and *Pinus koraiensis* forests, *Picea koraiensis* and *Picea jezoensis*, *Abies nephrolepis*, and *Pinus koraiensis* forests, *Tilia amurensis* and *Pinus koraiensis* forests, *Betula costata*, and *Pinus koraiensis* forests. However, these primary forests have degraded into secondary forests because of the historical excessive timber harvesting. Forests in this region have been divided into three types for classified management: the special ecological welfare forests, the general ecological welfare forests, and the commercial forests to restore forest resources and increase stocks since the Natural Forest Conservation Project in 1998. Timber harvest was conducted to increase incomes for the locals in the general ecological welfare forests and the commercial forests but was excluded in the special ecological welfare forests before the year 2014. All commercial timber harvesting has been prohibited since the Commercial Harvest Exclusion Policy since the year 2014, and now seed harvesting is an important anthropogenic disturbance.

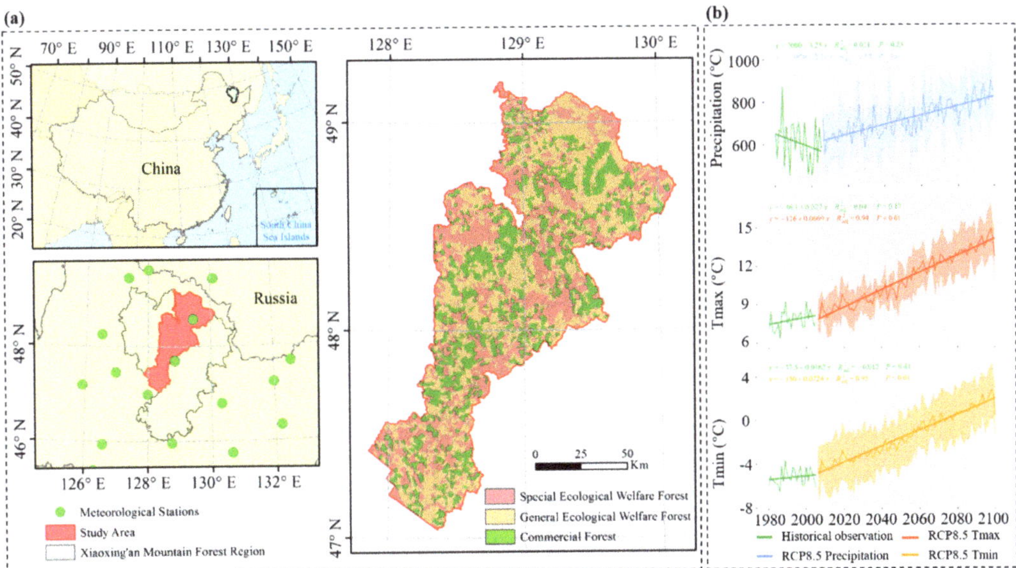

Figure 1. Geographical location and three categories of management zones of the study area (**a**), and the historical climate observation and future climates projected by the four General Circulation Models (GCMs) under the RCP8.5 scenario (**b**).

2.2. Climate Data

We obtained the maximum temperature, minimum temperature, precipitation, average wind speed, and average daily radiation from the China Meteorological Data Network from 1980 to 2009 as a baseline climate (http://data.cma.cn/, accessed on 24 February 2017). We selected the abovementioned climate variables under the representative concentration pathway 8.5 (RCP8.5) scenario predicted by the four GCMs from the Coupled Model Intercomparison Project phase 5 (CMIP5) as the climate change scenario (https://esgf-node.llnl.gov/search/cmip5/, accessed on 22 April 2016), since the RCP8.5 scenario with the highest emission provided the upper limit of climate change. The four selected GCMs included the Climate Model version 3 of the Geophysical Fluid Dynamics Laboratory (GFDL-CM3), the Hadley Center Global Environment Model version 2 Earth System (HadGEM2-ES), the Model for Interdisciplinary Research on Climate version 5 (MIROC5), and the Meteorological Research Institute Climate General Circulation Model version 3 (MRI-CGCM3). We chose the four GCMs since the temperature and precipitation predicted by them had large seasonal and annual differences which captured the uncertainties of future climate and they all output the climate variables used in our study [11].

2.3. LINKAGES 3.0 Model and Parameterization

LINKAGES, an ecosystem process model, incorporates succession and soil nutrient cycles to simulate forest dynamics at the stand scale with a one-year time step [21]. Individual regeneration is controlled by the growing degree per day, light, and water availability. Tree species growth and biomass accumulation are jointly regulated by growing degree per day, light and water availability, and soil available nitrogen. Mortality was determined probabilistically by annual tree growth of $\leq$10% of maximum possible diameter growth and by a background rate threshold defined as 4.605/maximum longevity against which a random number was compared. Random draws less than or equal to the threshold lead to mortality. LINKAGES principally requires the inputs of species' biological traits, daily climate data, and soil data. Since LINKAGES needs daily climate data, it can capture the effect of subtle changes in climate on forest dynamics. The outputs of LINKAGES mainly comprise species area and biomass, soil nitrogen availability, soil organic matter, and net primary productivity.

We derived fifteen major species' biological traits, mainly included in Table 1, from the Flora of China (http://frps.iplant.cn/, accessed on 4 October 2018) and the previous studies in this region [11]. We interpolated the above climate elements under historical and future scenarios into different sites using R package *"meteoland"* [22]. We obtained soil data from the Soil Database of China for Land Surface Modeling (http://globalchange.bnu.edu.cn/research/soil2.jsp, accessed on 4 November 2018). We calculated each soil layer's field capacity and wilting point with the Century Model soil calculator (https://www.nrel.colostate.edu/projects/century/soilCalculator.html, accessed on 10 November 2018).

We used species biomass and total biomass under historical and future climate conditions to evaluate the species establishment probability (SEP) and the maximum growing space occupied (MGSO) in different land types, respectively, which were two critical parameters for LANDIS. We divided forest landscapes into eight land types with relatively homogeneous interiors and heterogeneous exteriors based on elevation, aspect, and the $\geq$10 °C cumulative temperature line. We assumed the SEP and MGSO were homogeneous in one land type but heterogeneous among land types. We ran LINKAGES with each species from a bare plot until the year 30 for 20 replicates under alternative climate scenarios and calculated the ratio of individual species biomass and the sum of all species biomass as the SEP. We executed LINKAGES with all species from the bare plot until the year 100 for 20 replicates under alternative climate scenarios and estimated the maximum total biomass reaching on each land type among climate scenarios as the MGSO [16].

Table 1. The major species parameters in the LINKAGES 3.0 model.

Species	DMX	DMN	B3	B2	G	D3	FT	TL
Pinus koraiensis Sieb. & Zucc.	2500	900	0.29	62.96	104.53	0.28	−35	12
Picea jezoensis var. *microsperma* (Lindl.) W.C.Cheng & L.K.Fu and *Picea koraiensis* Nakai	2500	600	0.35	63.62	87.79	0.19	−34	11
Abies nephrolepis (Trautv.) Maxim.	1800	400	0.40	67.36	87.65	0.23	−34	10
Larix gmelinii (Rupr.) Kuzen.	1900	400	0.32	60.27	87.92	0.42	−38	12
Fraxinus mandschurica Rupr.	2800	1000	0.34	67.26	121.90	0.19	−31	2
Juglans mandshurica Maxim.	2650	850	0.23	41.40	72.89	0.28	−32	2
Phellodendron amurense Rupr.	3200	1000	0.23	43.43	79.53	0.23	−32	8
Quercus mongolica Fisch. ex Turcz.	3100	1100	0.32	60.27	87.92	0.51	−34	9
Ulmus davidiana var. *japonica* (Rehder) Nakai	2700	900	0.35	63.62	105.35	0.18	−33	5
Acer mono Maxim.	3200	1000	0.52	62.10	89.69	0.23	−32	2
Betula costata Trautv.	1900	700	0.35	63.62	105.35	0.19	−35	4
Betula dahurica Pall.	3100	600	0.75	74.52	118.67	0.49	−35	4
Tilia amurensis Rupr.	2400	800	0.40	67.36	87.65	0.23	−33	2
Betula platyphylla Suk.	3100	600	0.75	74.52	118.67	0.41	−38	4
Populus davidiana Dode	3000	700	0.66	78.77	146.52	0.33	−34	7

DMX, degree day maximum, which is counted above 5 °C (d·°C); DMN, degree day minimum, which is counted above 5 °C (d·°C); B2 and B3 are growth parameters of Richard function; G scales the growth rate; D3, the proportion of growing season species can withstand drought; FT, minimum January temperature species can withstand (°C); TL, leaf litter type.

2.4. LANDIS PRO Model and Parameterization

LANDIS, a raster-based forest landscape model, records species distribution (presence or absence) and abundance (density and basal area) by age cohorts on each pixel. LANDIS simulates population dynamics, seed dispersal, and anthropogenic disturbances (such as harvest) with flexible temporal and spatial resolutions of 1 to 10 years and 10 to 500 m [20,23]. Population dynamics are mainly driven by species' ecological traits and include growth, fecundity, competition, colonization, and mortality. Individual growth is represented by the relationship between age and diameter at the breast height. LANDIS simulates fecundity comprising seed germination and resprouting to consider species birth or recruitment. Competition is controlled by the growing space occupied (GSO), and the larger GSO indicates stronger competition. Once the GSO reaches the maximum, self-thinning will be initiated, which releases some growing space and alleviates competition in the stand. When species seeds or resprouting reach a pixel, it checks that species shade tolerance, the GSO, and the SEP to determine whether it can be colonized. LANDIS considers the mortality induced by competition, disturbance, and excessing the maximum longevity. Seed dispersal is simulated by the location and abundance of seed provenance and dispersal distance, which establishes spatial interaction among pixels. Disturbance such as harvest is conducted in specific management units based on different harvest strategies (e.g., harvest type and proportion, species preference).

We parameterized LANDIS with the temporal and spatial resolutions of 10 years and 100 m. We generated the species composition map based on the forest stand map detailing the species composition and abundance, and the inventory data provided the age information. We introduced how to divide forest landscapes into eight land types in Section 2.3. We derived the stand map and management area map based on the boundaries of forestry management and forest function type described by the forest stand map, respectively. Based on the previous study, we parameterized the species' biological traits in Table 2 [10,11,24]. The growth curve, the relationship between the number and width of the tree ring, was derived from the studies in this region [11,24]. The SEP and MGSO were calculated by the ecosystem process model LINKAGES in Section 2.3 [16,25]. Before executing all simulation scenarios, we also verified the total basal area predicted by LANDIS with forest inventory data and ensured no significant difference between them [10]. We considered the effect of seed harvesting on the natural regeneration of *Pinus koraiensis* by reducing its number of potential germination seeds (NPGS) and gained the outputs from LANDIS including its basal area and the abundance of juveniles with ages less than 10 years.

Table 2. The main species' biological traits in the LANDIS model.

Species	MT	LG	ST	MD	MDBH	MSDI	NPGS
Pinus koraiensis Sieb. & Zucc.	40	300	4	150	110	550	20
Picea jezoensis var. *microsperma* (Lindl.) W.C.Cheng & L.K.Fu and *Picea koraiensis* Nakai	30	300	4	150	90	600	20
Abies nephrolepis (Trautv.) Maxim.	30	300	4	150	85	650	20
Larix gmelinii (Rupr.) Kuzen.	20	300	2	300	95	650	30
Fraxinus mandschurica Rupr.	30	250	3	300	100	600	25
Juglans mandshurica Maxim.	20	250	2	200	90	650	25
Phellodendron amurense Rupr.	20	250	3	300	95	650	25
Quercus mongolica Fisch. ex Turcz.	20	300	2	200	95	600	20
Ulmus davidiana var. *japonica* (Rehder) Nakai	20	250	3	800	90	600	25
Acer mono Maxim.	20	200	3	200	60	700	25
Betula costata Trautv.	20	250	3	800	90	650	25
Betula dahurica Pall.	15	150	2	800	50	750	25
Tilia amurensis Rupr.	30	300	3	200	85	650	20
Betula platyphylla Suk.	15	150	1	2000	50	800	30
Populus davidiana Dode	15	150	1	2000	60	800	30

MT, species maturity age (year); LG, species maximum (year); ST, shade tolerance class ranging from 1 to 5, 1 and 5 denote the least and the most tolerance, respectively; MD, maximum dispersal distance (m); MDBH, maximum diameter at breast height (cm); MSDI, maximum stand density index (number of standard trees per hectare, which refers to 25.5 cm tree); NPGS, number of potential germination seeds per mature tree (number/one raster cell).

2.5. Experimental Design

We designed a two-factor experiment for seed harvesting (with or without harvesting) and climate change (historical climate and future climate) to evaluate how seed harvesting and climate change affected the natural regeneration of *Pinus koraiensis* (Table 3), resulting in four types of simulation scenarios, that is, historical climate without harvesting (S_{HN}), historical climate with harvesting (S_{HH}), future climate without harvesting (S_{FN}), and climate with harvesting (S_{FH}). Given the uncertainties of future climate change, we selected the predictions from four GCMs GFDL-CM3, HadGEM2-ES, MIROC5, and MRI-CGCM3 under the RCP8.5 scenario as the future climate scenarios to represent the ranges of climate variables such as temperature, precipitation and so on. The previous investigation found that seed harvesting decreased the seed bank of *Pinus koraiensis* by at least 80% in the mixed *Pinus koraiensis* forests [9,18]. Thus, we reduced its NPGS by 80% in the LANDIS simulations to consider the effect of seed harvesting. We simulated each scenario of seed harvesting and climate change in combination with five replicates to account for simulation stochasticity. Fifty simulations were conducted (2 seed harvesting scenarios × 5 climate scenarios × 5 replicates). We simulated changes in the natural regeneration of *Pinus koraiensis* under each scenario with the same initial stand condition by LANDIS in the next 150 years.

Table 3. Factorial experiment.

Climate Scenarios	Seed Harvesting Scenarios	
	Without Harvesting	**With Harvesting**
Historical climate observation	S_{HN}	S_{HH}
Future climate RCP8.5	S_{FN}	S_{FH}

S_{HN}, historical climate without harvesting; S_{HH}, historical climate with harvesting; S_{FN}, future climate without harvesting; S_{FH}, climate with harvesting.

2.6. Data Analysis

We used the abundance of *Pinus koraiensis* juveniles with ages less than 10 years to represent its natural regeneration since the abundance of seedlings or saplings in each size class strongly depended on the quantities of seedlings or saplings in the next smaller size class at the same site, which reflected demographic inertia [26]. We compared the differences in the natural regeneration by the ANOVA and the Duncan test to determine how its regeneration changed with the simulation in the short- (years 0–50), medium- (years 60–100), and long-term (years 110–150), respectively. We calculated the change rate of the abundance of juveniles and basal area of all age cohorts to represent the regeneration capacity of *Pinus koraiensis* at the landscape scale since the density of juveniles was significantly related to its basal area [27]. We derived the regeneration capacity from the abundance of juveniles and basal area predicted by seed harvesting and climate combination scenarios in 50, 100, and 150. We selected the spatial distribution of regeneration of *Pinus koraiensis* at the years 50, 100, and 150, and we calculated the mean values within each 1 km zones along the longitudinal direction to reflect how seed harvesting and climate change affected its migration into high-latitude regions. We quantified the seed harvesting effect (SHE), climate change effect (CCE), and their interactive effect (ITE) and combined effect (CBE) based on the relative change in the abundance of juveniles between the scenarios in the short-, medium-, and long-term by Equation (1) to evaluate how the four effects alone and in synergy impacted on the regeneration of *Pinus koraiensis*. We calculated the magnitudes of the SHE, the CCE, and the CBE by the relative change in the abundance of *Pinus koraiensis* juveniles between the scenario S_{HH} and S_{HN}, between the scenario S_{FN} and S_{HN}, and between the scenario S_{FH} and S_{HN}, respectively. We calculated the magnitude of the ITE by the CBE subtracting the additive effect (ADE) defined as the sum of the SHE and the CCE [16,28].

$$\frac{S_1 - S_2}{S_2} \times 100 \tag{1}$$

where S_1 and S_2 represented different scenarios of seed harvesting and climate change combination separately.

3. Results

3.1. Temporal Variations in the Natural Regeneration of Pinus koraiensis

The abundance of *Pinus koraiensis* juveniles increased under alternative scenarios of seed harvesting and climate change combination over the simulation (Figure 2). The differences in the abundance of *Pinus koraiensis* juveniles significantly increased with the simulation. The maximum difference of 1, 8, and 26 stems/ha occurred separately in the short-, medium-, and long-term. Seed harvesting significantly decreased the abundance of juveniles and the diminishing magnitude increased over the simulation, irrespective of climate change. Compared with the excluding harvesting scenario, seed harvesting reduced *Pinus koraiensis* juveniles by 1, 14, and 18 stems/ha under the historical climate (S_{HN}, S_{HH}), and diminished by 1, 17, and 24 stems/ha under the future climate (S_{FN}, S_{FH}) in the short-, medium-, and long-term, respectively. In comparison with the historical climate scenario, climate change increased *Pinus koraiensis* juveniles by 1, 4, and 8 stems/ha when excluding seed harvesting (S_{FN}, S_{HN}), and increased by 1, 1, and 2 stems/ha under seed harvesting (S_{FH}, S_{HH}) in the short-, medium-, and long-term separately.

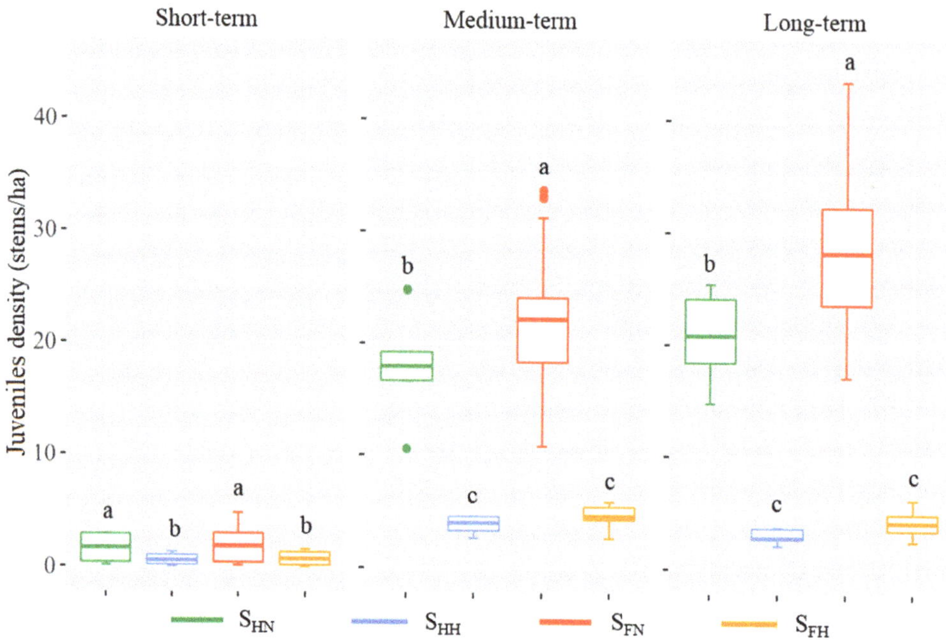

Figure 2. The abundance of *Pinus koraiensis* juveniles under different scenarios of climate and seed harvesting combinations at the whole region scale in the short, medium, and long term, respectively. Different letters displayed significant differences at the 0.05 level among scenarios as determined by the ANOVA and the Duncan test. S_{HN}, historical climate without harvesting; S_{HH}, historical climate with harvesting; S_{FN}, future climate without harvesting; S_{FH}, climate with harvesting.

3.2. Variations in the Natural Regeneration Capacity of Pinus koraiensis

Seed harvesting dramatically reduced the regeneration capacity of *Pinus koraiensis* (change rate between the abundance of *Pinus koraiensis* juveniles and the basal area of all age cohorts) under both historical climate and future climate, and the decreased magnitudes of regeneration capacity increased over time (Figure 3). Compared with excluding harvesting, seed harvesting diminished the regeneration capacity of *Pinus koraiensis* by 16.2, 27.7, and 29.3 stems/m^2 under the historical climate scenario and averagely decreased by 16.5, 29, and 31.8 stems/m^2 under the future climate scenarios at the years 50, 100, and 150, respectively. In contrast to historical climate, climate change increased the regeneration capacity of *Pinus koraiensis* by 0.2, 0.1, and 1.3 stems/m^2 under seed harvesting scenarios and increased by 0.5, 1.4, and 3.8 stems/m^2 under excluding harvesting scenarios at the years 50, 100, and 150 respectively.

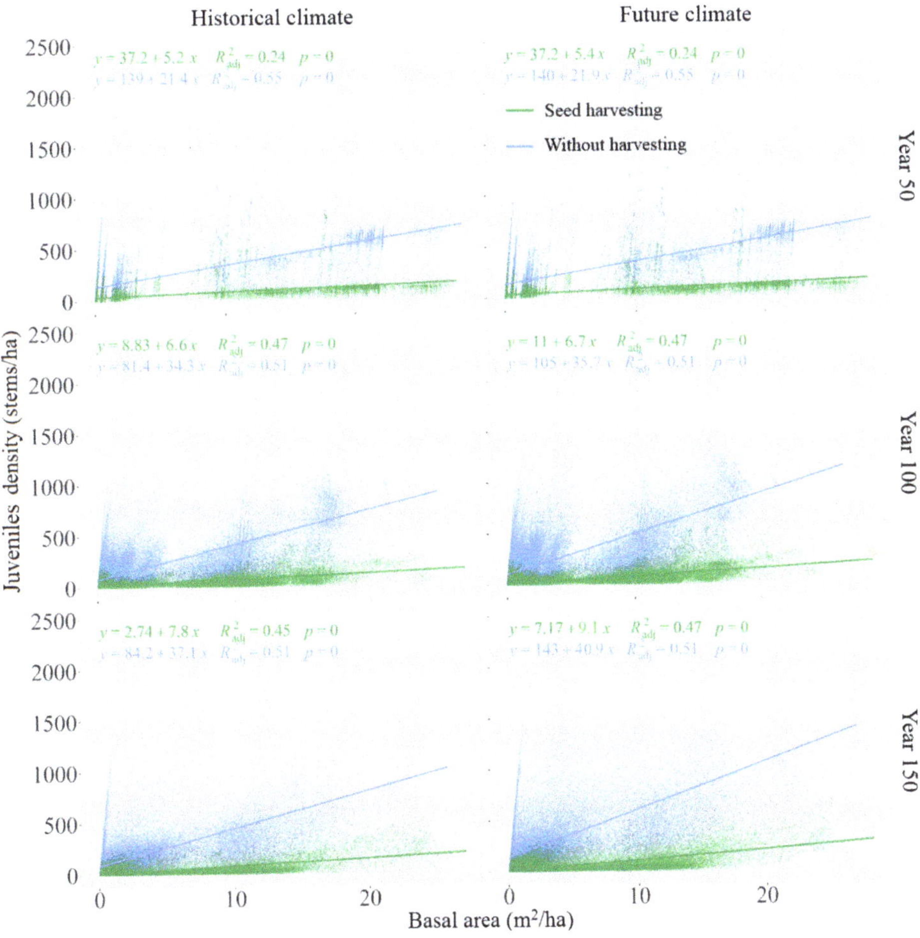

Figure 3. The relationship between the abundance of *Pinus koraiensis* juveniles and its basal area at the years 50, 100, and 150 under the different scenarios of seed harvesting and climate change combination.

3.3. Spatial Changes in the Natural Regeneration of Pinus koraiensis

The area and quantity of *Pinus koraiensis* juveniles increased with the simulation under the four scenarios of seed harvesting and climate change combination (Figure 4a,b). The leading edge of *Pinus koraiensis* juveniles migrated into high-latitude regions irrespective of combination scenarios. Compared with the seed harvesting scenario (S_{HH}) and the excluding harvesting scenario (S_{HN}) under historical climate, it was found that seed harvesting decreased the *Pinus koraiensis* juveniles by 142, 147, and 217 stems/ha on average in the whole region at the years 50, 100 and 150 separately. In contrast to the future climate scenario (S_{FN}) and the historical climate scenario (S_{HN}) excluding harvesting, climate change increased the *Pinus koraiensis* juveniles by 11, 35, and 75 stems/ha on average throughout the region at the years 50, 100 and 150, respectively (Figure 4a). Seed harvesting reduced the *Pinus koraiensis* juveniles throughout the study area regardless of climate change but had regional differences at different periods (Figure 4b). Seed harvesting had an unobvious impact on the juveniles in the *Pinus koraiensis* distribution with a latitude above 48.5° N at the year 50 but had evident effects at the years 100 and 150.

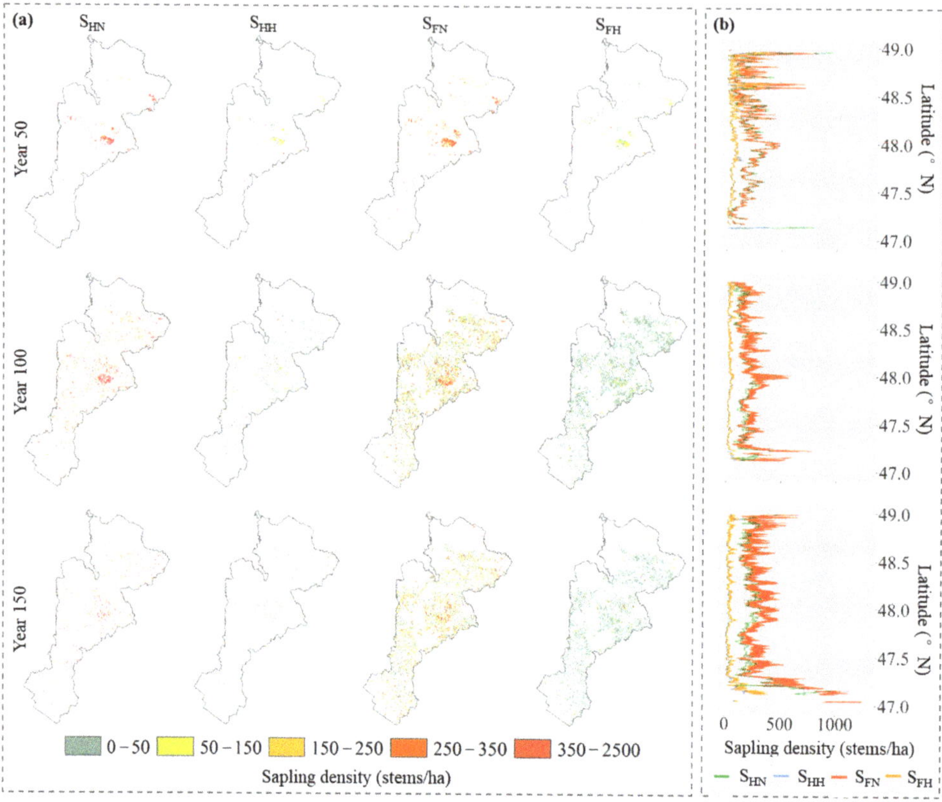

Figure 4. The spatial distribution of the abundance of *Pinus koraiensis* juveniles under the different scenarios of climate and seed harvesting combinations in 50, 100, and 150 (**a**); The spatial distribution of the abundance of *Pinus koraiensis* juveniles along the longitudinal direction under the different scenarios of seed harvesting and climate change combination in 50, 100 and 150 (**b**). Error bands represented one standard deviation of the abundance of *Pinus koraiensis* juveniles among scenarios of seed harvesting and climate change combinations. S_{HN}, historical climate without harvesting; S_{HH}, historical climate with harvesting; S_{FN}, future climate without harvesting; S_{FH}, climate with harvesting.

3.4. Effects of Seed Harvesting and Climate Change

The adverse effects of seed harvesting on the natural regeneration of *Pinus koraiensis* were always dominant, and the average effect magnitudes were −61.1%, −78.4%, and −85.7% in the short-, medium, and long-term, respectively (Figure 5). The favorable effects of climate change were 16.5%, 20.9%, and 38.2% in the short-, medium, and long term separately. There was an interactive effect between seed harvesting and climate change, and the interaction negatively impacted the regeneration of *Pinus koraiensis*. The interactive effect increased over the simulation and the average effect magnitudes were −10.1%, −16.2%, and −32.0% in the short-, medium, and long-term, respectively. Climate change and seed harvesting combinedly reduced the regeneration of *Pinus koraiensis*. The magnitudes of the combined effects were −54.7%, −73.8%, and −79.5% in the short-, medium- and long-term, respectively. Climate change failed to offset the unfavorable effect of seed harvesting. The effect sizes of climate change were 44.6%, 57.5%, and 47.5% lower than the absolute values of seed harvesting separately in the short-, medium- and long-term.

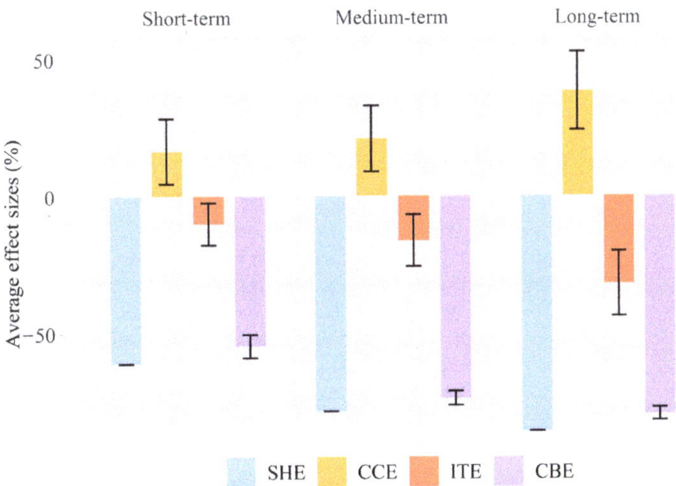

Figure 5. The average effect sizes of seed harvesting (SHE), climate change (CCE), and their interaction (ITE) and combination (CBE) on *Pinus koraiensis* regeneration in the short-, medium-, and long-term, respectively. Error bars of SHE represented one standard deviation of five replicates, and the remaining bars represented one standard deviation of the combination scenarios of four future climates projected by GCMs and five replicates (20 simulations, 4 climate scenarios × 5 replicates).

4. Discussion

Our results show that seed harvesting aggravates the poor natural regeneration of *Pinus koraiensis*, which generally is restricted by the light availability in Northeast China [5,29,30]. Seed harvesting has significantly decreased the natural regeneration of *Pinus koraiensis*, and the negative effect increases with forest succession. However, we find the differences in the short-term are small but large in the medium- and long-term between the scenarios with and without seed harvesting (Figure 2), which indicates seed harvesting has a lagging effect on the natural regeneration of *Pinus koraiensis* that is determined by its mature age. Although seed harvesting cannot cause the death of the existing *Pinus koraiensis*, it reduces the provenances and further decreases the abundance of *Pinus koraiensis* seedlings at the landscape scale, resulting in the loss of adult trees in the future. Decreases in the adult trees of *Pinus koraiensis* will further diminish its seed banks and exacerbate its poor regeneration. The previous studies at the plot or stand scale find that seed harvesting significantly reduces the seed and seedling banks of *Pinus koraiensis* [4,9,17]. Still, they fail to consider the effect of seed dispersal that may alleviate the adverse effect induced by seed harvesting. We apply a forest landscape model and simultaneously consider the effects of seed harvesting and dispersal in this study, which can provide more accurate results at the landscape scale from the methodology perspective. Our results indicate that the pressures from the continued predatory seed harvesting and the poor natural regeneration of *Pinus koraiensis* increase the challenges for restoring the secondary forests to the primary mixed *Pinus koraiensis* forests at the landscape scale [2].

Our results demonstrate climate change with the warming and wetting trends promotes the natural regeneration of *Pinus koraiensis*, and the favorable effect increases with the simulation. The previous study finds warm and wet climate combinations promote seed germination which supports our finding [31]. Still, a warm and dry climate combination has an inhibition effect which agrees with a dendrochronological study that drought induced by climate change restricts the growth of *Pinus koraiensis* at the trailing edge and its young age cohort [14]. There are regional differences in the response of *Pinus koraiensis* to a warming climate. The previous study has shown that rising temperature increases the growth of *Pinus koraiensis* at the leading edge, but decreases the growth in the midwestern regions and has little effect at the trailing edge [32]. Considering the differences in these responses and the uncertainties of future climate change, we suggest that multiple climate models should be integrated when investigating the responses of tree species to climate change in order to capture their diverse responses.

We find that seed harvesting decreases the regeneration capacity of *Pinus koraiensis* and hinders its migration at the leading edge irrespective of climate conditions. Generally, the stand with a more basal area of *Pinus koraiensis* can provide more seed provenance to germinate [27], consistent with our results that the abundance of juveniles increases with the rising basal area of *Pinus koraiensis* at the landscape scale (Figure 3). Regeneration dynamics is the key process that determines species shifts induced by climate change [33]. Seed harvesting reduces the opportunities for *Pinus koraiensis* recruitments, especially at the leading edge with low levels of dominance and the seed bank of *Pinus koraiensis*. Our results show the adverse effect of seed harvesting at the leading edge is unobvious in the short-term (year 50), but evident in the medium- and long-term (years 100 and 150) (Figure 4). We assume that the low dominance and seed bank of *Pinus koraiensis* results in a lagging effect of seed harvesting on its shift. Recent studies show that tree species migration fails to keep up with the step of climate change [13,16,34], while seed harvesting further impedes the migration of *Pinus koraiensis* to high-latitude regions, which is unfavorable for altering its distribution to adapt to the future climate.

Our results indicate the interactive effects between seed harvesting and climate change on the natural regeneration of *Pinus koraiensis*. The interaction has the same direction as seed harvesting and thereby exacerbates the contraction of recruitments induced by seed harvesting. It is since that climate change, under the circumstances that seed harvesting markedly obstructs the regeneration of *Pinus koraiensis*, benefits the recruitment of the other species, such as *Picea koraiensis* and *Picea jezoensis* and *Abies nephrolepis*, with similar ecological niche to *Pinus koraiensis* [10]. The effect of seed harvesting alone predominates throughout the simulation, followed by climate change alone and the interaction between them (Figure 5). We find that seed harvesting, climate change, and their interaction combinedly have adverse effects on the natural regeneration of *Pinus koraiensis*, and the positive effect of climate change fails to counteract the negative effect of seed harvesting. The previous study finds there is an interaction between climate change and timber harvesting on species distribution, emphasizing the interaction between multiple factors [16]. Our results also indicate the interactive effect between seed harvesting and climate change is comparable to climate change alone. Given the importance of interaction, we also suggest multiple drivers of recruitment changes should be assessed concurrently since the interactive effect is unable to be determined when the individual effect is considered independently.

Future forest management should consider silvicultural treatments, such as thinning from below and artificial plantation combinations, to protect the mother tree of *Pinus koraiensis* and promote its regeneration. A recent study finds that predatory seed harvesting has a long-lasting effect and decreases the dominance of *Pinus koraiensis* in the long term [10]. It is necessary to reduce the intensity and rotation of seed harvesting to ensure sufficient provenance to maintain the regeneration of *Pinus koraiensis* in the future. Although *Pinus koraiensis* seedlings are shade tolerant, their need for light increases with age [35]. Previous studies have shown that understory light availability limits the growth and survival of *Pinus koraiensis* seedlings [5,29,30] while thinning from below increases light availability by creating forest gaps, promoting the regeneration of shade-tolerant species such as *Pinus koraiensis* [5,36,37]. A previous study finds that the low thinning and slash burning treatment enhances initial Spanish black pine seedling recruitments [38]. The previous study shows thinning from below promotes *Pinus koraiensis* growth and increases the abundance of its juveniles at the landscape scale [11]. Artificial treatments such as direct seeding or seedling planting can be applied to promote the regeneration of *Pinus koraiensis* [39,40]. A recent study found that planting seedlings have a higher survival rate than direct seeding. However, the high costs of planting seedlings greatly restrict its application to a large spatial extent [39], while direct seeding may have more potential. Therefore, compositing these treatments may provide effective ways to promote the regeneration of *Pinus koraiensis*, which is of great significance for maintaining its dominance and restoring the primary mixed *Pinus koraiensis* forests.

Our study also has some limitations. *Pinus koraiensis* mainly relies on rodents that carry and bury seeds to achieve the process of seed dispersal [39]. Since seed harvesting directly reduces the food sources for these animals and may decrease their population size, which in turn will diminish the abundance of the buried seeds and thus adversely impact the natural regeneration of *Pinus koraiensis* [10]. We fail to take rodent population dynamics into account, which increases the uncertainties of the simulation. The seed yield of *Pinus koraiensis* has obvious inter-annual fluctuation [41], which alters the behavior of rodents and the seed dispersal process [37], and ignoring these impacts increases the uncertainties of *Pinus koraiensis* recruitment at the landscape scale. Generally, the natural regeneration of *Pinus koraiensis* occurs in forest gaps [42,43]. Wind disturbance naturally creates forest gaps in the forest landscape; we exclude the effects of wind on the regeneration of *Pinus koraiensis* in this study, which increases the uncertainties of prediction. Additionally, the variations indirectly induced by climate change in intensity, extent, and rotation of wind also contribute to the uncertainties [43]. Despite these shortcomings, our study still highlights the separate and interactive effects of seed harvesting and climate change on the regeneration of *Pinus koraiensis*, which provides an important reference for promoting the regeneration and dominance of *Pinus koraiensis* and restoring the mixed *Pinus koraiensis* forests from a landscape perspective.

5. Conclusions

We assessed and quantified how seed harvesting and climate change impacted the natural regeneration of *Pinus koraiensis* at the landscape scale by coupling an ecosystem process model with a forest landscape model in this study. We found that seed harvesting significantly exacerbated the poor natural regeneration of *Pinus koraiensis*. Seed harvesting evidently decreased the regeneration capacity of *Pinus koraiensis* and delayed its migration into the high latitude regions to adapt to future climate. We concluded seed harvesting always was dominant throughout the simulation, and climate change failed to neutralize the adverse effect of seed harvesting. Additionally, we found that the interactive effect between seed harvesting and climate change approached the magnitude of climate change alone, which indicated the interaction almost offset the positive effect of climate change. We suggest future management should decrease the intensity and rotation of seed harvesting and promote the regeneration and restoration of the primary mixed *Pinus koraiensis* forests through silvicultural treatments such as thinning from below, seedling planting, or direct seeding.

Author Contributions: Conceptualization, K.L. and H.S.H.; formal analysis, K.L. and H.S.; funding acquisition, K.L.; resources, K.L.; supervision, H.S.H.; writing—original draft, K.L.; writing—review and editing, K.L., H.S., H.S.H. and X.G. All authors have read and agreed to the published version of the manuscript.

Funding: This research was funded by the Natural Science Foundation of Jilin Province, China, grant number YDZJ202201ZYTS487, the National Natural Science Foundation of China, grant number 42101107, the Fundamental Research Funds for the Central Universities.

Data Availability Statement: The data presented in this study are available on request from the corresponding author.

Acknowledgments: We thank Muhammad Naveed for revising the language. We thank three anonymous reviewers for their useful recommendations on our manuscript.

Conflicts of Interest: The authors declare no conflict of interest.

References

1. Shao, G.; Schall, P.; Weishampel, J.F. Dynamic simulations of mixed broadleaved-*Pinus koraiensis* forests in the Changbaishan Biosphere Reserve of China. *For. Ecol. Manag.* **1994**, *70*, 169–181. [CrossRef]
2. Zhu, J.; Zhu, C.; Lu, D.; Wang, G.G.; Zheng, X.; Cao, J.; Zhang, J. Regeneration and succession: A 50-year gap dynamic in temperate secondary forests, Northeast China. *For. Ecol. Manag.* **2021**, *484*, 118943. [CrossRef]
3. Jin, X.; Pukkala, T.; Li, F.; Dong, L. Optimal management of Korean pine plantations in multifunctional forestry. *J. For. Res.* **2017**, *28*, 1027–1037. [CrossRef]
4. Li, Y.-B.; Mou, P.; Wang, T.-M.; Ge, J. Evaluation of regeneration potential of *Pinus koraiensis* in mixed pine-hardwood forests in the Xiao Xing'an Mountains, China. *J. For. Res.* **2012**, *23*, 543–551. [CrossRef]
5. Sun, Y.; Zhu, J.; Sun, O.J.; Yan, Q. Photosynthetic and growth responses of *Pinus koraiensis* seedlings to canopy openness: Implications for the restoration of mixed-broadleaved Korean pine forests. *Environ. Exp. Bot.* **2016**, *129*, 118–126. [CrossRef]
6. Liu, K.; Liang, Y.; He, H.S.; Wang, W.J.; Huang, C.; Zong, S.; Wang, L.; Xiao, J.; Du, H. Long-term impacts of China's new commercial harvest exclusion policy on ecosystem services and biodiversity in the temperate forests of Northeast China. *Sustainability* **2018**, *10*, 1071. [CrossRef]
7. Dai, L.; Li, S.; Zhou, W.; Qi, L.; Zhou, L.; Wei, Y.; Li, J.; Shao, G.; Yu, D. Opportunities and challenges for the protection and ecological functions promotion of natural forests in China. *For. Ecol. Manag.* **2018**, *410*, 187–192. [CrossRef]
8. Jin, X.; Li, F.; Pukkala, T.; Dong, L. Dong Modelling the cone yields of Korean pine. *For. Ecol. Manag.* **2020**, *464*, 118086. [CrossRef]
9. Piao, Z.; Tang, L.; Swihart, R.K.; Wang, S. Human–wildlife competition for Korean pine seeds: Vertebrate responses and implications for mixed forests on Changbai Mountain, China. *Ann. Forest Sci.* **2011**, *68*, 911–919. [CrossRef]
10. Liu, K.; He, H.S.; Sun, H.; Wang, J. Evaluating the Legacy Effects of the Historical Predatory Seed Harvesting on the Species Composition and Structure of the Mixed Korean Pine and Broadleaf Forest from a Landscape Perspective. *Forests* **2023**, *14*, 402. [CrossRef]
11. Liu, K.; He, H.; Xu, W.; Du, H.; Zong, S.; Huang, C.; Wu, M.; Tan, X.; Cong, Y. Responses of Korean pine to proactive managements under climate change. *Forests* **2020**, *11*, 263. [CrossRef]
12. Jiang, R.; Sun, L.; Sun, C.; Liang, X.-Z. CWRF downscaling and understanding of China precipitation projections. *Clim. Dynam.* **2021**, *57*, 1079–1096. [CrossRef]
13. Alexander, J.M.; Chalmandrier, L.; Lenoir, J.; Burgess, T.I.; Essl, F.; Haider, S.; Kueffer, C.; McDougall, K.; Milbau, A.; Nuñez, M.A. Lags in the response of mountain plant communities to climate change. *Glob. Change Biol.* **2018**, *24*, 563–579. [CrossRef]
14. Wang, X.; Pederson, N.; Chen, Z.; Lawton, K.; Zhu, C.; Han, S. Recent rising temperatures drive younger and southern Korean pine growth decline. *Sci. Total Environ.* **2019**, *649*, 1105–1116. [CrossRef]
15. Liang, Y.; Gustafson, E.J.; He, H.S.; Serra-Diaz, J.M.; Duveneck, M.J.; Thompson, J.R. What is the role of disturbance in catalyzing spatial shifts in forest composition and tree species biomass under climate change? *Glob. Change Biol.* **2023**, *29*, 1160–1177. [CrossRef]
16. Wang, W.J.; Thompson III, F.R.; He, H.S.; Fraser, J.S.; Dijak, W.D.; Jones-Farrand, T. Climate change and tree harvest interact to affect future tree species distribution changes. *J. Ecol.* **2019**, *107*, 1901–1917. [CrossRef]
17. Ji, L.; Liu, Z.; Hao, Z.; Wang, Q.; Wang, M. Effect of Cones Picking on Broad—leaved *Pinus koraiensis* Forest in Changbai Mountian. *Chin. J. Ecol.* **2002**, *21*, 39–42.
18. Jin, G.; Yang, G.; Ma, J.; Li, L.; Xu, Z.; Zhao, X.; Hong, M. Effect of Anthropogenic Cone-picking on Seed Bank and Seedling Bank of Korean Pine in the Major Forest Types in Lesser Hing'an Mountains. *J. Nat. Resour.* **2010**, *25*, 1845–1854.
19. He, H.S.; Hao, Z.; Mladenoff, D.J.; Shao, G.; Hu, Y.; Chang, Y. Simulating forest ecosystem response to climate warming incorporating spatial effects in north-eastern China. *J. Biogeogr.* **2005**, *32*, 2043–2056. [CrossRef]
20. He, H.S.; Gustafson, E.J.; Lischke, H. Modeling forest landscapes in a changing climate: Theory and application. *Landsc. Ecol.* **2017**, *32*, 1299–1305. [CrossRef]
21. Dijak, W.D.; Hanberry, B.B.; Fraser, J.S.; He, H.S.; Wang, W.J.; Thompson, F.R. Revision and application of the LINKAGES model to simulate forest growth in central hardwood landscapes in response to climate change. *Landsc. Ecol.* **2017**, *32*, 1365–1384. [CrossRef]
22. De Caceres, M.; Martin-StPaul, N.; Turco, M.; Cabon, A.; Granda, V. Estimating daily meteorological data and downscaling climate models over landscapes. *Environ. Modell. Softw.* **2018**, *108*, 186–196. [CrossRef]
23. Wang, W.J.; He, H.S.; Spetich, M.A.; Shifley, S.R.; Thompson III, F.R.; Dijak, W.D.; Wang, Q. A framework for evaluating forest landscape model predictions using empirical data and knowledge. *Environ. Modell. Softw.* **2014**, *62*, 230–239. [CrossRef]
24. Xiao, J.; Liang, Y.; He, H.S.; Thompson, J.R.; Wang, W.J.; Fraser, J.S.; Wu, Z. The formulations of site-scale processes affect landscape-scale forest change predictions: A comparison between LANDIS PRO and LANDIS-II forest landscape models. *Landsc. Ecol.* **2017**, *32*, 1347–1363. [CrossRef]
25. He, H.S.; Mladenoff, D.J.; Crow, T.R. Linking an ecosystem model and a landscape model to study forest species response to climate warming. *Ecol. Model.* **1999**, *114*, 213–233. [CrossRef]
26. Rooney, T.P.; McCormick, R.J.; Solheim, S.L.; Waller, D.M. Regional variation in recruitment of hemlock seedlings and saplings in the upper Great Lakes, USA. *Ecol. Appl.* **2000**, *10*, 1119–1132. [CrossRef]
27. Ge, X.; Zhu, J.; Lu, D.; Zhu, C.; Gao, P.; Yang, X. Effects of Korean Pine basal area in mixed broadleaved–Korean pine forest stands on its natural regeneration in northeast China. *Forest Sci.* **2021**, *67*, 179–191. [CrossRef]

28. García-Valdés, R.; Svenning, J.C.; Zavala, M.A.; Purves, D.W.; Araújo, M.B. Evaluating the combined effects of climate and land-use change on tree species distributions. *J. Appl. Ecol.* **2015**, *52*, 902–912. [CrossRef]

29. Zhu, J.; Wang, K.; Sun, Y.; Yan, Q. Response of *Pinus koraiensis* seedling growth to different light conditions based on the assessment of photosynthesis in current and one-year-old needles. *J. For. Res.* **2014**, *25*, 53–62. [CrossRef]

30. Zhou, G.; Liu, Q.; Xu, Z.; Du, W.; Yu, J.; Meng, S.; Zhou, H.; Qin, L.; Shah, S. How can the shade intolerant Korean pine survive under dense deciduous canopy? *For. Ecol. Manag.* **2020**, *457*, 117735. [CrossRef]

31. Zhao, J.; Song, Y.; Sun, T.; Mao, Z.; Liu, C.; Liu, L.; Liu, R.; Hou, L.; Li, X. Response of seed germination and seedling growth of *Pinus koraiensis* and Quercus mongolica to comprehensive action of warming and precipitation. *Acta Ecol. Sin.* **2012**, *32*, 7791–7800. [CrossRef]

32. Lyu, S.; Wang, X.; Zhang, Y.; Li, Z. Different responses of Korean pine (*Pinus koraiensis*) and Mongolia oak (*Quercus mongolica*) growth to recent climate warming in northeast China. *Dendrochronologia* **2017**, *45*, 113–122. [CrossRef]

33. Martínez-Vilalta, J.; Lloret, F. Drought-induced vegetation shifts in terrestrial ecosystems: The key role of regeneration dynamics. *Glob. Planet Change* **2016**, *144*, 94–108. [CrossRef]

34. Liang, Y.; Duveneck, M.J.; Gustafson, E.J.; Serra-Diaz, J.M.; Thompson, J.R. How disturbance, competition, and dispersal interact to prevent tree range boundaries from keeping pace with climate change. *Glob. Change Biol.* **2018**, *24*, e335–e351. [CrossRef]

35. Li, J.; Li, J. Regeneration and restoration of broad-leaved Korean pine forests in Lesser Xing'an Mountains of Northeast China. *Acta Ecol. Sin.* **2003**, *23*, 1268–1277.

36. Zhang, M.; Yan, Q.; Zhu, J. Optimum light transmittance for seed germination and early seedling recruitment of *Pinus koraiensis*: Implications for natural regeneration. *Iforest* **2015**, *8*, 853. [CrossRef]

37. Wang, J.; Yan, Q.; Lu, D.; Diao, M.; Yan, T.; Sun, Y.; Yu, L.; Zhu, J. Effects of microhabitat on rodent-mediated seed dispersal in monocultures with thinning treatment. *Agric. Forest Meteorol.* **2019**, *275*, 91–99. [CrossRef]

38. Tardós, P.; Lucas-Borja, M.; Beltrán, M.; Onkelinx, T.; Piqué, M. Composite low thinning and slash burning treatment enhances initial Spanish black pine seedling recruitment. *For. Ecol. Manag.* **2019**, *433*, 1–12. [CrossRef]

39. Wang, J.; Wang, G.G.; Zhang, T.; Yuan, J.; Yu, L.; Zhu, J.; Yan, Q. Use of direct seeding and seedling planting to restore Korean pine (*Pinus koraiensis* Sieb. Et Zucc.) in secondary forests of Northeast China. *For. Ecol. Manag.* **2021**, *493*, 119243. [CrossRef]

40. Zhao, F.; Yang, J.; He, H.S.; Dai, L. Effects of natural and human-assisted regeneration on landscape dynamics in a Korean pine forest in Northeast China. *PLoS ONE* **2013**, *8*, e82414. [CrossRef]

41. Li, B.; Hao, Z.; Bin, Y.; Zhang, J.; Wang, M. Seed rain dynamics reveals strong dispersal limitation, different reproductive strategies and responses to climate in a temperate forest in northeast China. *J. Veg. Sci.* **2012**, *23*, 271–279. [CrossRef]

42. Ishikawa, Y.; Krestov, P.V.; Namikawa, K. Disturbance history and tree establishment in old-growth *Pinus koraiensis*-hardwood forests in the Russian Far East. *J. Veg. Sci.* **1999**, *10*, 439–448. [CrossRef]

43. Zhang, Y.; Drobyshev, I.; Gao, L.; Zhao, X.; Bergeron, Y. Disturbance and regeneration dynamics of a mixed Korean pine dominated forest on Changbai Mountain, North-Eastern China. *Dendrochronologia* **2014**, *32*, 21–31. [CrossRef]

**forests**

Article

Forest Damage by Extra-Tropical Cyclone Klaus-Modeling and Prediction

Łukasz Pawlik [1,*], Janusz Godziek [1,2] and Łukasz Zawolik [1]

[1] Institute of Earth Sciences, Faculty of Natural Sciences, University of Silesia in Katowice, ul. Będzińska 60, 41-200 Sosnowiec, Poland
[2] International Environmental Doctoral School, University of Silesia, ul. Będzińska 60, 41-200 Sosnowiec, Poland
* Correspondence: lukasz.pawlik@us.edu.pl

Abstract: Windstorms may have negative consequences on forest ecosystems, industries, and societies. Extreme events related to extra-tropical cyclonic systems remind us that better recognition and understanding of the factors driving forest damage are needed for more efficient risk management and planning. In the present study, we statistically modelled forest damage caused by the windstorm Klaus in south-west France. This event occurred on 24 January 2009 and caused severe damage to maritime pine (*Pinus pinaster*) forest stands. We aimed at isolating the best potential predictors that can help to build better predictive models of forest damage. We applied the random forest (RF) technique to find the best classifiers of the forest damage binary response variable. Five-fold spatial block cross-validation, repeated five times, and forward feature selection (FFS) were applied to the control for model over-fitting. In addition, variable importance (VI) and accumulated local effect (ALE) plots were used as model performance metrics. The best RF model was used for spatial prediction and forest damage probability mapping. The ROC AUC of the best RF model was 0.895 and 0.899 for the training and test set, respectively, while the accuracy of the RF model was 0.820 for the training and 0.837 for the test set. The FFS allowed us to isolate the most important predictors, which were the distance from the windstorm trajectory, soil sand fraction content, the MODIS normalized difference vegetation index (NDVI), and the wind exposure index (WEI). In general, their influence on the forest damage probability was positive for a wide range of the observed values. The area of applicability (AOA) confirmed that the RF model can be used to construct a probability map for almost the entire study area.

Keywords: forest damage; windstorm; random forest; block cross-validation; forward feature selection

Citation: Pawlik, Ł.; Godziek, J.; Zawolik, Ł. Forest Damage by Extra-Tropical Cyclone Klaus-Modeling and Prediction. *Forests* **2022**, *13*, 1991. https://doi.org/10.3390/f13121991

Academic Editors: Any Mary Petritan and Mirela Beloiu

Received: 10 October 2022
Accepted: 23 November 2022
Published: 25 November 2022

Publisher's Note: MDPI stays neutral with regard to jurisdictional claims in published maps and institutional affiliations.

1. Introduction

Windstorms can bring devastating energy over Europe and a number of negative consequences to terrestrial ecosystems, soil cover, industry, and society [1,2]. It is estimated that annual losses to the EU and the UK are as high as 5 billion EUR, or approximately 0.04% of the total GDP [3]. Due to such a high level of losses, the problem of windstorm-related geohazards and disturbances requires serious consideration and new tools and methods, particularly in the context of the potential effects of climate change [4–6]. Extra-tropical windstorms have been studied for decades, and their impact on forests can be modelled and predicted to some extent [7–9]. To date, two types of models have been proposed: (1) hybrid-mechanistic and (2) statistical/machine learning models [10–12]. Hybrid-mechanistic models are based on information regarding the biomechanical behavior of trees under the forces generated by wind currents and information related to local and regional wind regimes [10,13]. Wind-related data can be obtained from models approximating the airflow over complex terrain, such as WindStation [14]. One example is the ForestGALES model, which is based on critical wind speed data and a windiness scoring system that allows the prediction of the probability of forest damage [13,15]. The best mechanistic models, such as ForestGALES or HWIND, are able to make predictions at a high level of accuracy if they have been parameterized for specific tree species [16,17]. Statistical/machine learning methods allow a certain degree of generalization based on large datasets and repeated evaluation of model training results, (e.g., in the case of k-fold cross-validation). Both approaches offer some advantages, but they also have limitations. Hybrid-mechanistic models can be sensitive to input data (such as expected wind speed or tree height), while statistical/machine learning models may be site-specific and sometimes do not work sufficiently well for new data [10].

Wind (speed, gustiness, direction) is the most difficult climate variable to model and predict due to its high dynamics (measured in seconds) and a wide range of factors influencing its properties. At the same time, wind is a key climatic agent affecting forest stand structures and dynamics, across the world, in both managed and old-growth forests. This has been reported in numerous scientific studies that have provided evidence of forest damage caused by strong winds [18–22].

It is commonly believed that the magnitude and frequency of catastrophic winter windstorms have recently increased, and this trend is predicted to continue, with a negative impact on forests [4,21,23,24]. However, there are contrasting opinions about the main drivers of the increase in storms and the scale of forest disturbance. For instance, Schelhaas et al. [23] concluded that the average volume of forest growing stock in Europe increased (along with forest area and stand age), thus increasing forest vulnerability to damage. This corresponds to the conclusions formulated by Seidl et al. [24], who also underlined the increasing standing timber volume that strongly influenced its susceptibility to disturbances. These findings agree with [9], who found that tree biomass was the most important predictor of forest damage in the Sudety Mountains in Poland [9]. However, other studies have reported increased storminess, which may have caused increased forest damage over the past decades [21]. This corresponds to the previous synthesis by Ranson et al. [4], who found, based on a literature review, that whilst the number of extra-tropical cyclones in the northern hemisphere may decrease, the number of extreme (very low pressure) winter cyclones may increase in certain regions. Nonetheless, Spinoni et al. [3] did not find any robust increasing trends in European windstorms. Assuming 3 °C warming (under RCP8.5; Representative Concentration Pathway), 16% of the land area in Europe may experience reduced maximum wind speeds. At the same time, the maximum wind speeds may increase in over 10% of Europe (including in the Alps), while remaining stable over the rest of the continent [3]. Under this climate scenario, wind losses may reach 4.6 billion EUR, annually. In the climate context, it seems the most dangerous are clusters of extreme events, which in terms of extra-tropical cyclones, mean several cyclones moving one after another in a short period of time [25].

Windstorm Klaus was a devastating event that caused severe damage to European forests, particularly in the Aquitaine region of France on 24 January 2009. It is estimated that it damaged 42 Mm^3 of trees, mainly maritime pine (*Pinus pinaster*) [26]. The total financial loss for these stands was estimated at ca 1.5 billion EUR in France alone [26]. Liberato et al. [27], based on a report by Aon-Benfield [28], claimed that the windstorm was the costliest weather hazard event, globally, in 2009, causing damage in France, Spain, and Portugal that was estimated between 4.0 and 6.0 billion USD. The Extreme Wind Storms Catalogue (http://www.europeanwindstorms.org/, accessed on 1 October 2022) showed that windstorm Klaus was the sixth most damaging event between 1987 and 2013 in terms of insured loss (3.5 billion USD). This event also killed at least 26 people [28].

With the increasing uncertainty regarding the magnitude of changes in wind properties and storm occurrence, and their severity under different climate change scenarios, a better understanding of the vulnerability of forests is needed. From the viewpoint of forest management ecology, this is a key issue in many European countries. In the present study, we aimed to contribute to the understanding of the negative consequences of extra-tropical cyclones on forest ecosystems. To explain the spatial distribution of wind-related forest damage, we applied one of the most efficient machine learning techniques, the random forest (RF) method [9]. We used it for the first time to model forest damage caused by the windstorm Klaus in 2009 in southwest France. We selected windstorm Klaus due to the relatively large database, with detailed information on forest damage during the event that allowed us to construct more robust prediction models. We aimed to isolate and calculate the impact of the most important predictors to support the construction of wind damage probability maps.

We describe the study area (Section 2.1) and explain how we define the response variable (Section 2.3.1). The method of selection and preparation of the set of potential predictors is presented in Sections 2.3.2 and 2.3.3. All steps of our modelling approach are elaborated in Section 2.4. The results are included in Section 3, and thoroughly discussed in Section 4, followed by our recommendations and conclusions in Section 5.

2. Materials and Methods

2.1. Study Area—Physical Settings and Climate

The study area is in southwest France, in the *Nouvelle Aquitaine* administrative region and the Gascony historical region. It is bordered, to the west, by the Bay of Biscay, which is part of the Atlantic Ocean (Figure 1). The relief has a lowland character, with small differences in the relative altitude (up to 250 m). Limestone and marl plateaus comprise the bedrock in the north, while sandstone and conglomerates form the so-called molasse in the south. A major part of the study area is dominated by podzols underlined by an impermeable layer of iron pan or bedrock [29].

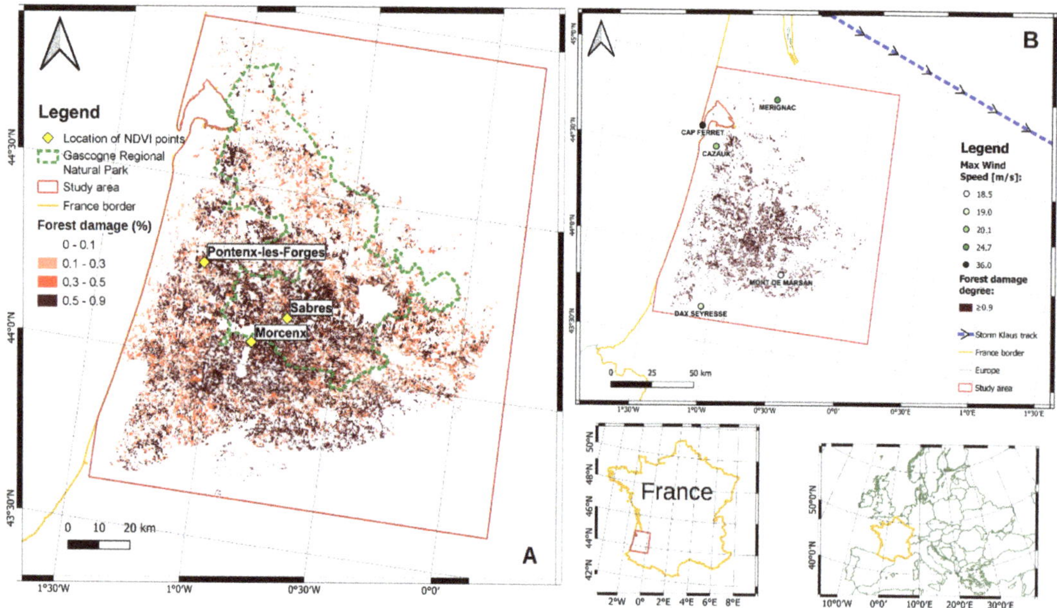

Figure 1. Study area with damage indicated according to the FORWIND database Forzieri et al. [30]. Map (**A**): the degree of forest damage (%) detected after windstorm Klaus in SW France. Yellow squares indicate places for which NDVI and LAI timeseries were constructed and showed in Figure 2. Map (**B**): the maximum wind speed for selected meteorological stations showed against the highest class of forest damage degree ($\geq$0.9), and the windstorm Klaus trajectory (dashed blue line).

According to the Koppen-Geier climate classification [31], the climate of the area under study is temperate and oceanic, with hot summers and no dry season (Cfb). The mean annual temperature in Cazaux is +13.4 °C (+6.6 °C in January, +20.6 °C in July), while in Mont de Marsan the mean annual temperature is +13.2 °C (+5.7 °C in January, +21.1 °C in

July). The mean total annual precipitation is 862 mm and 844 mm in Cazaux and Mont de Marsan, respectively [32]. The mean annual wind speed in Cazaux is 3.2 m s^{-1} (3.2 m s^{-1} in January, 3.3 m s^{-1} in July); in Mont de Marsan the mean annual wind speed is 2.4 m s^{-1} (2.2 m s^{-1} in January, 2.5 m s^{-1} in July).

Windstorm Klaus developed from a small wave perturbation on 21 January and intensified on 23 January 2009, moving rapidly towards the Bay of Biscay [27]. The system underwent explosive development when crossing the polar jet at an unusually low latitude position [27]. According to the data provided by the NOAA (National Oceanic and Atmospheric Administration) [32], the maximum wind speed during windstorm Klaus reached 36 m s^{-1}, on 24 January 2009. However, other sources reported that the wind speed was over 44 m s^{-1} [27].

2.2. Forest Conditions before and after Windstorm Klaus

The area of interest (AOI) was originally occupied by heath and marshes. Currently, most of the area is made up by the *Landes forest*, which is considered to be one of the largest man-made woodlands in Western Europe. The forests were planted in the 18th and 19th centuries to combat erosion and rehabilitate landscapes, as well as to provide wood for industrial purposes. The dominant species was the maritime pine (*Pinus pinaster*). The central part of the *Landes forest* area is protected by the Landes de Gascogne Regional Natural Park, established in 1970. Part of the study area is also used for agriculture, mainly for grain farming [29,33]. The data in the Forest Global Change web service (https://glad.earthengine.app/view/global-forest-change#dl=1;old=off;bl=off;lon=20;lat=10;zoom=3; accessed on 1 October 2022) did not show any spectacular damage in that part of France between 2000 and 2008 [34]. However, other data sources indicate that forests in the area were damaged during a storm on 27 December 1999 [35]. The affected region was between Bordeaux and the Atlantic coastline, with a damage level exceeding 30% of the forest area in each forest inspectorate [35]. The event resulted in estimated losses of 26.0 Mm^3 of wood [36]. It is likely that because of the 1999 event, damage caused by windstorm Klaus was much lower in that region than in the southern part of the AOI. The Leaf Area Index (LAI) and NDVI time series were inspected to obtain a better understanding of the forest stand conditions before and after windstorm Klaus (Figure 2). The NDVI decreased after the Klaus windstorm, and its values were lower than those in the same period during 2008. For the same points, the LAI showed only a sudden drop in the area close to Pontenx-les-Forges. Essentially, there are no clearly visible negative outcomes of the windstorm Klaus in the LAI graphs. Two plots (Sabres and Pontenx-les-Forges) showed even higher LAI values for July 2009 than a year before. Based on this exploratory analysis, it was decided that NDVI would likely be a better metric for exploring the impacts of windstorm Klaus on the region's forests. This was tested in the training module of the present project.

2.3. Data Sources and Pre-Processing

2.3.1. Response Variable

Information concerning the forest damage caused by windstorm Klaus was obtained from the dataset compiled by Forzieri et al. [30]. The FORWIND dataset consists of information on wind disturbances in European forests from 2000 to 2018. It includes almost 90,000 records of areas disturbed by strong winds. The response variable was defined in the database as the degree of damage (D). The D has five classes and single values ascribed to those classes: 0.1 (D $\leq$ 20%), 0.3 (20% < D $\leq$ 40%), 0.5 (40% < D $\leq$ 60%), 0.7 (60% < D $\leq$ 80%), and 0.9 (80% < D $\leq$ 100%) (Table S1, Supplementary Materials). For data related to windstorm Klaus, this classification was based on the interpretation of aerial images obtained by the Institut National de information geographique et forestiere. In total, the FORWIND database consists of 21,691 records related to windstorm Klaus (polygons with determined degree of damage). Most of the damage was concentrated along the

Atlantic Ocean coastline between Bordeaux, Mont de Marsan, and Dax. The damage area included the Landes de Gascogne Regional Natural Park (Figure 1).

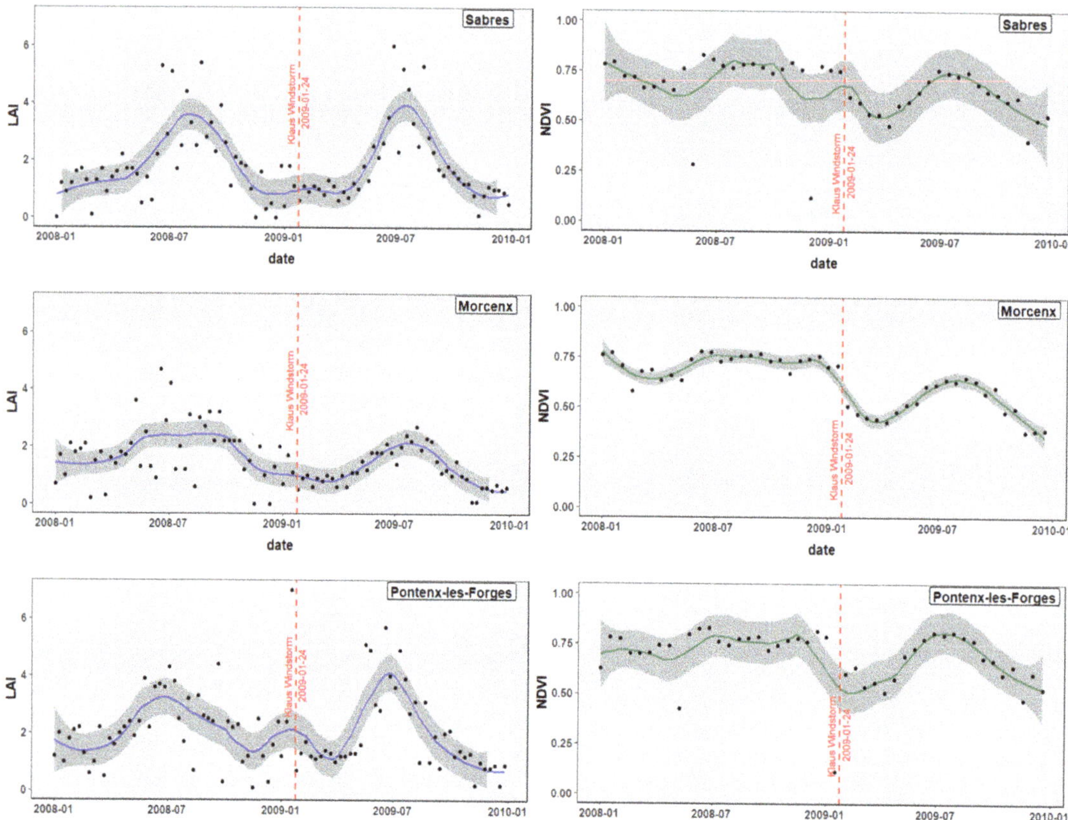

Figure 2. Timeseries of MODIS-based NDVI and LAI showing the negative impact of windstorm Klaus. The timeseries are for three localities shown in Figure 1 with a damage degree ≥ 0.9. Authors' own compilation based on the MODIS web service (https://modis.ornl.gov/data/modis_webservice. html; accessed on 1 October 2022). The data was imported using the *MODISTools* 1.1.2 R package [37].

2.3.2. Potential Environmental Predictors

In this study, we used four types of potential predictors: (1) climate; (2) geomorphic; (3) soil properties; and (4) vegetation indices (Table 1).

The climate predictors included information on wind speed. The first wind speed layer used is a product offered by the Windstorm Information Service (WISC) that contains spatial data on the most significant European windstorms since 1940. In that service, storm tracks, footprints, summary data and loss estimates are based on the ERA5 (ECMWF Reanalysis 5th Generation, ECMWF stands for European Centre for Medium-Range Weather Forecasts) reanalysis data sets. The windstorm Klaus footprint was a gridded dataset showing a maximum three-second wind gust speed (in m s^{-1}) at each grid point during the entire 72 h storm period. The wind speed layer used was available as a NC (NetCDF–Network Common Data Form) file.

Table 1. List of potential predictors used in the present study.

Name	Format	Original Resolution	Source	Direct URL
Elevation (m asl)	Raster layer	25 m	EU-DEM 1.1	https://land.copernicus.eu/imagery-in-situ/eu-dem/eu-dem-v1.1?tab=mapview (accessed on 1 October 2022)
Terrain slope	Raster layer	25 m	Calculated from EU-DEM	
Terrain roughness	Raster layer	25 m	Calculated from EU-DEM	
Profile curvature	Raster layer	25 m	Calculated from EU-DEM	
Planform curvature	Raster layer	25 m	Calculated from EU-DEM	
Topographic Wetness Index (TWI)	Raster layer	25 m	Calculated from EU-DEM	
Wind Exposition Index (WEI)	Raster layer	25 m	Calculated from EU-DEM	
Wind speed ($m·s^{-1}$)	Raster layer	$0.04° \times 0.04°$	Windstorm Information Service	https://climate.copernicus.eu/windstorm-information-service (accessed on 1 October 2022)
January mean wind speed (for based period 1999–2008) in $m·s^{-1}$	Raster layer	~4 × 4 km	TERRACLIMATE	https://www.climatologylab.org/terraclimate.html (accessed on 1 October 2022)
January 2009 wind speed ($m·s^{-1}$)	Raster layer	~4 × 4 km	TERRACLIMATE	https://www.climatologylab.org/terraclimate.html (accessed on 1 October 2022)
Sand fraction content in % (kg/kg) at 10 cm depth	Raster layer	250 m	OpenLandMap	https://openlandmap.org (accessed on 1 October 2022)
Clay fraction content in % at 10 cm depth	Raster layer	250 m	OpenLandMap	https://openlandmap.org (accessed on 1 October 2022)
Soil pH in H_2O at 10 cm depth	Raster layer	250 m	OpenLandMap	https://openlandmap.org (accessed on 1 October 2022)
Soil bulk density (fine earth) $10 \times kg/m^3$ at 10 cm depth	Raster layer	250 m	OpenLandMap	https://openlandmap.org (accessed on 1 October 2022)
Soil organic carbon content in × 5 g/kg at 10 cm depth	Raster layer	250 m	OpenLandMap	https://openlandmap.org (accessed on 1 October 2022)
MODIS Leaf Area Index	Raster layer	500 m	NASA MODIS/VIIRS Subsets	https://modis.ornl.gov/data/modis_webservice.html (accessed on 1 October 2022)
MODIS NDVI (Normalized Difference Vegetation Index)	Raster layer	250 m	NASA MODIS/VIIRS Subsets	https://modis.ornl.gov/data/modis_webservice.html (accessed on 1 October 2022)
Coastline buffer zones	Rasterized vector layer	1 km step buffer zones	geoBoundaries	https://www.geoboundaries.org/ (accessed on 1 October 2022)
Windstorm track line buffer zones	Rasterized vector layer	1 km step buffer zones	Windstorm Information Service	https://climate.copernicus.eu/windstorm-information-service (accessed on 1 October 2022)

The second wind-related predictor was the mean wind speed in January, calculated from gridded data for the based period 1999–2008, i.e., prior to windstorm Klaus in 2009.

In addition, we used the January 2009 wind speed. The data were downloaded from the Climatology Lab web service (https://www.climatologylab.org/terraclimate.html, accessed on 1 October 2022).

In machine learning models, the most commonly used geomorphic variable is elevation, which is used for the calculation of other terrain properties, e.g., slope, terrain roughness, etc. We used the EU-DEM v1.1 elevation data, downloaded from the Copernicus web service (Table 1). The data can be downloaded in the 1000×1000 km tiles as zipped GeoTIFF files. The initial resolution was 25 m (vertical accuracy ± 7 m RMSE), and the layer is available in ETRS89-LAEA (European Terrestrial Reference System 1989—Lambert Azimuthal Equal-Area) projection (EPSG:3035). The digital elevation model was used to calculate the terrain roughness, profile and planform curvature, topographic wetness index (TWI), and wind exposure index (WEI). The terrain slope, roughness, profile, and planform curvature were calculated using the *raster* R package [38]. For the slope calculation, the package follows Horn's [39] method. Terrain roughness is defined as the difference between the maximum and minimum cell values of a cell and the eight surrounding cells. The profile and planform curvatures were calculated using the *spatialEco* R package [40]. The planform and profile curvatures are the second derivatives of terrain elevation, or slope of the slope. The TWI and WEI were calculated using the SAGA GIS software 7.9.1 [41]. TWI is a dimensionless index that describes the potential surface water flow and accumulation [42]. The WEI provides information about terrain exposure to potential airflow where places with WEI < 1 are shadowed from the potential influence of airflow, and places with WEI > 1 are exposed to wind impact. The TWI and WEI were previously used in ecological and geomorphic modelling [9,43].

Soil properties are commonly considered an important type of information for environmental modelling, including the modelling of wind-related damage [12,20]. We used five types of soil feature data, measured at 10 cm depth: (1) sand; (2) clay fraction; (3) soil pH; (4) soil bulk density; and (5) soil organic carbon content. Raster layers containing information on soil properties were the result of ensemble machine learning modelling on a large set of reference points [44] and were made available as open-source data through the OpenLandMap web service (https://openlandmap.org/, accessed on 1 October 2022).

We applied two vegetation indices: the Leaf Area Index (LAI) [45] and the normalized difference vegetation index (NDVI), calculated from the Moderate Resolution Imaging Spectroradiometer (MODIS) satellite images. These two products were accessed using the *MODISTools* 1.1.2 R package [37]. The LAI is defined as the one-sided green leaf area (m^2) per unit ground area (m^2) in broadleaf canopies, and one-half of the total needle surface area per unit ground area in coniferous canopies (https://lpdaac.usgs.gov/products/mcd1 5a2hv006/, accessed on 1 October 2022). These data are available in an eight day composite dataset with a pixel size of 500 m. For this project, we used images from 1 June 2008. The other plant-related predictor was NDVI, generated every 16 days at a spatial resolution of 250 m (https://lpdaac.usgs.gov/products/mod13q1v006/, accessed on 1 October 2022). The NDVI was calculated as follows: $(NIR - R)/(NIR + R)$, where NIR stands for the near infrared spectrum of the electromagnetic radiation (ER), and R stands for the red spectrum of the ER. NDVI is a product of satellite image applications to detect changes in vegetation cover. This indicates chlorophyll absorption and NIR reflectance, which can be interpreted in terms of vegetation content and vigor [46]. For this study, we chose one image from 9 June 2008, which offers information on the forest stand condition before the windstorm Klaus.

The last two variables were created based on the distance from the windstorm Klaus track and from the Atlantic Ocean coastline. The separation distance between each belt (buffer zone) was 1 km. The windstorm Klaus track was downloaded from WISC, and the buffer zones were calculated using QGIS 3.16.4 [47]. The windstorm tracks were identified as the maximum relative vorticity ($\times 10^{-5}$ s^{-1}) at 850 hPa at three hourly locations of the extra tropical cyclone (ETC) [48–51].

2.3.3. Presence-Absence Data and the Data Set Preparation

The raster stack was built and used for subsequent sampling based on the points indicating places with damage (presence data) and without damage (absence data). The presence data were generated using the FORWIND vector layer with polygons that have attributes of the damage degree. Based on our initial results of the model training, we chose only polygons with the forest damage rate ≥ 0.7 (D $\geq 60\%$), which guaranteed a stronger signal from the variables controlling the response variable. For each polygon, the central point was generated using the *sf::st_point_on_surface* function [50]. To obtain an equal number of random points indicating absence data, we first extracted areas without damage from the FORWIND shapefile using *the rgeos::gDifference* function [51]. Subsequently, the *sf::st_sample* function was used to obtain randomly generated points with absence data. After combining the points with presence and absence data, they were used to extract spatial information from the raster layers (Figure 3). The spatial information inherited from the initial resolution of the raster layers was preserved (see similar approach by Bonannella et al. [52]).

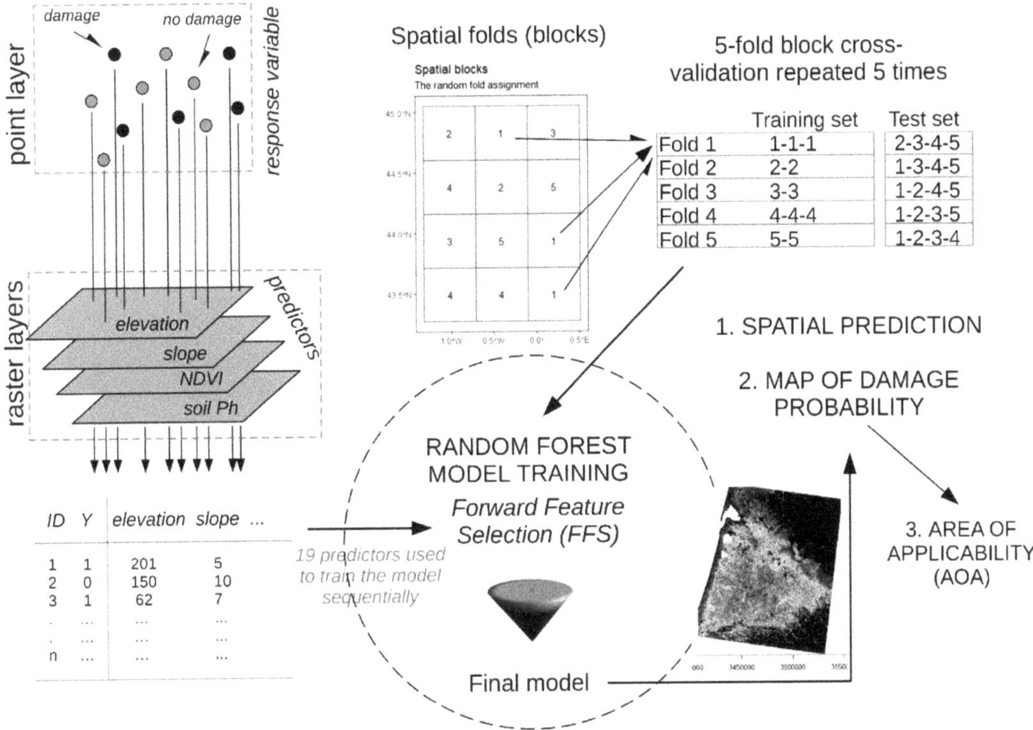

Figure 3. Modelling pattern adopted in the present study. The subsequent stages show raster layers sampling based on points, block cross-validation, random forest model training, the construction of map of forest damage probability, and final validation with the area of applicability (AOA) technique.

2.4. Model Training and Evaluation

2.4.1. Data Pre-Processing and Model Training

The *caret* R package was used for model training [53,54]. We applied RF to solve the classification problem. The RF model is a modern decision tree based ensemble learning technique that can be used for classification and regression [55,56]. The method builds multiple independent trees using bagging and averaging their results [57]. This is known to increase model stability and accuracy. The RF allows hyperparameters tuning such as *mtry*, which is a random sample of the *m* predictors, to split at in each node, and *min_n*, which is a minimum terminal node size. The number of trees can be tuned but it is also recommended to set it to a sufficiently and computationally effective number [58]. In the present project, the number of trees was set to 500.

Before the training process, highly correlated (|r| > 0.75) variables were excluded to avoid multicollinearity between independent variables. When two variables were correlated, the program analyzed how they were correlated with other variables and excluded the one which had a higher level of correlation with them. In order to avoid over-fitting, instead of random cross-validation, we applied spatial block cross-validation [59,60], the procedure that takes into account the spatial domain of the data structure. For that task we used the *blockCV* [60] and the *caretSDM* [61] R package. Five-fold block cross-validation repeated five times was applied (Figure 3). Such an approach is based on a Leave-Location-Out (LLO) Cross-Validation principle [62]. Ignoring the spatial scale and selecting the training and test set randomly from the entire study area could lead to over-fitting because the training and test points were sampled from the same area [62,63]. In other words, the random selection of the test points does not guarantee independence from training points due to spatial autocorrelation [63].

Before the training procedure, the data set was split into training and test sets (60 and 40%, respectively). During the model training stage, we used forward feature selection (FFS) to automatically identify and remove variables that could lead to over-fitting [62]. This method was implemented in the *CAST* R package [62].

2.4.2. Model Evaluation

For model evaluation, we used the following metrics: (1) the area under the receiver operating characteristics curve (ROC AUC); (2) accuracy; (3) variable importance (VI); and (4) accumulation local effects (ALE) plots.

Accuracy is the ratio between the correct predictions and all predictions. The ROC curve computes the sensitivity (i.e., recall, true positive class divided by the total number of positive results) and specificity (true negative divided by the sum of false positives and true negatives) over a continuum of different event thresholds [64,65]. The ROC AUC is a measure of the probability of correctly identifying a positive signal (response 1 in binary classification) above the noise [66]. In the present study, we adopted the common threshold level of probability used to discriminate between a positive and negative class of 0.5. The classes of ROC AUC were defined as follows: AUC = 0.5 (random discrimination between binary classes), AUC = (0.5, 0.7] (weak discrimination), AUC = (0.7, 0.8] (acceptable discrimination), AUC = (0.8, 0.9] (excellent discrimination), AUC = (0.9, 1.0) (outstanding discrimination), and AUC = 1.0 (perfect discrimination between classes) [67,68]. Variable importance (VI) was calculated to evaluate each predictor's contribution to the model. VI is a metric that describes accuracy and relies on the information in each predictor [69]. A higher VI indicates that a predictor has a stronger impact on the response variable and accumulation local effects (ALE) is a model diagnostic tool that explains the average influence of the features on the prediction [70]. It works in a similar way to partial dependence plots (PDP), but it is considered a better measure because it is less sensitive to correlations between predictors.

2.4.3. Prediction and Probability Maps

Spatial prediction was performed using the *raster::predict* function in R. To speed up the computation time, all rasters were aggregated to 100 m of spatial resolution. We applied the Area of Applicability (AOA) technique to test whether the RF model performs equally well across the entire study area [71]. AOA is the area where the model can learn relationships between variables based on the training data, and where the estimated cross-validation performance holds [71].

3. Results

Our results indicated that the RF model correctly discriminated between binary classes (damage vs. no damage) and can be a useful tool in forest damage modelling and prediction. From nineteen potential predictors, nine were selected during the FFS procedure as important and did not lead to model over-fitting. The four most important predictors were the distance from the windstorm trajectory, sand fraction content, NDVI, and WEI (Figure 4). Unexpectedly, the January mean wind speed (mean_ws) was less important, as were the other abiotic predictors (soil carbon content, elevation, soil pH, and slope). The model performance of the optimal model measured by ROC_AUC was 0.895, and the accuracy was 0.82. The model was trained on 7275 observations. After applying the model to the test set ($n = 4844$) the ROC_AUC was 0.899, and the accuracy reached 0.837 (balanced accuracy = 0.835).

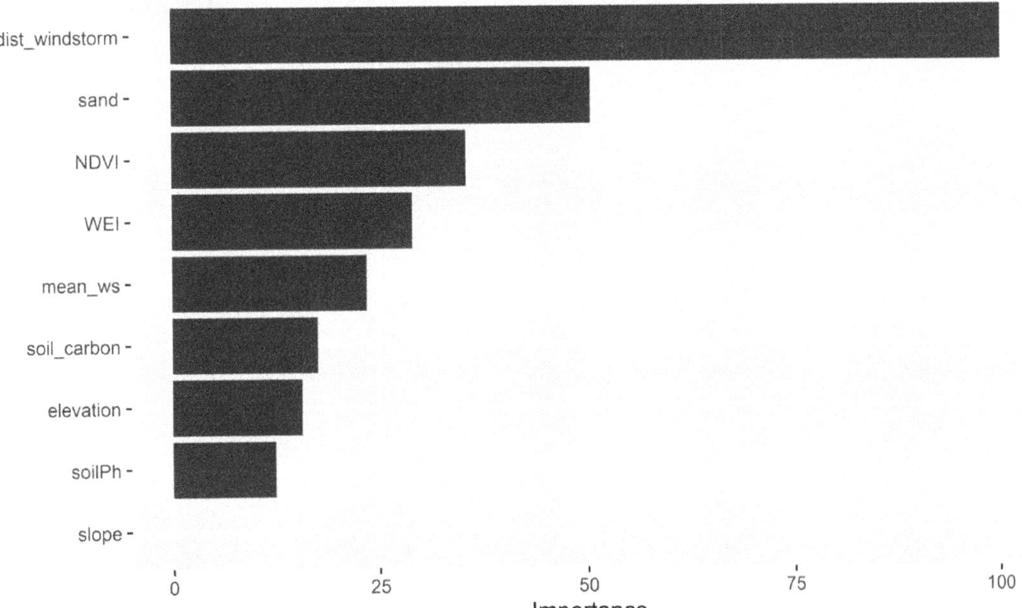

Figure 4. Feature importance plots for the best RF model. Abbreviations: dist_windstorm—distance from the windstorm Klaus trajectory, sand—sand fraction content, NDVI—Normalized Difference Vegetation Index, WEI—Wind Exposition Index, mean_ws—January mean wind speed (m s^{-1}) for 1999–2008, soil_carbon—soil organic carbon content, elevation—terrain elevation in m asl, soilPh—soil pH, slope—surface slope (in radians) (see also Table 1).

In terms of the individual effect of each predictor, the four most important features (distance from the windstorm track, sand content, NDVI, and WEI) had a positive effect; with their increasing values, the damage probability also increased (Figure 5). The probability of damage increased at a distance between 75 and 150 km from the windstorm trajectory (dist_windstorm). The soil sand fraction content had a positive impact on forest damage when reaching between 35 and 60%. In addition, the forest damage probability increased when the WEI increased over 0.95 and remained almost constant over WEI 1.05. The values of NDVI over 0.5 also positively impacted the forest damage probability. Five other variables (less important), such as mean wind speed in January, soil carbon content, elevation, soil pH, and slope, had different effects on damage probability (only partially positive). The mean wind speed had a positive effect when reaching up to 2.0 ms^{-1} and for 2.75–3.25 ms^{-1}. A higher mean wind speed did not affect the damage probability. The soil carbon content, which is a measure of soil fertility and edaphic conditions of forest growth, had a positive impact up to 4 units. After that point, the level of damage probability decreased and remained almost constant for soil carbon over 6 units. The terrain elevation had positive impact on the damage probability up to ca. 60 m asl. Soil pH, which is also a measure of soil fertility, has a minor positive impact on forest damage probability up to ca. 6.2 units. The slope had a negligible effect on forest vulnerability to damage (Figure 5).

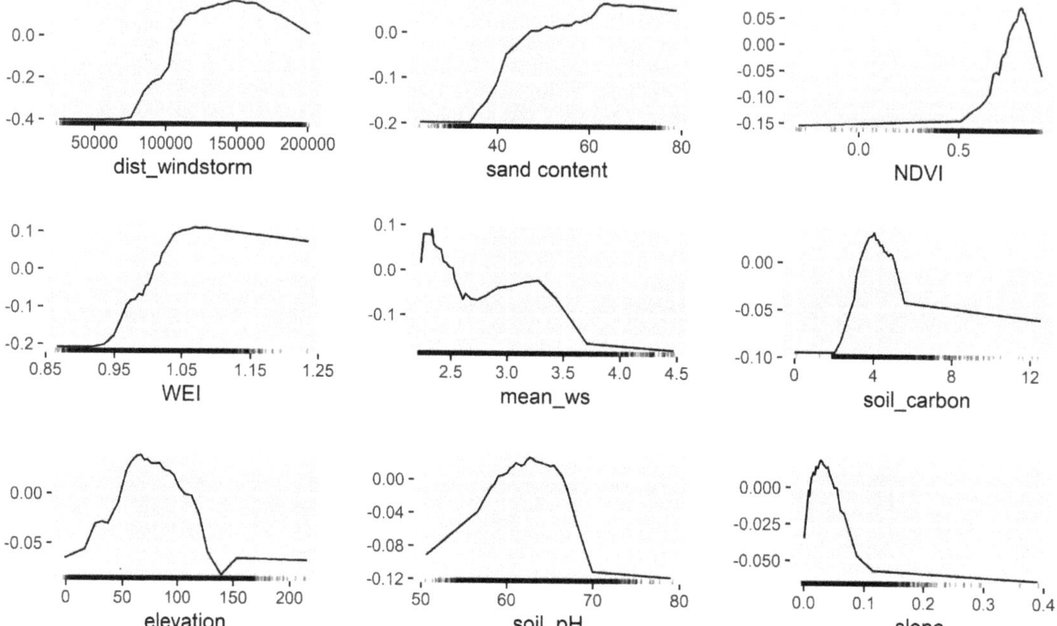

Figure 5. Accumulated Local Effects (ALE) plots for the most important features of the RF model. The X-axis represents predictors values while the Y-axis represents probability. Abbreviations: dist_windstorm—distance from the windstorm Klaus trajectory, sand—sand fraction content, NDVI—Normalized Difference Vegetation Index, WEI—Wind Exposition Index, mean_ws—January mean wind speed (m s^{-1}) for 1999–2008, soil_carbon—soil organic carbon content, elevation—terrain elevation in m asl, soilPh—soil pH, slope—surface slope (in radians) (see also Table 1).

After applying the RF model to the entire AOI, we were able to obtain maps of damage probability. Clusters of high forest damage probability were computed primarily for the central part of the study area, some distance from the Atlantic Ocean coastline, and the windstorm trajectory, which was North of the area (Figure 6). The Area of Applicability (AOA) confirmed the RF model could be successfully applied to the entire area, with some minor areas of model inapplicability along the coastline.

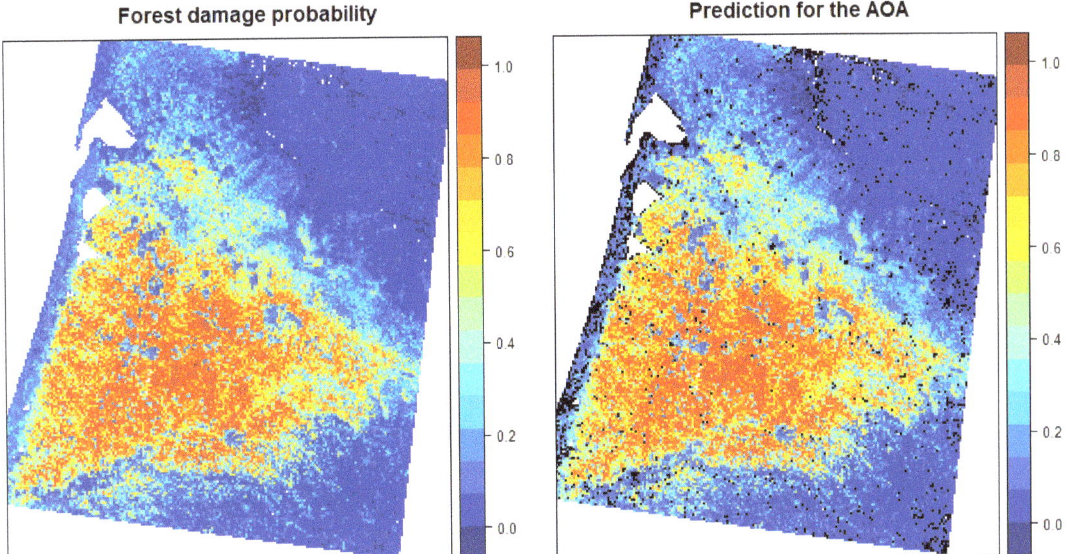

Figure 6. Map of forest damage probability and prediction of the Area of Applicability (AOA). Areas outside of the AOA are shown in black.

4. Discussion

We found a greater forest damage probability with increasing distance from the windstorm track, up to 150 km. A similar tendency was reported in New England (USA) forests, which were impacted by hurricanes in 1944 [72]. It was found that the sites that experienced higher rates of damage were located further east of the storm track and closer to the area of maximum estimated wind speed [73]. This is an important, and yet understudied, finding because it allows for the approximate evaluation of windstorm damage and probability of damage based on predicted windstorm trajectory, even in near real-time projections. Other studies found the highest importance of forest biomass (volume in m^3) and tree age predictors [9]. However, in many instances, forest feature data were not available or not up to date, thus favoring the application of other predictors that were easier to access and implement. We found a higher damage probability with increasing NDVI. High NDVI values can be interpreted as mature forests with a better health condition [74]. Some studies have found weak correlations between NDVI and forest biomass [74]. That agrees with our previous results, where the tree biomass was found to be the most important predictor [9].

In the past, several approaches were tested to isolate the best predictive models of forest damage triggered by strong winds [11,68,74]. However, the results differed in the level of model predictive power [68,75–77], and in most cases our RF model outperformed other models. We were able to reach a higher precision of the RF model, which was 0.899 (ROC AUC) and 0.837 (accuracy) after applying the model to the test set. Fridman and Valinger [75] obtained an accuracy of 0.8 using tree, stand, and site characteristics from study plots in Sweden as an input data for the logistic regression model. Schindler et al. [76] analyzed the damage caused in Germany by the windstorm Lothar in 1999. After using the logistic regression modelling the model precision measured by ROC AUC reached 0.79. The highest probability of damage was predicted for coniferous forest stands growing on acidic, fresh, and moist soils [76]. Klaus et al. [77] obtained an accuracy level of 0.7, based on GLM modelling of damage caused by the windstorm Kyrill in Germany in 2007. They highlighted the sensitivity of forest stands to windthrow as the effect of a high proportion of coniferous trees, a complex topography, and immature soils. Suvanto et al. [68] obtained

ROC AUC close to 0.73 for damage probability in Finland and were able to demonstrate a level of precision for a large database over the entire country. Pawlik and Harrison [9] presented the RF model, of which the precision measured by ROC AUC was 0.71. They analyzed data for the Sudety region in Poland and found that forest volume and tree age were the most important predictors. There might be several reasons behind the differences in the models' precision obtained by various authors in comparison to the results presented here. The most likely are: (1) different machine learning methods applied (here RF which is a decision tree-based method); (2) a different set and number of potential predictors (19 predictors reduced to 9); (3) application of spatial block cross-validation; (4) use of the forward feature selection.

For the same study area, Kamimura et al. [10] used tree census data for the evaluation and modelling of the damage caused by windstorm Klaus. They used data from 235 plots, collected in 2007 and 2008 by the French foresters during national inventories. Additionally, trees damaged by windstorm Klaus were identified. This unique database consisted of several tree features, including height, diameter at breast height, and tree age. With as many as 25 predictors, Kamimura et al. [10] were able to train the logistic regression model, which reached AUC = 0.791 and an accuracy of 0.717. Although such a level of accuracy is acceptable, it can also indicate that the data set used in their study was too small, or that the modelling based on individual tree features was not able to capture any significant signal from the data. Despite the statistical modelling of windstorm Klaus, the authors had some difficulties finding the optimal logistic regression model. Nevertheless, they were able to isolate the most important predictors, which were soil type and the year of forest establishment [10]. In our study, we considered five soil features (Table 1) that we thought could support a better explanation of the spatial structure of the forest damage. Soil clay and sand fraction content is critical for the development and architecture of the root systems, which are important for tree stability during high wind events [10,20]. The root systems are weakly developed in soils that are immature, too shallow, or containing too high content of gravels. In such soils, root systems may not provide secure anchorage in the ground or overall stability. Additionally, hydromorphic soils (saturated or waterlogged soils) can prevent trees from developing deep root systems. Maritime pine trees were found to have less anchorage in wet soils [78]. We found that the damage probability was positively influenced by a higher sand content. The soil type in the area of investigation primarily contained podzols of more than 55 cm in depth [10]. Sandy soils, such as podzols, cannot secure good tree stability. Further, data on Sitka spruce (*Picea sitchensis*) collected during tree-pulling experiments in the UK suggested better anchorage on peat than on gleyed mineral soils [79]. While applying the tree-pulling experiments in the same study area in southwest France, Cucchi et al. [37] found evidence of higher stability (better anchorage) for maritime pine on dry podzols with a deeper groundwater table, and a broken or absent hard pan. However, such conditions frequently cause stem failure. In addition, we found important, albeit minor, impact of soil pH and carbon content on the level of forest damage probability. Both indicators provide information on soil fertility and quality. It was also found that the risk of windstorm damage increased with a growing deterioration of the humus form and higher soil acidity (i.e., lower soil pH) [80]. This was partly supported by our study, as we found that the forest stands growing on alkaline soils had a very low damage probability.

We concluded that the WEI increased the probability of damage. When >1.0, this dimensionless index indicates areas exposed to wind impact. The RF model correctly discriminated between exposed (>1.0) and unexposed (<1.0) places where the increase in damage probability was always over zero (Figure 4). The study area was rather flat; however, at a resolution of 25 m the WEI algorithm applied in SAGA GIS was able to find significant differences in the terrain exposure to wind. This is in line with our previous results [9]. Rapid damage propagation can be expected close to the coastline, irrespective of tree features, as pointed out by Kamimura et al. [10]; however, we did not find the distance from the coastline as an important predictor. Although difficult to measure, wind speed is a key predictor in forest damage modelling [9,81]. Of the three wind-related predictors used in our study (Table 1), only mean January wind speed was important, but not as significant as expected. According to our results, the forest damage probability increased at short intervals when the mean wind speed was >0.25 ms^{-1}, and between 2.75 and 3.25 ms^{-1}. In the Sudety Mountains, SW Poland, it was found that the forest damage probability increased proportionally to the mean June wind speed, between 3 and 6 m s^{-1} [9]. Several historical studies reviewed by Gardiner [12] indicated an increasing damage probability with increasing wind speed. In Finland, it was found that the relationship between wind gust speed and volume of forest damage followed a power function, with a power of approximately 10 [81]. However, there is a common agreement that such a relationship is highly dependent on regional wind conditions and tree features (tree species, height, age, and health status). Due to the managed nature of the forest and past forest management in our AOI, we cannot assume that maritime pines had natural resistance to wind impact [10]. The raster layer containing wind speed data from the Windstorm Information Service (Table 1) indicated rapid wind speed decline, from 47 to 27 m s^{-1}, with increasing distance from the coastline, 20 to 50 km, particularly in the northern part of the region (Figure S1, Supplementary Materials). The highest damage density for the class of damage ≥ 0.7 was between 15 and 50 km from the coastline. However, we presume that most of the trees in some heavily impacted forest inspectorates were damaged by wind speeds below the level of 30 m s^{-1} (which is still high). Past reports suggest that even lower, but sustained, wind speeds and gusts can cause widespread damage in forests [12].

The biggest limitation of our modelling approach was the lack of detailed information on tree and forest properties before and after the event. One of the ways to overcome that limitation was to use vegetation indices based on satellite images, i.e., NDVI. Data on individual tree features are difficult to obtain, especially after windstorm events, when all trees in managed forests are subject to sanitary removal for safety and economic reasons. Therefore, we focused on forest characteristics that are easily accessible via the Internet and are also high-quality products (in terms of classification and spatial resolution precision) based on the reanalysis of satellite images. Furthermore, in some cases, building a model using information on individual tree characteristics cannot provide satisfactory prediction accuracy [10]. Considering these advantages and limitations, this study used general information about forest conditions, namely NDVI and LAI. Our results suggest that only NDVI can accommodate a strong signal of forest damage in specific conditions of the maritime pine forest.

5. Conclusions

Windstorm Klaus was among the most devastating extreme climate events in the history of European meteorological records. The windstorm left a significant imprint in forest stands, particularly in France, as well as high financial losses and many casualties. After applying machine learning techniques, we were able to identify some underlying relationships that may help improve forest damage risk management due to strong wind events. This can be a valuable starting point for other analyses and planning in regions with similar natural conditions as our study area in southwest France affected by windstorm Klaus. We were able to map the probability of forest damage using the RF method. RF can be used for a quick evaluation of the most important features behind the high rate of forest damage during extra-tropical cyclones. These events may occur more frequently in the near future, and for that reason, their thorough understanding is a key prerequisite for better planning. This can also ensure a basic level of warning system functionality. In the present study, we found, for the first time, that the distance from the windstorm track, wind exposure of the terrain, and soil properties, should always be considered as important predictors. When detailed information on forest structure and features is not directly accessible, NDVI can be used as a good approximation of forest conditions and history. The information can be easily obtained using various web services. Additionally, prediction maps offer a general overview of forest damage probability in geographic regions and will aid future steps in forest management, ecology, and geomorphology.

Supplementary Materials: The following supporting information can be downloaded at: https://www.mdpi.com/article/10.3390/f13121991/s1, Figure S1: Density function of the distance from the Atlantic coastline displayed for the two classes of the binary response variable. Abbreviations: dam—damage class, no_dam—places with no damage class; Table S1: Basic properties of the damage classes adopted in the study.

Author Contributions: Conceptualization, Ł.P.; methodology, Ł.P.; validation, Ł.P.; formal analysis, Ł.P. and J.G.; investigation, Ł.P. and J.G.; data curation, Ł.P.; writing—original draft preparation, Ł.P. and J.G.; writing—review and editing, Ł.P. and J.G.; visualization, Ł.P., J.G. and Ł.Z.; supervision, Ł.P.; project administration, Ł.P.; funding acquisition, Ł.P. All authors have read and agreed to the published version of the manuscript.

Funding: This study was supported by the Polish National Center—Narodowe Centrum Nauki (grant no 2019/35/O/ST10/00032).

Data Availability Statement: Data will be shared on request to the corresponding author.

Acknowledgments: Two anonymous reviewers are thanked for their insightful comments and suggestions that allowed us to improve the final version of the manuscript. In addition, we want to thank Lukasz Longosz for proofreading the final draft of the manuscript.

Conflicts of Interest: The authors declare no conflict of interest.

References

1. Walz, M.A.; Kruschke, T.; Rust, H.W.; Ulbrich, U.; Leckebusch, G.C. Quantifying the extremity of windstorms for regions featuring infrequent events. *Atmos. Sci. Lett.* **2017**, *18*, 315–322. [CrossRef]
2. Romeiro, J.M.N.; Eid, T.; Antón-Fernández, C.; Kangas, A.; Trømborg, E. Natural disturbances risks in European Boreal and Temperate forests and their links to climate change—A review of modelling approaches. *For. Ecol. Manag.* **2022**, *509*, 120071. [CrossRef]
3. Spinoni, J.; Formetta, G.; Mentaschi, L.; Forzieri, G.; Feyen, L. *Global Warming and Windstorm Impacts in the EU*; JRC Technical Report; EUR 29960 EN; Publications Office of the European Union: Luxembourg, 2020. [CrossRef]
4. Ranson, M.; Kousky, C.; Ruth, M.; Jantarasami, L.; Crimmins, A.; Tarquinio, L. Tropical and extratropical cyclone damages under climate change. *Clim. Chang.* **2014**, *127*, 227–241. [CrossRef]
5. Senf, C.; Seidl, R. Storm and fire disturbances in Europe: Distribution and trends. *Glob. Chang. Biol.* **2021**, *27*, 3605–3619. [CrossRef]
6. Masson-Delmotte, V.; Zhai, P.; Pirani, A.; Connors, S.L.; Péan, C.; Berger, S.; Caud, N.; Chen, Y.; Goldfarb, L.; Gomis, M.I.; et al. (Eds.) IPCC Climate Change 2021: The Physical Science Basis. In *Contribution of Working Group I to the Sixth Assessment Report of the Intergovernmental Panel on Climate Change*; Cambridge University Press: Cambridge, UK, 2021. Available online: https://www.ipcc.ch/report/ar6/wg1/ (accessed on 1 October 2022).
7. Jahani, A.; Saffariha, M. modelling of trees failure under windstorm in harvested Hercynian forests using machine learning techniques. *Sci. Rep.* **2021**, *11*, 1124. [CrossRef] [PubMed]
8. Pettit, J.L.; Pettit, J.M.; Janda, P.; Rydval, M.; Čada, V.; Schurman, J.S.; Nagel, T.A.; Bače, R.; Saulnier, M.; Hofmeister, J.; et al. Both cyclone-induced and convective storms drive disturbance patterns in European primary beech forests. *J. Geophys. Res. Atmos.* **2021**, *126*, e2020JD033929. [CrossRef]
9. Pawlik, Ł.; Harrison, S.P. modelling and prediction of wind damage in forest ecosystems of the Sudety Mountains, SW Poland. *Sci. Total Environ.* **2022**, *815*, 151972. [CrossRef] [PubMed]
10. Kamimura, K.; Gardiner, B.; Dupont, S.; Guyon, D.; Meredieu, C. Mechanistic and statistical approaches to predicting wind damage to individual maritime pine (*Pinus pinaster*) trees in forests. *Can. J. For. Res.* **2016**, *46*, 1. [CrossRef]
11. Hart, E.; Sim, K.; Kamimura, K.; Meredieu, C.; Guyon, D.; Gardiner, B. Use of machine learning techniques to model wind damage to forests. *Agric. For. Meteorol.* **2019**, *265*, 16–29. [CrossRef]
12. Gardiner, B. Wind damage to forets and trees: A review with an emphasis on planted and managed forests. *J. For. Res.* **2021**, *26*, 248–266. [CrossRef]
13. Gardiner, B.; Byrne, K.; Hale, S.; Kamimura, K.; Mitchell, S.J.; Peltola, H.; Ruel, J.C. A review of mechanistic modelling of wind damage risk to forests. *Forestry* **2008**, *81*, 447–463. [CrossRef]
14. Lopes, A.M.G. WindStation—A software for the simulation of atmospheric flows over complex topography. *Environ. Model. Softw.* **2003**, *18*, 81–96. [CrossRef]
15. Hale, S.; Gardiner, B.; Peace, A.; Nicoll, B.; Taylor, P.; Pizzirani, S. Comparison and validation of three versions of a forest wind risk model. *Environ. Model. Softw.* **2015**, *68*, 27–41. [CrossRef]

16. Peltola, H.; Kellomäki, S.; Vaisanen, H.; Ikonene, V.-P. A mechanistic model for assessing the risk of wind and snow damage to single trees and stands of Scots pine, Norway spruce, and birch. *Can. J. For. Res.* **1999**, *29*, 647–661. [CrossRef]

17. Gardiner, G.; Suárez, J.; Achim, A.; Hale, S.; Nicoll, B. *ForestGALES: A PC-Based Wind Risk Model for British Forests*; User's Guide Version 2.0; Forestry Commission: Edinburgh, UK, 2004.

18. Everham, E.M.; Brokaw, N.V. Forest damage and recovery from catastrophic wind. *Bot. Rev.* **1996**, *62*, 113–185. [CrossRef]

19. Mitchell, S. Wind as a natural disturbance agent in forests: A synthesis. *Forestry* **2013**, *86*, 147–157. [CrossRef]

20. Gardiner, B.; Schuck, A.; Schelhaas, M.-J.; Orazio, C.; Blennow, K.; Nicoll, B. *Living with Storm Damage to Forests: What Science Can Tell Us*; European Forest Institute: Joensuu, Finland, 2013.

21. Gregow, H.; Laaksonen, A.; Alper, M.E. Increasing large scale windstorm damage in western, central and northern European forests, 1951–2010. *Sci. Rep.* **2017**, *7*, 46397. [CrossRef] [PubMed]

22. Negrón-Juárez, R.I.; Jenkins, H.S.; Raupp, C.F.M.; Riley, W.J.; Kueppers, L.M.; Magnabosco Marra, D.; Ribeiro, G.H.P.M.; Monteiro, M.T.F.; Candido, L.A.; Chambers, J.Q.; et al. Windthrow variability in Central Amazonia. *Atmosphere* **2017**, *8*, 28. [CrossRef]

23. Schelhaas, M.-J.; Nabuurs, G.-J.; Schuck, A. Natural disturbances in the European forests in the 19th and 20th centuries. *Glob. Chang. Biol.* **2003**, *9*, 1620–1633. [CrossRef]

24. Seidl, R.; Schelhaas, M.-J.; Lexer, M.J. Unraveling the drivers of intensifying forest disturbance regimes in Europe. *Glob. Chang. Biol.* **2011**, *17*, 2842–2852. [CrossRef]

25. Dacre, H.F.; Pinto, J.G. Serial clustering of extratropical cyclones: A review of where, when and why it occurs. *NPJ Clim. Atmos. Sci.* **2020**, *3*, 48. [CrossRef]

26. Caurla, S.; Garcia, S.; Niedzwiedz, A. Store or export? An economic evaluation of financial compensation to forest sector after windstorm. The case of Hurricane Klaus. *For. Policy Econ.* **2015**, *61*, 30–38. [CrossRef]

27. Liberato, M.L.R.; Pinto, J.G.; Trigo, I.F.; Trigo, R.M. Klaus—An exceptional winter storm over northern Iberia and southern France. *Weather* **2011**, *66*, 330–334. [CrossRef]

28. Aon-Benfield, Annual Global Climate and Catastrophe Report IF 2009. 2010. Available online: https://www.aon.com/attachments/reinsurance/200912_ab_if_impact_forecasting_2009_report.pdf (accessed on 1 October 2022).

29. Tuppen, J.N.; Bachrach, B.S.; Higonnet, P.L.-R.; Flower, J.E.; Popkin, J.D.; Wright, G.; Bisson, T.N.; Shennan, J.H.; Fournier, G.; Elkins, T.H.; et al. "France". Encyclopedia Britannica. 3 November 2021. Available online: https://www.britannica.com/place/France (accessed on 1 October 2022).

30. Forzieri, G.; Pecchi, M.; Girardello, M.; Mauri, A.; Klaus, M.; Nikolov, C.; Rüetschi, M.; Gardiner, B.; Tomaštík, J.; Small, D.; et al. A spatially explicit database of wind disturbances in European forests over the period 2000–2018. *Earth Syst. Sci. Data* **2020**, *12*, 257–276. [CrossRef]

31. Beck, H.E.; Zimmermann, N.E.; McVicar, T.R.; Vergopolan, N.; Berg, A.; Wood, E.F. Present and future Köppen-Geiger climate classification maps at 1-km resolution. *Sci. Data* **2018**, *5*, 180214. [CrossRef]

32. NOAA GSOD, National Oceanic and Atmospheric Administration, Global Summary of the Day, U.S. Department of Commerce. 2021. Available online: https://www7.ncdc.noaa.gov/CDO/cdoselect.cmd?datasetabbv=GSOD (accessed on 6 November 2021).

33. Alison, C. (Ed.) *Michelin Green Guide: French Atlantic Coast*; Michelin Apa Publications: London, UK, 2010; Volume 8, pp. 258–263, ISBN 1-906261-79-2.

34. Hansen, M.C.; Potapov, P.V.; Moore, R.; Hancher, M.; Turubanova, S.A.; Tyukavina, A.; Thau, D.; Stehman, S.V.; Goetz, S.J.; Loveland, T.R.; et al. High-resolution global maps of 21st-century forest cover change. *Science* **2013**, *342*, 850–853. [CrossRef] [PubMed]

35. Cucchi, V.; Bert, D. Wind-firmness in *Pinus pinaster* Aït. Stands in Southwest France: Influence of stand density, fertilisation and breeding in two experimental stands damaged during the 1999 storm. *Ann. For. Sci.* **2003**, *60*, 209–226. [CrossRef]

36. Cucchi, V.; Meredieu, C.; Stokes, A.; Barthier, S.; Bert, D.; Najar, M.; Denis, A.; Lastennet, R. Root anchorage of inner and edge trees in stands of Maritime pine (*Pinus pinaster* Ait.) growing in different podzolic soil conditions. *Trees* **2004**, *18*, 460–466. [CrossRef]

37. Tuck, S.L.; Phillips, H.R.P.; Hintzen, R.E.; Scharlemann, J.P.W.; Purvis, A.; Hudson, L.N. MODISTools—Downloading and processing MODIS remotely sensed data in R. *Ecol. Evol.* **2014**, *4*, 4658–4668. [CrossRef]

38. Hijmans, R.J. Raster: Geographic Data Analysis and Modelling. R Package Version 3.3-13. 2020. Available online: https://CRAN.R-project.org/package=raster (accessed on 1 October 2022).

39. Horn, B.K.P. Hill shading and the reflectance map. *Proc. IEEE* **1981**, *69*, 14–47. [CrossRef]

40. Evans, J.S. _spatialEco_. R Package Version 1.3-6. 2021. Available online: https://github.com/jeffreyevans/spatialEco (accessed on 1 October 2022).

41. Conrad, O.; Bechtel, B.; Bock, M.; Dietrich, H.; Fischer, E.; Gerlitz, L.; Wehberg, J.; Wichmann, V.; Böhner, J. System for Automated Geoscientific Analyses (SAGA) v. 2.1.4. *Geosci. Model Dev.* **2015**, *8*, 1991–2007. [CrossRef]

42. Moore, I.D.; Grayson, R.B.; Ladson, A.R. Digital terrain modelling: A review of hydrological, geomorphological, and biological applications. *Hydrol. Processes* **1991**, *5*, 3–30. [CrossRef]

43. Dyderski, M.K.; Pawlik, Ł. Drivers of forest aboveground biomass and its increments in the Tatra Mountains after 15 years. *Catena* **2021**, *205*, 105468. [CrossRef]

44. Hengl, T.; Mendes de Jesus, J.; Heuvelink, G.B.M.; Ruiperez Gonzalez, M.; Kilibarda, M.; Blagotić, A.; Shangguan, W.; Wright, M.N.; Geng, X.; Bauer-Marschallinger, B.; et al. SoilGrids250m: Global gridded soil information based on machine learning. *PLoS ONE* **2017**, *12*, e0169748. [CrossRef] [PubMed]

45. Watson, D.J. The estimation of leaf area in field crops. *J. Agric. Sci.* **1937**, *27*, 474–483. [CrossRef]
46. Clark, J.; Bobbe, T. Using remote sensing to map and monitor fire damage in forest ecosystems. In *Understanding Forest Disturbance and Spatial Pattern*; Michael, A.W., Steven, E.F., Eds.; Remote Sensing and GIS Approaches; Taylor and Francis: New York, NY, USA, 2007; pp. 113–131.
47. QGIS.org, QGIS Geographic Information System. QGIS Association. 2021. Available online: https://www.qgis.org (accessed on 1 October 2022).
48. Hodges, K.I. Feature tracking on the unit sphere. *Mon. Weather Rev.* **1995**, *123*, 3458–3465. [CrossRef]
49. Whitelaw, A.; Shaffrey, L.; Hodges, K. WISC Storm Tracks Description. Copernicus Climate Change Service. 2017. Available online: https://wisc.climate.copernicus.eu/wisc/documents/shared/C3S_WISC_Storm%20Track_Description_v1.0.pdf (accessed on 1 October 2022).
50. Pebesma, E. Simple Features for R: Standardized Support for Spatial Vector Data. *R J.* **2018**, *10*, 439–446. [CrossRef]
51. Bivand, R.; Rundel, C. rgeos: Interface to Geometry Engine—Open Source ('GEOS'). R Package Version 0.5-7. 2021. Available online: https://CRAN.R-project.org/package=rgeos (accessed on 1 October 2022).
52. Bonannella, C.; Hengl, T.; Heisig, J.; Parent, L.; Wright, M.N.; Herold, M.; de Bruin, S. Forest tree species distribution for Europe 2000-2020: Mapping potential and realized distributions using spatiotemporal machine learning. *PeerJ* **2022**, *10*, e13728. [CrossRef]
53. Kuhn, M. Building predictive models in R using the caret package. *J. Stat. Softw.* **2008**, *28*, 1–26. [CrossRef]
54. Kuhn, M. Caret: Classification and Regression Training. R Package Version 6.0-86. 2020. Available online: https://CRAN.R-project.org/package=caret (accessed on 1 October 2022).
55. Breiman, L. Random forests. *Mach. Learn.* **2001**, *45*, 5–32. [CrossRef]
56. Probst, P.; Boulesteix, A.-L. To tune or not to tune the number of trees in random forest? *arXiv* **2017**, arXiv:1705.05654.
57. Lesmeister, C.; Chinnamgari, S.K. *Advanced Machine Learning with R*; Packt: Birmingham, UK; Mumbai, India, 2019; 649p.
58. Hengl, T.; Nussbaum, M.; Wright, M.N.; Heuvelink, G.B.M.; Gräler, B. Random forest as a generic framework for predictive modeling of spatial and spatio-temporal variables. *PeerJ* **2018**, *6*, e5518. [CrossRef]
59. Roberts, D.R.; Bahn, V.; Ciuti, S.; Boyce, M.S.; Elith, J.; Guillera-Arroita, G.; Hauenstein, S.; Lahoz-Monfort, J.J.; Schröder, B.; Thuiller, W.; et al. Cross-validation strategies for data with temporal, spatial, hierarchical, or phylogenetic structure. *Ecography* **2017**, *40*, 913–929. [CrossRef]
60. Valavi, R.; Elith, J.; Lahoz-Monfort, J. BlockCV: An R package for generating spatially or environmentally separated folds for k-fold cross-validation of species distribution models. *Methods Ecol. Evol.* **2019**, *10*, 225–232. [CrossRef]
61. Corrêa, P. caretSDM—Species Distribution Models Using Caret, v.0.2.0. 2021. Available online: https://github.com/correapvf/caretSDM (accessed on 1 October 2022).
62. Meyer, H.; Reudenbach, C.; Hengl, T.; Katurji, M.; Nauss, T. Improving performance of spatio-temporal machine learning models using forward feature selection and target-oriented validation. *Environ. Model. Softw.* **2018**, *101*, 1–9. [CrossRef]
63. Ploton, P.; Mortier, F.; Réjou-Méchain, M.; Barbier, N.; Picard, N.; Rossi, V.; Dormann, C.; Cornu, G.; Viennois, G.; Bayol, N.; et al. Spatial validation reveals poor predictive performance of large-scale ecological mapping models. *Nat. Commun.* **2020**, *11*, 4540. [CrossRef] [PubMed]
64. Fawcett, T. An introduction to ROC analysis. *Pattern Recognit. Lett.* **2006**, *27*, 861–874. [CrossRef]
65. Kuhn, M.; Silge, J. Tidy Modelling with R. 2020. Available online: https://www.tmwr.org/ (accessed on 1 October 2022).
66. Hanley, J.A.; McNeil, B.J. The meaning and use of the area under a receiver operating characteristic (ROC) curve. *Radiology* **1982**, *143*, 29–36. [CrossRef] [PubMed]
67. Hosmer, D.W.; Lemeshow, S.; Strudivant, R.X. *Applied Logistic Regression*, 3rd ed.; John Wiley and Sons: New York, NY, USA, 2013.
68. Suvanto, S.; Peltoniemi, M.; Tuominen, S.; Strandström, M.; Lehtonen, A. High-resolution mapping of forest vulnerability to wind for disturbance-aware forestry. *For. Ecol. Manag.* **2019**, *453*, 117619. [CrossRef]
69. Fisher, A.; Rudin, C.; Dominici, F. All models are wrong, but many are useful: Learning a variable's importance by studying an entire class of prediction models simultaneously. *J. Mach. Learn. Res.* **2019**, *20*, 1–81.
70. Molnar, C. Interpretable Machine Learning. A Guide for Making Black Box Models Explainable. 2020. Available online: https://christophm.github.io/interpretable-ml-book/ (accessed on 1 October 2022).
71. Meyer, H.; Pebesma, E. Predicting into unknown space? Estimating the area of applicability of spatial prediction models. *Methods Ecol. Evol.* **2021**, *12*, 1620–1633. [CrossRef]
72. Busby, P.E.; Motzkin, G.; Boose, E.R. Landscape-level variation in forest response to hurricane disturbance across a storm track. *Can. J. For. Res.* **2008**, *38*, 2942–2950. [CrossRef]
73. Ogaya, R.; Barbeta, A.; Başnou, C.; Penuelas, J. Satellite data as indicators of tree biomass growth and forest dieback in a Mediterranean holm oak forest. *Ann. For. Sci.* **2015**, *72*, 135–144. [CrossRef]
74. Hanewinkel, M.; Zhou, W.; Schill, C. A neural network approach to identify forest stands susceptible to wind damage. *For. Ecol. Manag.* **2004**, *196*, 227–243. [CrossRef]
75. Fridman, J.; Valinger, E. Modelling probability of snow and wind damage using tree, stand, and site characteristics from *Pinus sylvestris* sample plots. *Scand. J. For. Res.* **1998**, *13*, 348–356. [CrossRef]
76. Schindler, D.; Grebhan, K.; Albrecht, A.; Schönborn, J. Modelling the wind damage probability in forests in Southwestern Germany for the 1999 winter storm 'Lothar'. *Int. J. Biometeorol.* **2009**, *53*, 543–554. [CrossRef]

77. Klaus, M.; Holsten, A.; Hostert, P.; Kropp, J.P. Integrated methodology to assess windthrow impacts on forest stands under climate change. *For. Ecol. Manag.* **2011**, *261*, 1799–1810. [CrossRef]
78. Danjon, F.; Fourcaud, T.; Bert, D. Root architecture and wind-firmness of mature *Pinus pinaster*. *New Phytol.* **2005**, *168*, 387–400. [CrossRef]
79. Nicoll, B.C.; Gardiner, B.A.; Rayner, B.; Peace, A.J. Anchorage of coniferous trees in relation to species, soil type and rooting depth. *Can. J. For. Res.* **2006**, *36*, 1871–1883. [CrossRef]
80. Hanewinkel, M.; Albrecht, A.; Schmidt, M. Influence of stand characteristics and landscape structure on wind damage. In *What Science Can Tell Us. Living with Storm Damage to Forests*; Gardiner, B., Schuck, A., Schelhaas, M.-J., Orazio, C., Blennow, K., Nicoll, B., Eds.; European Forest Institute: Joensuu, Finland, 2013; pp. 207–224.
81. Valta, H.; Lehtonen, I.; Laurila, T.K.; Venäläinen, A.; Laapas, M.; Gregow, H. Communicating the amount of windstorm induced forest damage by the maximum wind gust speed in Finland. *Adv. Sci. Res.* **2019**, *16*, 31–37. [CrossRef]

forests

MDPI

Article

Impact of Forest Fires on Air Quality in Wolgan Valley, New South Wales, Australia—A Mapping and Monitoring Study Using Google Earth Engine

Sachchidanand Singh [1,2], Harikesh Singh [1,2], Vishal Sharma [1,2], Vaibhav Shrivastava [1,2], Pankaj Kumar [3,*], Shruti Kanga [4], Netrananda Sahu [5], Gowhar Meraj [6], Majid Farooq [6] and Suraj Kumar Singh [7]

[1] RBased Services Private Limited, Delhi 110086, India; sachin.iirs@gmail.com (S.S.); harikeshsngh77@gmail.com (H.S.); vishal.iirs@gmail.com (V.S.); iirs.vaibhav@gmail.com (V.S.)
[2] Indian Institute of Remote Sensing, Dehradun 248001, India
[3] Institute for Global Environmental Strategies, Hayama, Miura-gun 240-0115, Kanagawa, Japan
[4] Centre for Climate Change & Water Research, Suresh Gyan Vihar University, Jaipur 302017, India; shruti.kanga@mygyanvihar.com
[5] Department of Geography, Delhi School of Economics, University of Delhi, Delhi 110007, India; nsahu@geography.du.ac.in
[6] Department of Ecology, Environment and Remote Sensing, Government of Jammu and Kashmir, Srinagar 190018, India; gowharmeraj@gmail.com (G.M.); majid_rsgis@yahoo.com (M.F.)
[7] Centre for Sustainable Development, Suresh Gyan Vihar University, Jaipur 302017, India; suraj.kumar@mygyanvihar.com
* Correspondence: kumar@iges.or.jp

Citation: Singh, S.; Singh, H.; Sharma, V.; Shrivastava, V.; Kumar, P.; Kanga, S.; Sahu, N.; Meraj, G.; Farooq, M.; Singh, S.K. Impact of Forest Fires on Air Quality in Wolgan Valley, New South Wales, Australia—A Mapping and Monitoring Study Using Google Earth Engine. *Forests* **2022**, *13*, 4. https://doi.org/10.3390/f13010004

Academic Editors: Any Mary Petritan, Mirela Beloiu and Timothy A. Martin

Received: 28 September 2021
Accepted: 19 December 2021
Published: 21 December 2021

Publisher's Note: MDPI stays neutral with regard to jurisdictional claims in published maps and institutional affiliations.

Abstract: Forests are an important natural resource and are instrumental in sustaining environmental sustainability. Burning biomass in forests results in greenhouse gas emissions, many of which are long-lived. Precise and consistent broad-scale monitoring of fire intensity is a valuable tool for analyzing climate and ecological changes related to fire. Remote sensing and geographic information systems provide an opportunity to improve current practice's accuracy and performance. Spectral indices techniques such as normalized burn ratio (NBR) have been used to identify burned areas utilizing satellite data, which aid in distinguishing burnt areas using their standard spectral responses. For this research, we created a split-panel web-based Google Earth Engine app for the geo-visualization of the region severely affected by forest fire using Sentinel 2 weekly composites. Then, we classified the burn severity in areas affected by forest fires in Wolgan Valley, New South Wales, Australia, and the surrounding area through Difference Normalized Burn Ratio (dNBR). The result revealed that the region's burnt area increased to 6731 sq. km in December. We also assessed the impact of long-term rainfall and land surface temperature (LST) trends over the study region to justify such incidents. We further estimated the effect of such incidents on air quality by analyzing the changes in the column number density of carbon monoxide and nitrogen oxides. The result showed a significant increase of about 272% for Carbon monoxide and 45% for nitrogen oxides. We conclude that, despite fieldwork constraints, the usage of different NBR and web-based application platforms may be highly useful for forest management to consider the propagation of fire regimes.

Keywords: forest fire (FF); Google Earth Engine (GEE); burnt vegetation; difference normalized burn ratio (dNBR); normalized burn ratio (NBR)

1. Introduction

Forests are an essential natural resource that plays a crucial role in sustaining environmental sustainability. Forest health is a true predictor of the predominant ecological condition in the region. The frequent occurrence of forest fires (FFs) is one of the main reasons why most of our valuable flora have been depleted and distressed [1]. Further, the devastation from these deadly fires directly or indirectly impacts human beings [2]. Forest fires are also viewed as a potential human, ecological, economic, and environmental threat.

Fire causes partial or complete forest canopy loss, altering radiation balance by increasing the surface albedo, water drainage, and increased soil erosion [3].

Using remote sensing to monitor, analyze, and restore burned regions has become an essential part of postfire mitigation efforts on a global and regional scale. It provides reliable and speedy data and enables rapid diagnosis across burned areas. This approach further demands innovative technologies in the prompt, cost-effective collection, encoding, and proper visualization of spatial information. Space technology benefits from the ability of a computer to store and process enormous amounts of data. In this regard, the Google Earth Engine—a remote sensing datasets processing cloud-based web platform [4–7]—plays a vital role. It offers an efficient assessment of the fire status and the corresponding environmental impact of wildfire through geo-visualization. Change identification (e.g., amid postfire and pre-fire images) models commonly utilize remote sensing in fire intensity mapping [8–10].

Satellite imagery identifies forest fires using special techniques, including the fire intensity measurement by the normalized burn ratio (NBR), intended to distinguish areas burnt. Numerous indices, including the difference Normalized Burn Ratio (dNBR), and the soil-adjusted vegetation index burned zone index, have been derived and compared in the past [11–13]. The most widely used index, dNBR, provides a fair description of various vegetation populations (e.g., 60–70% precision as opposed to field validation) of the spatial disparity in intensity within a single fire [10,14]. Even the Landsat satellite data's short-wave infrared (SWIR) band is appropriate for detecting moisture both in vegetation and soils, and the near-infrared reflection (NIR) band is sensitive to green pigment, i.e., chlorophyll levels in leafy vegetation [15]. Therefore, satellite imagery analysis proves to be a powerful method for measuring and evaluating the intensity of fires since they have a sufficient temporal and spatial resolution [16].

Forest fires have environmental consequences, as they release carbon that contributes to global warming and can eventually alter biodiversity [17]. Fire activity has a significant influence on air pollution, atmospheric composition, and the climate. The toxic and chemical reactive gases, such as methane, carbon dioxide (CO_2), carbon monoxide (CO), hydrocarbons, nitrogen oxides (NO_x), methyl chloride, and particulate matter are released. The emissions from forest fires are a significant source of carbon dioxide (CO_2), affecting interannual variability and biogeochemical processes with atmospheric impact. Carbon monoxide (CO), emitted by incomplete combustion, affects the national and global air quality. Nitrogen oxides (NO_x), radical OH, volatile organic compounds, and black carbon can create troposphere ozone depletion while particulate matter affects human health and negatively impacts climate [18–20]. Various tests to assess the air quality of forest fires have since been conducted in the past [18,21,22]. The variability in weather can have a considerable impact on the fire regimes. Past experiments have shown that short-term (seasonal to annual) precipitation shifts that affect the humidity content of the fuel are related to the volatility of wildfires [23–25]. Further, due to climate change, the world's mean temperature is increasing, which may increase the chances of forest fires [4,24,26].

There have been many forest fires in the world recently, but the Australian forest fire has been the most powerful. By 9 March 2020, about 18.6 million hectares [27] had been burnt, at least 34 people killed [28–30], and more than 5900 buildings damaged [31]. About 1 billion animals were killed, and other endangered species could have been made extinct [32,33]. The objectives of this study were to estimate the forest fire footprints by utilizing Landsat-8 satellite imageries, analyze the fire intensity classification efficiency using standard spectral indices, estimate the impact of long-term temperature and precipitation on forest fire incidents, evaluate the impact of forest fires on air quality, and development of an interactive visualization web-application for quick geo-visualization of the burn severity. This application has been implemented to raise awareness of Australia's forest fire issues using an advanced Google Earth Engine (GEE) cloud-based platform.

2. Materials and Methods

2.1. Study Area

Australia is the sixth-largest nation in the world total-area-wise and Oceania's largest country. Twenty-six million people are mostly urbanized on the eastern seaboard. In Australia, semiarid and arid areas cover 50%–75% of the land [34]. The study region lies in the vicinity of Wolgan Valley near Lidsdale, Eastern Australia (Figure 1), areas hit by fires in 2019–2020. Wolgan Valley is a small valley on the Wolgan River in the New South Wales (NSW) region of Lithgow, Australia.

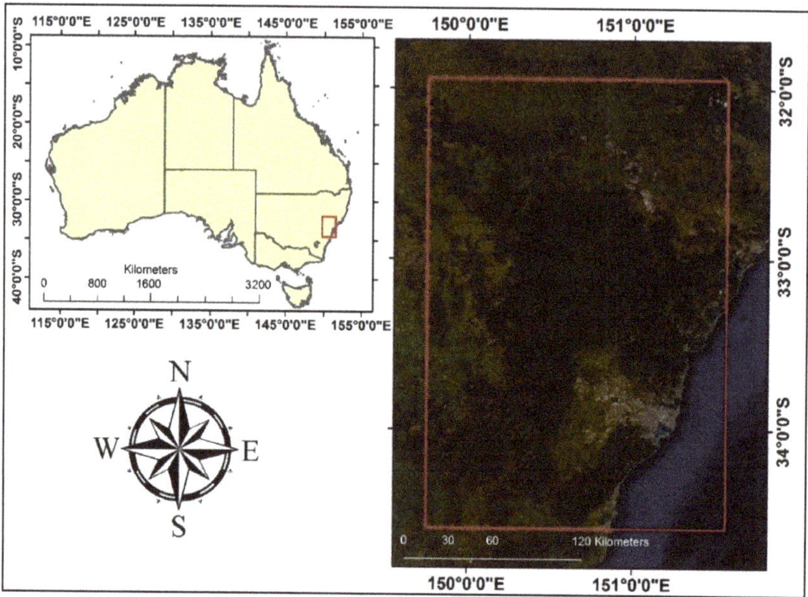

Figure 1. Location of the study areas. Wolgan Valley, near Lidsdale, Eastern Australia, is represented in the red box.

The valley lies about 32 km north of Lithgow and 150 km northwest of Sydney. Accessible from Castlereagh Highway via the Wolgan Valley Discovery Path (Wolgan Route), the route crosses the valley leading to Newnes historical village with its extensive industrial ruins. It runs wide east until it reaches the Capertee River and then the Colo River. The Wollemi Wilderness is the largest protected area in NSW and the largest in eastern Australia. The Wollemi protected area is 361,000 hectares east. Wolgan Valley comprises Wollemi National Park, Stone Gardens National Park, and the UNESCO World Heritage Region of Blue Mountains. Recently, in 2019, the region suffered from the hazardous incident of forest fire, which distressed flora and fauna of the region and deteriorated the air quality.

2.2. Materials and Methods

This study used the images from the Operational Land Imager (OLI) sensor of Landsat-8 Satellite to map the burn severity. The images were acquired using an image export algorithm in Google Earth Engine. The temporal filter was applied for March, and October to December 2019. The minimum cloud cover imageries were chosen for the research area to help classify burned areas during the 2019 fire. Using the principles and elements of image interpretation (pattern, situation, association, size, shape, tone, NBR, and dNBR), we identified the burnt patches and area, and the geospatial layer of the burnt area was generated. The overall methodology of this research is shown in Figure 2. For air quality monitoring, we used the Sentinel-5P dataset. CHIRPS (Climate Hazards Group Infrared Precipitation with Station) was used for rainfall time series estimation. The MODIS (Moderate Resolution Imaging Spectroradiometer) Terra Land Surface Temperature (LST) and Daily Emissivity dataset (MOD11A1.006) were used for validation. The details of the datasets are displayed in Table 1.

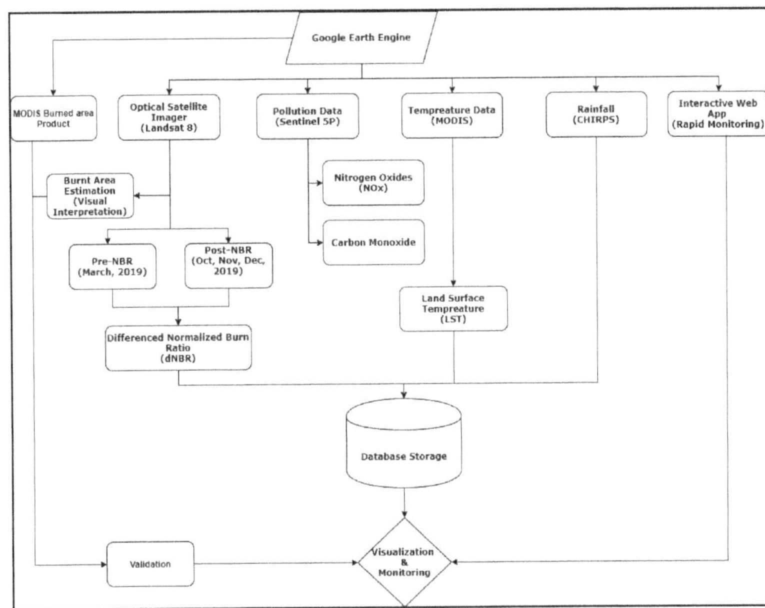

Figure 2. Overall methodology for this research.

Table 1. Various datasets employed in this research for burnt area estimation and pollution monitoring.

S. No.	Purpose	Data	Duration	Resolution/Scale	Source
1	Burned Area Mapping	LANDSAT-8 Operational Land Imager	March, October, November, December 2019	30 m	
2	Web App Visualization	Sentinel 2	March, October, November, December 2019	10 m	Google Earth Engine https://code.earthengine.google.com/ (Accessed on 20 August 2021).
3	Rainfall	CHIRPS daily	1981–2019	5000 m	
4	Land Surface Temperature	MODIS Terra LST daily	2001–2019	1000 m	
5	Pollution Mapping and Monitoring	Sentinel 5P	March, October, November, December 2019	1000 m	
6	Validation	MODIS Burned Area Monthly	October, November, December 2019	500 m	

2.2.1. Spectral Indices

The fire spectral index determines the edge of burning areas by normalization of burn ratio. The NBR is a normalized Burn Ratio index (see Equation (1)), based on the OLI sensor bands 5 and 7 reflectance data classifying burnt areas. Band 5 has a 30 m spatial resolution corresponding to Near-Infrared (NIR), equal to 0.85- and 0.88-μm spectral range; band 8 corresponds to SWIR, which has a 30 m spatial resolution equal to a 2.11–2.29 μm spectral range. In determining multitemporal identification of change, we considered the bispectral SWIR–NIR bands [16,35].

$$NBR = ((NIR - SWIR))/((NIR + SWIR)) \tag{1}$$

where NIR represents Landsat-8 OLI sensor Band 5 and SWIR denotes Landsat-8 OLI sensor Band 7.

We also measured the dNBR using the bitemporal difference of the NBR images (see Equation (2)) [36]. Teobaldo and Baptista (2016) found that the dNBR strengthens the differences between the NBR scenes and emphasizes fire [37].

$$dNBR = NBRpre - NBRpost \tag{2}$$

where NBRpre—pre-fire data, NBRpost—postfire data, and dNBR—difference NBR.

2.2.2. Rainfall and Temperature Retrieval

GEE makes fast analysis possible by using Google's machine infrastructure. For this study, CHIRPS, a quasi-global rainfall dataset, was coded to obtain Wolgan Valley's precipitation information from 1981 to 2019. We used GEE to determine the precipitation values around the study region with a $0.05° \times 0.05°$ daily temporal and spatial resolution. For the estimation of land surface temperature (LST), the MOD11A1.006 Terra LST dataset was processed in GEE to obtain the day surface temperature from 2001 to 2019. This method uses JavaScript coding on the GEE platform. The resulting chart was saved in 'csv' format.

2.2.3. Pollution Monitoring

The Nitrogen oxides and carbon monoxide datasets of Sentinel 5P satellite were retrieved using a GEE algorithm for pollution monitoring. Both Carbon monoxide (CO) and the Nitrogen oxides (NO_2 and NO) are important trace gases in the atmosphere to understand tropospheric chemistry. The primary sources of CO and NO_x include the combustion of fossil fuels, biomass burning in the atmosphere, and natural processes (wildfires, lightning, and microbiological mechanisms in soils). TROPOMI on the satellite Sentinel 5P measures CO global abundance using clear-sky and cloud-sky Earth's radiation parameters in the 2.3 μm-SWIR components of the solar spectrum. The TROPOMI NO_2 processing method for OMI is based on the algorithm innovations for the DOMINO-2 software, and the EU QA4ECV NO_2 reprocessed dataset has been modified for TROPOMI. The datasets for March, October, November, and December 2019, were used to evaluate the change in air quality due to the forest fire.

2.2.4. Web-App Development

We have created a split-panel interactive geo-visualization app using GEE's create app feature. The weekly composites using Sentinel 2 Multispectral Instrument (MSI) Level-1C images were also created. Sentinel-2 is a wide-ranging, multispectral, high-resolution imaging initiative promoting Copernicus Land Monitoring studies involving vegetation mapping, soil and waters cover, and the study of inland and coastal waterways. Filter metadata function was used to filter out the images with a cloud pixel percentage of less than 30%. For the proper visualization of the Fire event, False Color Composite was created using Bands 12 (Short-wave Infrared-2, 2202.4 nm), 8 (Near-Infrared, 835.1 nm), and 3 (Green, 560 nm) over the median images. The min and max band values for visualization were set to 0 and 5000, respectively. For the pre-fire event, March's second week composites were taken; for the postfire event, the weeks of October, November, and December 2019 were taken. The desired weekly composites could be selected from the dropdown list provided in the corners of each panel for the users.

Various inbuilt functions of GEE such as ui.Map(), ui.Label(), ui.Select(), ui.Panel(), ui.SplitPanel(), and ui.Map.Linker() were used judiciously to this interactive app. The output window was split into two parts to visualize the changes in the fire events easily. This app also has a search toolbar at the top to directly examine the affected places. The created app could be accessed from https://bit.ly/fires-aus (Accessed on 20 August 2021). This created app was used to visually analyze the zones deeply affected by forest fires over all of Australia.

3. Results and Discussion

This section may be divided by subheadings. It should provide a concise and precise description of the experimental results, their interpretation, as well as the experimental conclusions that can be drawn.

3.1. Web App Visualization

Numerous places in Australia are severely affected by the forest fire. We developed the split-panel web-based application to analyze the change in the forest fire event visually. Some of the places severely affected were Wolgan Valley, NSW; Crawney, NSW; Nowendoc, NSW; and Nullo Mountains, NSW, as shown in Figures 3–6. The weekly composites of the images of Sentinel 2 were effective in visualizing forest fire growth.

Among all, Wolgan valley was taken as a study area from the most severely affected region, as shown by the web app.

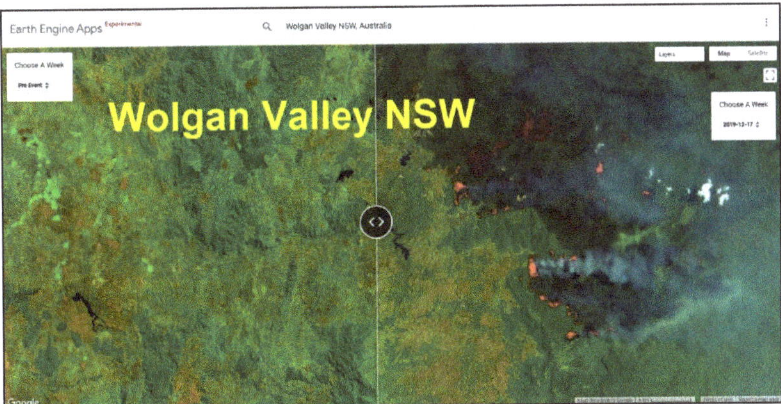

Figure 3. An overview of the Forest Fire web applications in Wolgan Valley, NSW, Australia.

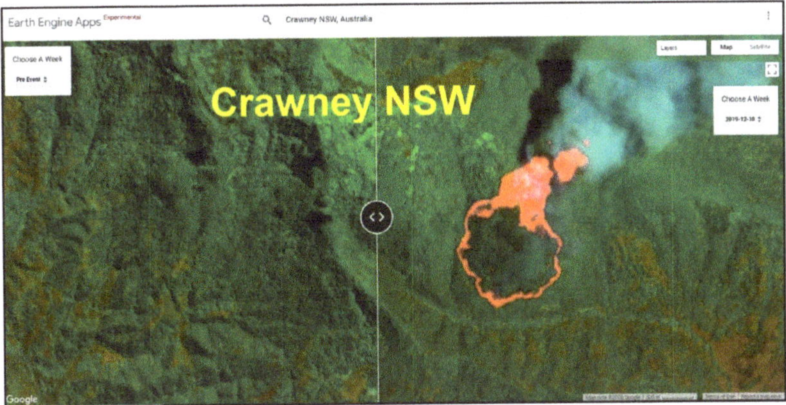

Figure 4. An example from the web app for a forest fire in Crawney, NSW, Australia.

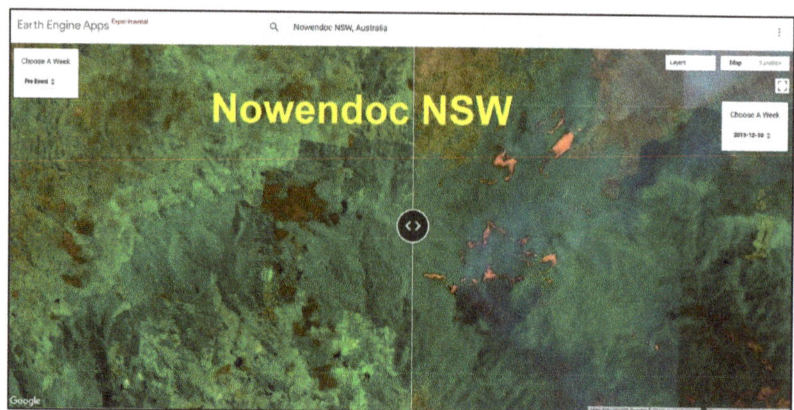

Figure 5. An example of the web app for visualizing forest fires in Nowendoc, NSW, Australia.

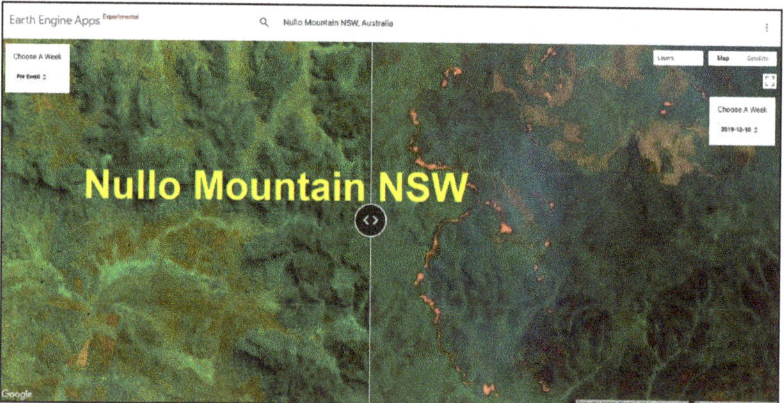

Figure 6. An example of the Forest Fire web app in Nullo Mountains, NSW, Australia.

3.2. Rainfall and Temperature Variation

Due to the warming climate, precipitation patterns and moisture levels change, leading to more dryness. The cooler air enhances the evaporation, leaving the soil drier by the atmosphere. Prolonged dry conditions are one of the driving factors behind forest fires. Dryness and high-temperature conditions cause the fire to occur frequently and exacerbate the length and severity of fires.

The annual rainfall time series analysis from 1981 to 2019 for Wolgan Valley was analyzed using the satellite-derived CHIRPS datasets in GEE (Figure 7). The region showed a decreasing trend (although insignificant) with a slope of −2.5. However, for 2019, the region received an annual rainfall of 596 mm, which is approximately 30% lower than the mean annual rainfall of 825 mm of the study area, which could be a major reason for the occurrence and enlargement of the forest fire. The Mann–Kendall Trend Test was also performed in order to obtain the trend behavior of the series; the test was performed based on yearly and yearly moving average. The test result is shown below in Table 2.

As shown in Table 2, the time series shows no trend when yearly data are considered, but it starts showing the decreasing trend as we move for a more realistic approach of moving average in the test. All parameters obtained by performing the Mann–Kendall trend test are shown in Table 2.

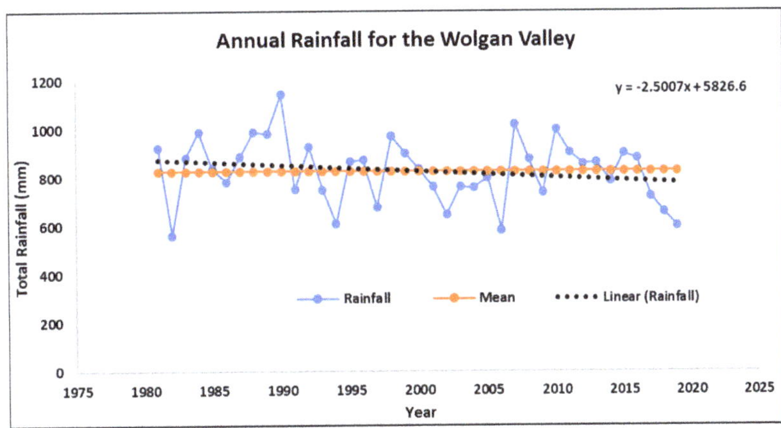

Figure 7. Annual rainfall time series from 1981–2019 for Wolgan Valley estimated from CHIRPS datasets in GEE.

The mean annual land surface temperature (LST) time series analysis during 2001 to 2019 for Wolgan Valley was performed using the satellite derived MOD11A1.006 Terra datasets in GEE (Figure 8), which found that the region showed an increasing trend with a slope of 0.0236. Moreover, for the year of 2019, the region experienced the mean annual LST of 23.7 °C, which is approximately 8% higher than the mean annual LST of 21.9 °C. The increase in temperature has a drying effect on the flora that could be one of the reasons for the forest fire events in the region.

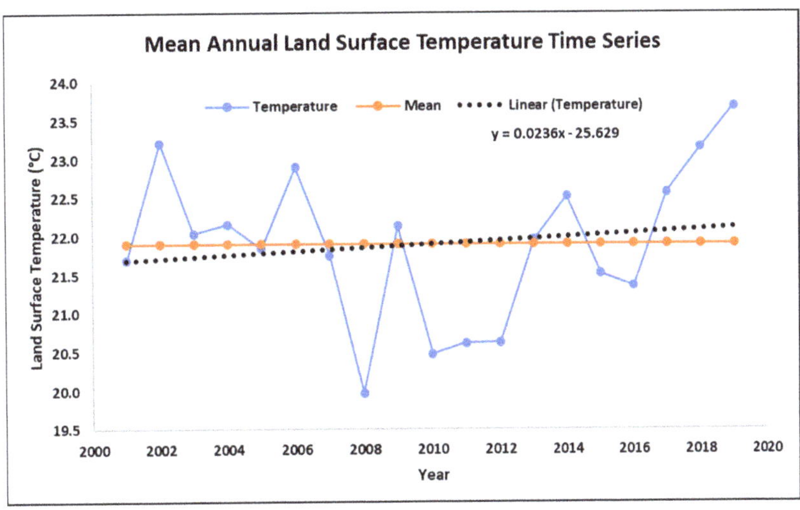

Figure 8. Annual mean land surface temperature (LST) from 2001–2019 for Wolgan Valley estimated from MODIS LST product datasets in GEE.

Table 2. Trend analysis of CHIRPS rainfall data using Mann–Kendall moving average method.

Time Period	Trend	h	p	z	Tau	s	var_s	Slope	Intercept
Yearly	no trend	FALSE	0.110313	−1.59679	−0.17949	−133	6833.667	−0.55886	48.80874
2-yearly moving average	no trend	FALSE	0.056014	−1.91093	−0.21764	−153	6327	−0.48733	48.69761
3-yearly moving average	decreasing	TRUE	0.008567	−2.62886	−0.3033	−202	5846	−0.56612	53.42137
4-yearly moving average	decreasing	TRUE	0.001132	−3.25539	−0.38095	−240	5390	−0.66736	58.51139
5-yearly moving average	decreasing	TRUE	0.000222	−3.69237	−0.43866	−261	4958.333	−0.72488	61.15642

3.3. Burnt Area Estimation

The altitude for the bounding box ranges from 47 to 1599 m and is situated in the east of New South Wales. By analyzing the periodic increase in the geospatial boundary of burnt regions, we found it was 21 sq. km in October 2019, which increased to 912 sq. km in November 2019 primarily in the eastern region; later in December 2019, it further increased to 6731 sq. km with the growth in the east, north, and south directions (refer to Figure 9). So, it could be easily determined that the region suffered a severe overall increase of about 6710 sq. km within two months of the forest fire.

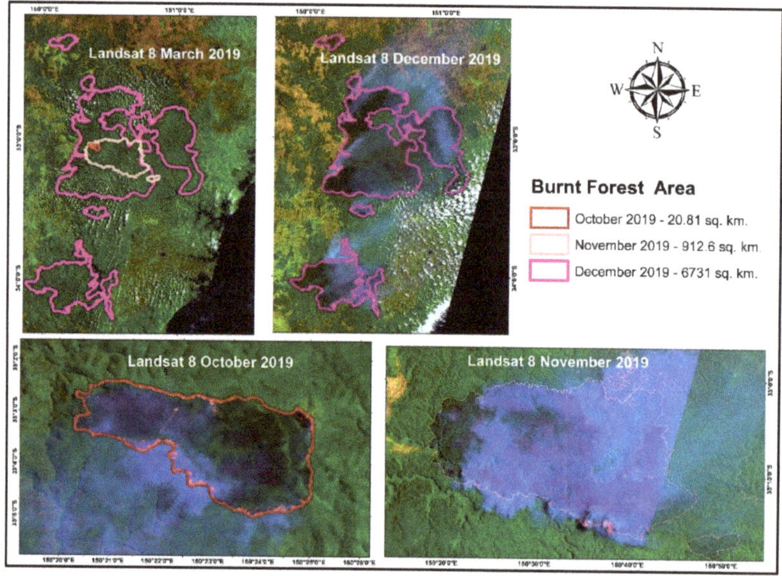

Figure 9. Map displaying the burned area footprints as estimated from visual interpretation.

The NBR multitemporal difference, i.e., due to their ease of application, the dNBR index has recently become the standard for fire intensity metrics using Landsat satellite data, as they typically provide a broad spectral range that can be accomplished between SWIR and NIR bands. The NBR pre-, NBR postestimation allows for the evaluation of change detection by multiple passes of thorough observation. From Equation (1), NBRpre and NBRpost values for the processed satellite imageries were observed. Key and Benson, (2006) who created the NBR, conceptualized the data-slicing index and suggested that the theoretical range spans from −1.0 to 1.0. An NBR near "0" means that clouds, grasses, exposed soil, or rocky outcrops can occur, and if pixels have a negative NBR, this implies extreme water stress on plants and the negative trace of a fire [38]. Thus, it is important to note that recent fire results usually vary from "0" to strongly negative. The dNBR (Equation (2)) combines multitemporal datasets of the NBR in one gradient, so the dNBR has a theoretical range from −2 to +2. Positive dNBR values indicate a reduction in vegetation, while negative values are a rise in vegetation cover [10,14,39].

Table 3 shows NBR data from the Landsat-8 image estimation for March, October, November, and December 2019. Concerning the NBR data for March 2019, it was found that pixel values ranged between −0.68 and +1.0; for October, the values ranged between −0.77 to 1; for November 2019, values ranged from −0.86 to +1.0, while the NBR data of December 2019 was found between −0.87 and +0.88.

Table 3. Theoretical and obtained bands, considering the pixels of Wolgan Valley, NSW NBR indices.

Month	Theoretical Range	NBR
Mar-19	[−1 to 1]	[−0.68 to +1]
Oct-19	[−1 to 1]	[−0.77 to +1]
Nov-19	[−1 to 1]	[−0.86 to +1]
Dec-19	[−1 to 1]	[−0.87 to +0.88]

The dNBR index for the severity groups was obtained using Equation (2). As such, Figure 10 shows the severity class for the fire burning in Wolgan Valley in the images used in this analysis. It is important to emphasize that the results of intensity display differences within the same burned area [36] and that determination and fire distribution perimeter, as well as the intensity within the fire, are useful for unit management managers seeking to grasp fire impacts on forests, such as the recovery of vegetation and postfire sequence for example [16,39].

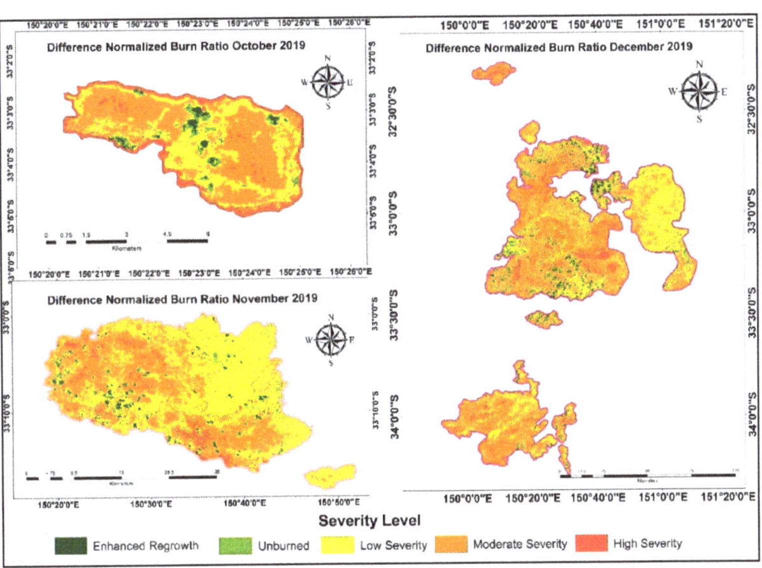

Figure 10. The dNBR indices for the identification of the severity level of the burnt area for Wolgan Valley, NSW.

The severity levels (Table 4) were used to classify the severity of indices. Moreover, we used a five-layer configuration, which has proven useful in several aspects. The dNBR value classes can differ between paired scenes. Values below −0.1 or greater than +0.66 may also exist, which may not be classified as burned. Alternatively, they are disguised as phenomena caused by lack of monitoring, clouds, or other causes not linked to actual ground cover variations.

Table 4. Severity levels and dNBR interval.

Severity Level	dNBR Range
Enhanced Regrowth	[>−0.1]
Unburned	[−0.1 to +0.1]
Low Severity	[+0.1 to +0.27]
Moderate Severity	[+0.27 to +0.66]
High Severity	[>+0.66]

The first level of severity reflects areas in which vegetation is present and can identify vegetation patches for postfire productivity. They occur almost entirely in vegetation where dNBR can be strongly negative, suggesting areas of improved after-fire efficiency (postfire NBR is much higher than pre-fire). Regular unburned pixels are located below zero. The last three stages include all other burned areas of which dNBR is explicitly constructive (postfire NBR is much less than pre-fire), including what is known as burned recently.

3.4. Validation

To validate the derived output of the burned area, the fire footprint for October and December 2019 were compared spatially with MCD64A1.006 MODIS Monthly Burned Area product obtained from GEE at 500 m-resolution as shown in Figure 11. On spatial interpretation, it was seen that the MODIS product overpredicted the burnt area derived from visual interpretation estimates for October and November. For October, MODIS product estimated the area to be 58 sq. km while, from Landsat-8 OLI visual interpretation, the area obtained was 22 sq. km. For December, MODIS product estimated the area to be 6205 sq. km while, from visual interpretation, the area obtained was 6731 sq. km. It could be stated that the visually interpreted region overlaps with the MODIS-derived burned area region. However, this variability could be due to the difference in their spatial resolution.

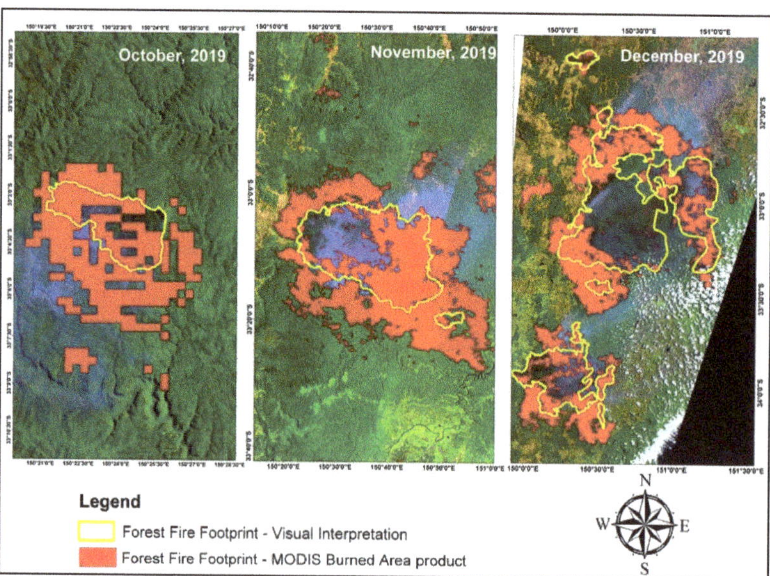

Figure 11. Forest fire footprints obtained by visual interpretation of Landsat-8 OLI and MODIS burned area products.

3.5. Impact on Air Pollution Due to the Forest Fire

A significant fire event recently took place in the woods of Australia from October to December 2019. A total of 5,595,739 hectares have been burned, and 2475 houses and 25 lives have been lost in 10,520 bushfires in NSW. These incidents further resulted in heavy pollution, in the form of CO and NO, in the vicinity of the event. For comparison, the air pollutants datasets were fetched for pre-fire and during-fire events.

It was found that the actual (satellite-based) mean spatial distribution of Carbon Monoxides Column Number Densities (COCND) (Figure 12) increased from 0.0177 mol/sq. m. to 0.066 mol/sq. m. during March–December 2019 with 272% change. Further, the periodic observation exhibits that the minimum and maximum COCND were 0.014 and 0.021 mol/sq. m, respectively, in March 2019, which increased to 0.022 and 0.035 mol/sq. m, respectively, in October 2019 with a 47% increase. In November 2019, it increased to a minimum and maximum value of 0.021 and 0.11 mol/sq. m, respectively, with 23% successive amplification mostly in the eastern direction. Later, in December 2019, the densities enlarged to a minimum and maximum value of 0.029 and 0.15 mol/sq. m, respectively, with 103% incremental increase in almost every direction. The stats obtained for the COCND are shown in Figure 13.

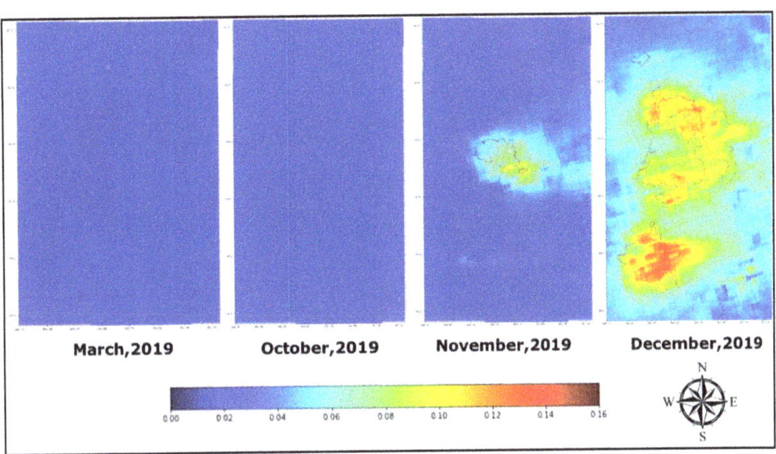

Figure 12. Maps showing the variation of COCND during the event of forest fire.

The obtained result could also be visualized from the histogram (Figure 14a) and violin graph (Figure 14b) of the COCND images for March, October, November, and December 2019. It is found that most of the pixel counts lie close to the value of 0.01 mol/sq. m for March; for the October month, it shifts towards 0.025 mol/sq. m. For November, the pixel spans with max counts close to 0.03 mol/sq. m and, finally, for December 2019, the pixels become very distributed with maximum counts close to the value of 0.05 mol/sq. m.

Similarly, the overall trend of mean Nitrogen Oxides Column Number Densities (NOCND) (Figure 15), for the forest area was found to increase from 0.0000287 mol/sq. m to 0.0000416 mol/sq. m during March–December 2019 with 45% change. It was observed that the minimum and maximum NOCND were 0.0000027 and 0.0002496 mol/sq. m, respectively, in March 2019, which increased to 0.0000041 and 0.0002627 mol/sq. m, respectively, in October 2019 with a 10% increase. In November 2019, it decreased to a minimum and maximum 0.0000033 and 0.0002478 mol/sq. m, respectively, with a 6% decrease. However, again in December 2019, the densities increased to a minimum and maximum value of 0.0000033 and 0.0004636 mol/sq. m, respectively, with 40% incremental increase. The stats obtained for the NOCND is shown in Figure 16.

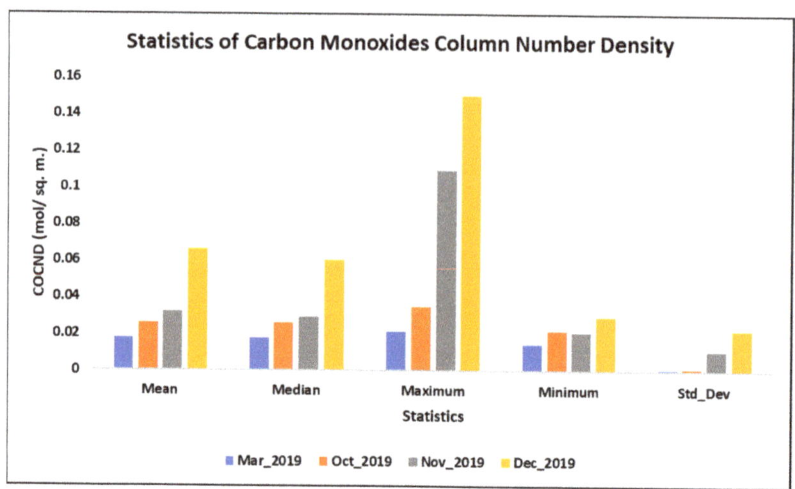

Figure 13. Statistics of COCND variation for the forest fire event.

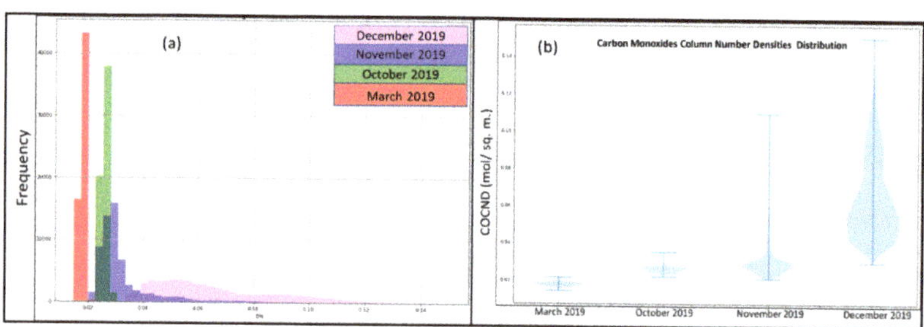

Figure 14. (**a**) Histogram and (**b**) Violin plot obtained for the COCND values for the forest fire incident.

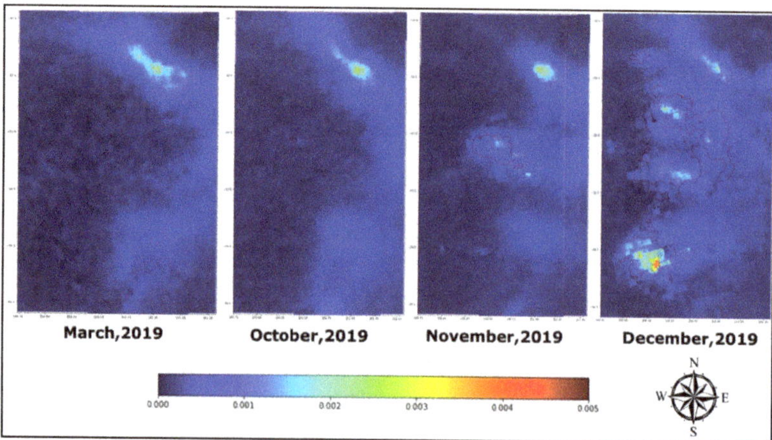

Figure 15. Maps showing the variation of NOCND during the event of forest fire.

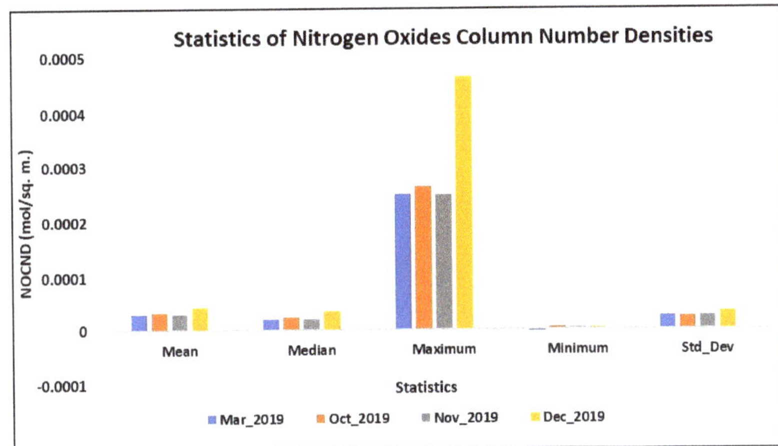

Figure 16. Statistics of NOCND variation for the forest fire event.

The obtained result could also be visualized from the histogram (Figure 17a) and violin graph (Figure 17b) of the NOCND images for March, October, November, and December 2019. It is found that most of the pixel counts lie close to the value of 0.0000116 mol/sq. m for March and, for October, it shifts towards 0.0000155 mol/sq. m. For November, the pixel value reduces to 0.000012 mol/sq. m and, finally, for December 2019, the pixels get distributed with maximum counts close to the value of 0.00002 mol/sq. m.

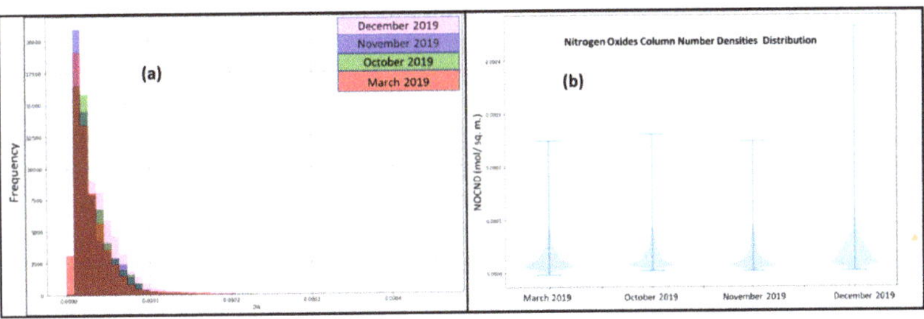

Figure 17. (**a**) Histogram and (**b**) Violin plot obtained for the NOCND values for the forest fire incident.

The increase in CO is attributed to incomplete biomass combustion. NO emissions are due to the high nitrogen content of biomass and new leaves. The analysis reveals that forest fire does not just affect the ecosystem in which it occurs, but also has effects on the surrounding regions.

4. Conclusions

Forest fires are one of the major sources of forest loss and air pollution. The NBR index was sensitive to pre- and postfire images of fire pixels in the SWIR–NIR region. Based on the methodology used in this work, the pre- and postfire difference index (dNBR) proved adequate to identify fire-affected pixels. The use of the dNBR index proved to be a valuable tool for the classification of burned areas in the region. It was found that the rate of loss of forest was very large (approx. 6730 sq. km) at the time of the fire. It is necessary to consider that the temporal information obtained from Landsat satellite images can provide valuable data for efficient management of natural resources. Moreover, in the limitations of fieldwork, the use of GEE-based Web-application, the geo-visualization of the fire events became quick and easy. This application could be extremely beneficial for forest management to understand the spread of fire regimes, as it could contribute to the execution of effective environmental training actions and restoration.

Burning biomass also has a significant effect on air quality in nearby cities. During the fire period, major contaminants CO and NOx were found to increase compared with the pre-fire time-period. The long-term time series analysis of rainfall and temperature revealed their increasing trend in the region. It is hypothesized that, due to climate change, forest fire incidences may further increase in the coming years. The joint management from the forest department and local village communities is important in this context. Further, there is an urgent need for effective management practices and a better weather forecasting system to ensure proper monitoring of such incidents.

Author Contributions: Conceptualization, S.S., H.S., V.S. (Vaibhav Shrivastava), V.S. (Vishal Sharma) and G.M.; methodology S.S., H.S., V.S. (Vaibhav Shrivastava), V.S. (Vishal Sharma) and G.M.; software, S.S., H.S., V.S. (Vaibhav Shrivastava) and V.S. (Vishal Sharma); validation, S.S., H.S., V.S. (Vaibhav Shrivastava), V.S. (Vishal Sharma) and G.M.; formal analysis, S.S., H.S., V.S. (Vaibhav Shrivastava), V.S. (Vishal Sharma) and G.M.; investigation, S.S., H.S., V.S. (Vaibhav Shrivastava), V.S. (Vishal Sharma), P.K., S.K.S., N.S., M.F., S.K. and G.M.; resources, S.S., H.S., V.S. (Vaibhav Shrivastava), V.S. (Vishal Sharma), P.K., S.K.S., N.S., G.M., M.F. and S.K.; data curation, S.S., H.S., V.S. (Vaibhav Shrivastava), V.S. (Vishal Sharma), P.K., S.K.S., N.S., G.M., M.F. and S.K.; writing—original draft preparation, S.S., H.S., V.S. (Vaibhav Shrivastava) and V.S. (Vishal Sharma); writing—review and editing, S.S., H.S., V.S. (Vaibhav Shrivastava), V.S. (Vishal Sharma), P.K., S.K.S., N.S., G.M., M.F. and S.K.; visualization, S.S., H.S., V.S. (Vaibhav Shrivastava), V.S. (Vishal Sharma), P.K., S.K.S., N.S., G.M., M.F. and S.K.; supervision, S.S., H.S., S.S., V.S. (Vaibhav Shrivastava), V.S. (Vishal Sharma), S.K.S. and S.K.; project administration, P.K.; funding acquisition, P.K. All authors have read and agreed to the published version of the manuscript.

Funding: Publication fund for this work is supported by the Strategy Research Fund 2021 (WHN-Planetary-Health), an in-house grant from Institute for Global Environmental Strategies (IGES).

Institutional Review Board Statement: Not Applicable.

Informed Consent Statement: Not Applicable.

Data Availability Statement: Data are available on request from the corresponding author.

Acknowledgments: The authors S.S., H.S., V.S. (Vaibhav Shrivastava) and V.S. (Vishal Sharma) are thankful to R-Based Services Private Limited, Delhi, India, for providing the necessary setup for conducting this research. The author G.M. is thankful to the Department of Science and Technology, Government of India (DST-GoI) for providing the Fellowship under Scheme for Young Scientists and Technology (SYST-SEED) [Grant no. SP/YO/2019/1362(G) & (C)].

Conflicts of Interest: The authors declare no conflict of interest.

References

1. Prasad, A.S.; Pandey, B.W.; Leimgruber, W.; Kunwar, R.M. Mountain hazard susceptibility and livelihood security in the upper catchment area of the river Beas, Kullu Valley, Himachal Pradesh, India. *Geoenviron. Disasters* **2016**, *3*, 1. [CrossRef]
2. Yao, J.; Raffuse, S.M.; Brauer, M.; Williamson, G.J.; Bowman, D.M.; Johnston, F.H.; Henderson, S.B. Predicting the minimum height of forest fire smoke within the atmosphere using machine learning and data from the CALIPSO satellite. *Remote Sens. Environ.* **2018**, *206*, 98–106. [CrossRef]
3. Beck, K.; Mariani, M.; Fletcher, M.-S.; Schneider, L.; Aquino-López, M.; Gadd, P.; Heijnis, H.; Saunders, K.; Zawadzki, A. The impacts of intensive mining on terrestrial and aquatic ecosystems: A case of sediment pollution and calcium decline in cool temperate Tasmania, Australia. *Environ. Pollut.* **2020**, *265*, 114695. [CrossRef]
4. Gibson, R.; Danaher, T.; Hehir, W.; Collins, L. A remote sensing approach to mapping fire severity in south-eastern Australia using sentinel 2 and random forest. *Remote Sens. Environ.* **2020**, *240*, 111702. [CrossRef]
5. Gorelick, N.; Hancher, M.; Dixon, M.; Ilyushchenko, S.; Thau, D.; Moore, R. Google Earth Engine: Planetary-scale geospatial analysis for everyone. *Remote Sens. Environ.* **2017**, *202*, 18–27. [CrossRef]
6. Mab, P.; Ly, S.; Chompuchan, C.; Kositsakulchai, E. Evaluation of Satellite Precipitation from Google Earth Engine in Tonle Sap Basin, Cambodia. In Proceedings of the THA 2019 International Conference on Water Management and Climate Change towards Asia's Water-Energy-Food Nexus and SDGs, Bangkok, Thailand, 23–25 January 2019.

7. Singh, S.; Dhasmana, M.K.; Shrivastava, V.; Sharma, V.; Pokhriyal, N.; Thakur, P.K.; Aggarwal, S.P.; Nikam, B.; Garg, V.; Chouksey, A.; et al. Estimation of revised capacity in Gobind Sagar reservoir using Google Earth Engine and GIS. *ISPRS—Int. Arch. Photogramm. Remote Sens. Spat. Inf. Sci.* **2018**, *XLII-5*, 589–595. [CrossRef]

8. Chafer, C.J.; Noonan, M.; Macnaught, E. The post-fire measurement of fire severity and intensity in the Christmas 2001 Sydney wildfires. *Int. J. Wildland Fire* **2004**, *13*, 227–240. [CrossRef]

9. Lentile, L.B.; Holden, Z.A.; Smith, A.; Falkowski, M.J.; Hudak, A.T.; Morgan, P.; Lewis, S.A.; Gessler, P.E.; Benson, N.C. Remote sensing techniques to assess active fire characteristics and post-fire effects. *Int. J. Wildland Fire* **2006**, *15*, 319–345. [CrossRef]

10. Miller, J.D.; Thode, A.E. Quantifying burn severity in a heterogeneous landscape with a relative version of the delta Normalized Burn Ratio (dNBR). *Remote Sens. Environ.* **2007**, *109*, 66–80. [CrossRef]

11. Kasischke, E.S.; Turetsky, M.R.; Ottmar, R.D.; French, N.H.F.; Hoy, E.E.; Kane, E.S. Evaluation of the composite burn index for assessing fire severity in Alaskan black spruce forests. *Int. J. Wildland Fire* **2008**, *17*, 515–526. [CrossRef]

12. Smith, A.M.; Eitel, J.U.; Hudak, A.T. Spectral analysis of charcoal on soils: Implications for wildland fire severity mapping methods. *Int. J. Wildland Fire* **2010**, *19*, 976–983. [CrossRef]

13. Marino, E.; Guillen-Climent, M.; Ranz Vega, P.; Tomé, J. Fire severity mapping in Garajonay National Park: Comparison between spectral indices. *FLAMMA* **2016**, *7*, 22–28.

14. Soverel, N.O.; Perrakis, D.D.; Coops, N.C. Estimating burn severity from Landsat dNBR and RdNBR indices across western Canada. *Remote Sens. Environ.* **2010**, *114*, 1896–1909. [CrossRef]

15. Meng, Q.; Meentemeyer, R.K. Modeling of multi-strata forest fire severity using Landsat TM Data. *Int. J. Appl. Earth Obs. Geoinf.* **2011**, *13*, 120–126. [CrossRef]

16. Escuin, S.; Navarro, R.; Fernández, P. Fire severity assessment by using NBR (Normalized Burn Ratio) and NDVI (Normalized Difference Vegetation Index) derived from LANDSAT TM/ETM images. *Int. J. Remote Sens.* **2007**, *29*, 1053–1073. [CrossRef]

17. Kollanus, V.; Tiittanen, P.; Niemi, J.; Lanki, T. Effects of long-range transported air pollution from vegetation fires on daily mortality and hospital admissions in the Helsinki metropolitan area, Finland. *Environ. Res.* **2016**, *151*, 351–358. [CrossRef]

18. McMahon, C.K. Forest fires and smoke—Impacts on air quality and human health in the USA. In Proceedings of the TAPPI International Environmental Conference, Nashville, TN, USA, 18–21 April 1999; TAPPI Press: Nashville, TN, USA, 1999; Volume 2, pp. 443–453.

19. Yin, S.; Wang, X.; Guo, M.; Santoso, H.; Guan, H. The abnormal change of air quality and air pollutants induced by the forest fire in Sumatra and Borneo in 2015. *Atmos. Res.* **2020**, *243*, 105027. [CrossRef]

20. Yuchi, W.; Yao, J.; McLean, K.E.; Stull, R.; Pavlovic, R.; Davignon, D.; Moran, M.D.; Henderson, S.B. Blending forest fire smoke forecasts with observed data can improve their utility for public health applications. *Atmos. Environ.* **2016**, *145*, 308–317. [CrossRef]

21. Adetona, O.; Reinhardt, T.E.; Domitrovich, J.; Broyles, G.; Adetona, A.; Kleinman, M.T.; Ottmar, R.D.; Naeher, L.P. Review of the health effects of wildland fire smoke on wildland firefighters and the public. *Inhal. Toxicol.* **2016**, *28*, 95–139. [CrossRef]

22. Langmann, B.; Duncan, B.; Textor, C.; Trentmann, J.; van der Werf, G. Vegetation fire emissions and their impact on air pollution and climate. *Atmos. Environ.* **2009**, *43*, 107–116. [CrossRef]

23. Bernard, M.; Bernard, L. Correlation between Wildfire Statistical Data, Weather and Climate. 2007. Available online: https://ams.confex.com/ams/pdfpapers/126829.pdf (accessed on 20 August 2021).

24. Trauernicht, C. Vegetation—Rainfall interactions reveal how climate variability and climate change alter spatial patterns of wildland fire probability on Big Island, Hawaii. *Sci. Total Environ.* **2019**, *650*, 459–469. [CrossRef] [PubMed]

25. Whitworth-Hulse, J.I.; Magliano, P.N.; Zeballos, S.R.; Gurvich, D.E.; Spalazzi, F.; Kowaljow, E. Advantages of rainfall partitioning by the global invader Ligustrum lucidum over the dominant native Lithraea molleoides in a dry forest. *Agric. For. Meteorol.* **2020**, *290*, 108013. [CrossRef]

26. Bar, S.; Parida, B.R.; Pandey, A.C. Landsat-8 and Sentinel-2 based Forest fire burn area mapping using machine learning algorithms on GEE cloud platform over Uttarakhand, Western Himalaya. *Remote Sens. Appl. Soc. Environ.* **2020**, *18*, 100324. [CrossRef]

27. Burton, J. It Was a Line of Fire Coming at Us': South West Firefighters Return Home. 2020. Available online: https://www.busseltonmail.com.au/story/6620313/it-was-a-line-of-fire-coming-at-us-firefighters-return-home/ (accessed on 20 August 2021).

28. ABCNews. Victorian Bushfires Death Toll Rises as Authorities Confirm Contractor's Death Was Fire-Related. 2020. Available online: https://www.abc.net.au/news/2020-01-15/fires-death-toll-rises-to-five-in-victoria/11869596 (accessed on 20 August 2021).

29. Green, M. Australia's Massive Fires Could Become Routine, Climate Scientists Warn. 2020. Available online: https://www.reuters.com/article/us-climate-change-australia-report/australias-massive-fires-could-become-routine-climate-scientists-warn-idUSKBN1ZD06W (accessed on 20 August 2021).

30. SBSNews. The Numbers Behind Australia's Catastrophic Bushfire Season. 2020. Available online: https://www.sbs.com.au/news/the-numbers-behind-australia-s-catastrophic-bushfire-season (accessed on 20 August 2021).

31. O'Mallon, F.; Tiernan, E. Australia's 2019–2020 Bushfire Season. 2020. Available online: https://www.canberratimes.com.au/story/6574563/australias-2019-20-bushfire-season/ (accessed on 20 August 2021).

32. Readfearn, G. "Silent Death": Australia's Bushfires Push Countless Species to Extinction. 2020. Available online: https://www.theguardian.com/environment/2020/jan/04/ecologists-warn-silent-death-australia-bushfires-endangered-species-extinction (accessed on 20 August 2021).

33. The University of Sydney. More Than One Billion Animals Killed in Australian Bushfires. 2020. Available online: https://www.sydney.edu.au/news-opinion/news/2020/01/08/australian-bushfires-more-than-one-billion-animals-impacted.html (accessed on 20 August 2021).
34. Zaman, M.A.; Rahman, A.; Haddad, K. Regional flood frequency analysis in arid regions: A case study for Australia. *J. Hydrol.* **2012**, *475*, 74–83. [CrossRef]
35. Veraverbeke, S.; Lhermitte, S.; Verstraeten, W.; Goossens, R. The temporal dimension of differenced Normalized Burn Ratio (dNBR) fire/burn severity studies: The case of the large 2007 Peloponnese wildfires in Greece. *Remote Sens. Environ.* **2010**, *114*, 2548–2563. [CrossRef]
36. Roy, D.P.; Boschetti, L.; Trigg, S.N. Remote sensing of fire severity: Assessing the performance of the normalized burn ratio. *IEEE Geosci. Remote Sens. Lett.* **2006**, *3*, 112–116. [CrossRef]
37. Teobaldo, D.; Baptista, G.M.D.E. Measurement of severity of fires and loss of carbon forest sink in the conservation units at Distrito Federal. *Rev. Bras. Geogr. Fis.* **2016**, *9*, 250–264. [CrossRef]
38. Key, C.H.; Benson, N.C. Landscape Assessment: Ground Measure of Severity, the Composite Burn Index; and Remote Sensing of Severity, the Normalized Burn Ratio. 2006. Available online: http://pubs.er.usgs.gov/publication/2002085 (accessed on 20 August 2021).
39. Dos Santos, S.M.B.; Bento-Gonçalves, A.; Franca-Rocha, W.; Baptista, G. Assessment of burned forest area severity and postfire regrowth in chapada diamantina national park (Bahia, Brazil) using dnbr and rdnbr spectral indices. *Geosciences* **2020**, *10*, 106. [CrossRef]

Article

Short-Term Effects of Anthropogenic Disturbances on Stand Structure, Soil Properties, and Vegetation Diversity in a Former Virgin Mixed Forest

Cosmin Ion Braga [1], Vlad Emil Crisan [1], Ion Catalin Petritan [2], Virgil Scarlatescu [3], Diana Vasile [1], Gabriel Lazar [1] and Any Mary Petritan [1,*]

[1] National Institute for Research and Development in Forestry "Marin Dracea", 13 Cloşca Str., 500040 Braşov, Romania

[2] Department of Forest Engineering, Forest Management Planning and Terrestrial Measurements, Faculty of Silviculture and Forest Engineering, Transylvania University, 1 Ludwig van Beethoven Str., 500123 Brasov, Romania

[3] National Institute for Research and Development in Forestry "Marin Dracea", Principala Str., 117470 Mihaesti, Romania

* Correspondence: apetritan@gmail.com or any_mary.petritan@icas.ro

Citation: Braga, C.I.; Crisan, V.E.; Petritan, I.C.; Scarlatescu, V.; Vasile, D.; Lazar, G.; Petritan, A.M. Short-Term Effects of Anthropogenic Disturbances on Stand Structure, Soil Properties, and Vegetation Diversity in a Former Virgin Mixed Forest. *Forests* **2023**, *14*, 742. https://doi.org/10.3390/f14040742

Academic Editor: Annemarie M. Bastrup-Birk

Received: 9 February 2023
Revised: 27 March 2023
Accepted: 31 March 2023
Published: 4 April 2023

Abstract: Despite the sharply growing interest in the disturbances occurring in primary forests, little is known about the response of European virgin forests to anthropogenic disturbance. The present study investigated the effect of the first silvicultural interventions that took place nine years earlier in a former virgin forest (FVF). Changes in the stand structure, environmental characteristics, and diversity of ground vegetation were studied in comparison with a nearby virgin forest (VF), both consisting of a mixture of European beech and silver fir. While the tree density did not differ significantly between the two forests, the number of large trees, the basal area, and the stand volume were significantly reduced in the FVF. The deadwood volume was twice as great in the VF as in the FVF and was found in both forests, particularly from silver fir. Despite significantly better light conditions in the FVF, natural regeneration was not significantly higher than in the VF. However, a slight improvement in the proportion of silver fir and other tree species into total regeneration was reported. The soil temperature was significantly higher in the FVF, independent of the measurement season, while the soil moisture showed a higher value in the VF only in spring. The FVF is characterized by a greater soil CO_2 emission, which is especially significant in summer and fall. The diversity of the ground vegetation did not yet react significantly to the silvicultural intervention. These preliminary findings are important in drawing suitable forest management practices that need to be applied in mixed beech–silver fir stands, especially in terms of maintaining species diversity. However, the short time frame since the intervention obliges further research on this VF–FVF pair over the next 10–20 years, at least regarding silver fir dynamics.

Keywords: former virgin forest; structure; regeneration; deadwood; soil respiration; ground vegetation diversity

1. Introduction

As a result of intense human exploitation, mainly for the creation of agricultural land or for wood consumption, only a few remnants of virgin forests remain in the contemporary landscape of Europe [1–3]. Furthermore, according to a recent study [4], virgin forests are scarce and continue to disappear. Therefore, one of the most important objectives of the "EU Biodiversity Strategy for 2030" agenda is to define, map, and monitor all of these remnants' forests. Despite the fact that forest covers 35% of Europe's total land area, forest undisturbed by man (i.e., virgin forest) covers only 2.2% of the European forest area. Furthermore, Romania hosts some of the largest remnants of virgin forest in Central Europe, covering about 59,110 ha according to [4] and 71,077 ha according to the last edition of the Romanian virgin forest catalogue [5]. It should be mentioned that the updating of both sources [4,5] is an ongoing process.

These increasingly rare untouched forests have attracted the interest of scientists since the middle of the 19th century, especially after the Second World War at the initiative of the IUFRO forestry section [6,7]. From the point of view of the landscape, this unique and critical forest ecosystem (i.e., primary forest) represents the highest range of naturalness and can be found particularly in the northern and eastern parts of Europe [4,8,9]. Still present in the Carpathian Mountains, such intact ecosystems have a key role in biodiversity conservation [6,10–12], but also in carbon sequestration [13,14], biomass accumulation [15], and riparian functionality [16]. Characterized by high structural complexity driven by natural disturbance processes and natural regeneration dynamics [17,18], virgin forests are an important reference for managed forests, especially for close-to-nature silviculture [19–22], providing a rare opportunity for understanding natural ecosystem processes and functions [11,23].

In the context of climatic change, an increase in the disturbance regime (i.e., in the frequency and severity) has been recorded in the first decade of the current century, with a potential negative impact on the carbon storage capacity of European forests and many implications for other ecosystem services [18,24]. The importance of the disturbance regime in modifying the structure and dynamics of virgin forests has already been shown [7,25]. In addition, the disturbance regime affects forest ecosystems from the functional plant level to the stand level [24,26], as it can change the natural cycle of forest development [7,27]. Forest disturbance can generally be related to natural causes (i.e., windthrow events, insect attacks, drought stress, forest fires) but can also include forest management activities (i.e., silvicultural interventions) with a potential influence on biodiversity maintenance [25,28]. Aside from the devaluation of timber, the consequences of disturbance will amplify the impact of climate change, and forest ecosystem provisioning services (i.e., drinking water) and regulating and supporting services (recreation value, air quality, etc.) will all be seriously affected [18,24]. Therefore, understanding the main effects of the disturbance regime on forest ecosystems constitutes a key challenge for ecological research.

Disturbances are described by the size, distribution, and structure of the resulting canopy gaps. The main natural disturbance regime of European primary forests, especially those consisting of shade-tolerant species [7,29], but not limited to them [27,30,31], is characterized by the frequent occurrence of small canopy gaps (i.e., less than 150 m^2), generated by the mortality of an individual tree or small group of trees and referred to as a low-severity disturbance regime [7]. However, moderate severity disturbance has been found as a key driver of tree species coexistence in primary temperate forests [32]. High-severity stand-replaced disturbances can also occur [7,30], with this type of disturbance being very important in the establishment of light-demanding tree species.

While the study of the natural disturbance regime of virgin forests has been intensified over the last decades [7,27,30,31], there is a lack of information about the short-term impact of first silvicultural interventions on a former virgin forest in Europe. In the Southern Carpathian Mountain range in Romania, in the immediate vicinity of the virgin forest Sinca (a forest included in the UNESCO World Heritage Site of "Primeval Beech Forests of the Carpathians and the Ancient Beech Forests of Germany" in 2017), there is a forest with the same composition and stand characteristics, where the first silvicultural (human) interventions (i.e., group shelterwood system) occurred in 2009. This rare situation gave us the opportunity to study the short-term impacts of the first management interventions on the main traits of the former virgin forest. A better understanding of how forest management (i.e., silvicultural interventions) reshapes the main characteristics of stand diversity is crucial to anticipating forest ecosystem responses to the intensity and frequency of disturbance events [33], and in our case, will contribute to understanding the main consequences of the loss of these unique and diverse forests.

Unlike the few comparative studies that have investigated stand characteristics in virgin and managed forests [6,11,22,34,35], our study provides valuable insights into the changes induced by the anthropic interventions within the first decade. The specific objectives of our study were the following: (i) to assess the short-term effects of silvicultural intervention on the stand structural characteristics, amount of deadwood, and natural regeneration; (ii) to describe the possible consequences of light condition changes on ground vegetation diversity; and (iii) to test whether the soil microclimate and soil respiration are affected by the reduction in stand stocking volume.

2. Materials and Methods

2.1. Study Site

The study area (Figure 1A) is located in the Southern Carpathians, in the northern part of the Fagaras Mountains, in central Romania (45°40′0.420″ N and 25°10′14.359″ E), at an altitude ranging from 850 to 1350 m (a.s.l.). The experimental sites of our study are homogeneous in terms of ecological and vegetation characteristics. They are part of the Pulmonario rubrae-Fagetum (Soo, 1964) association [36] and are in accordance with the Natura 2000 classification of the 91V0 Dacian beech forest (Symphylo-Fagion) [37]. The forest composition is largely dominated by European beech (*Fagus sylvatica* L.) and silver fir (*Abies alba* Mill.) trees in different proportions. The soils are derived from recent moraines of the Weichsel glacial period. The main soil type is Cambisol, developed on a parental material of crystalline schists. The main humus type is mull-moder. According to the Koppen classification, the climate is temperate continental, with a mean annual temperature of 6.1 °C and mean annual precipitation of 1100 mm [38].

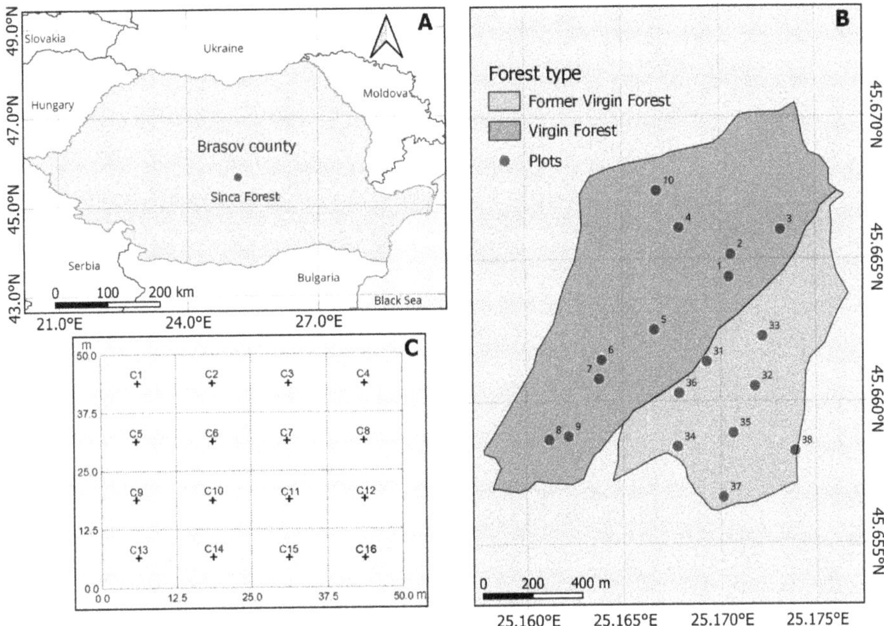

Figure 1. Study area maps: (**A**)—Location of Sinca Forest, (**B**)—Distribution of plots by each forest type (VF and FVF), (**C**)—Experimental design for soil properties and light availability inside of each plot (50 m × 50 m) where C1–C16 represent the center of each subplot.

In the former virgin forest, the initial felling of the group shelterwood system was applied in 2009, with a harvesting intensity of around 30% stand volume. The group shelterwood system is the most widespread management practice applied in Romania, currently used in 81% of Romanian forests [39]. This forest management consists of repeated and successive fellings. The trees are extracted irregularly and concentrated in gaps, their size and distribution of which depends on the species' composition and light requirements, with the goal to firstly provoke natural seedings and to secondly provide enough light to advance growth [40]. Therefore, this kind of silvicultural intervention is formed by three kinds of fellings, starting with gap opening, followed by extending the gaps to provide light conditions for seedlings, and finishing with removing the last old trees. Consequently, a stand with this type of forest management can simultaneously include areas with all of these three felling types with successive stages of regeneration. For shade-tolerant species (i.e., European beech and silver fir), in our case, the gaps were generally small, up to 0.5–0.75 of the mean stand height [40]. Thus, in this beech–silver fir mixed forest, the initial felling could be taken as a uniform intervention (i.e., only thinning around the seed trees). The regeneration period (including all three felling types) for shade-tolerant species is about 30 years.

2.2. Field Sampling

During the year 2017, eighteen permanent, randomly distributed 50 m × 50 m (2500 m^2) sample plots, ten in the virgin forest (VF) and eight in the former virgin forest (FVF), were established (Figure 1B). In each plot, the standing trees with a diameter at breast height (dbh) greater than 6 cm were recorded by their species and vitality attributes (live/dead). Their dbh and total height (H) were measured using tape and a Vertex IV Hypsometer (Haglof, Sweden), respectively. For the broken standing trees, the diameter at the breakage point was measured with an optical caliper. All lying deadwood pieces with a length $\geq$ 3 m and thick-end diameter $\geq$ 15 cm were registered with species, and their diameter at both ends and the total length within the plot were measured. According to Keller [41], each piece of deadwood (standing and lying) was classified within one of the five decay classes, namely, fresh (A), hard (B), rotten (C), moldering (D), or mull (E) wood. For the FVF site, we noted if the deadwood was caused by silvicultural intervention or natural disturbance. Following the methodology described by Petritan et al. [42], the natural regeneration (all samplings taller than 10 cm and smaller than 6 cm in dbh) was sampled in four circular subplots, the centers of which were located 17.68 m apart from each corner on the diagonals of the plot. Each subplot included three concentric circles: a 5 m^2 circle (RC1) for saplings between 10 and 39.9 cm tall, a 10 m^2 circle (RC2) for saplings between 40 and 129.9 cm height, and a 20 m^2 circle (RC3) for saplings with a height $\geq$1.3 m and dbh <6 cm. Saplings smaller than 130 cm were evaluated for ungulate browsing to the leading shoot. The ground vegetation composition was determined through phytosociological relevés conducted in each plot. The cover of each vascular plant species was estimated using the Braun-Blanquet scale [43]. For data analysis, scale values were transformed into percentage values as follows: r = 0.1; + = 0.5; 1 = 2.5; 2m = 5; 2a = 10; 2b = 20; 3 = 37.5; 4 = 62.5; 5 = 87.5 [44]. Twelve of the eighteen permanent sample plots were randomly selected, with six in each forest, to study the environmental conditions. These plots were divided into sixteen equal subplots of 12.5 m × 12.5 m (Figure 1C). Light availability, microclimate, and soil respiration were determined at the center of each subplot, i.e., at 96 points in each forest. In summer, a hemispherical photograph was taken of the center of each square using a digital camera coupled with a fisheye lens. The light availability conditions, expressed by diffuse light (indirect site factor (ISF)) and direct light (direct site factor (DSF)), as a percentage of the full light, were determined for each photo using the WinScanopy Pro B software 2012 (Regent Instruments Inc., Quebec City, QC, Canada). The soil temperature (Ts) was measured using a soil temperature sensor (STP; PP Systems) that was inserted at a depth of 5 cm into the soil, and the soil moisture (Us) was measured at a depth of 14 cm into the soil using a Trime Pico 64 sensor coupled with an HD2 data logger (IMKO GmbH, Germany). Two weeks prior to the start of soil respiration (Rs) measurements, at the center of each subplot, a PVC collar (10 cm in diameter and 8 cm in height) was inserted into the soil at approximatively 3 cm, as recommended by Curiel Yuste et al. [45]. A portable infrared gas analyzer (IRGA) connected to a soil respiration chamber (EGM-4 and SRC-1; PP Systems, USA) was used to carry out the soil respiration measurements. The volume of the collars above the soil was added to the chamber volume. The duration of each measurement was 2 min. Soil respiration measurements were taken between 9 a.m. and 18 a.m. (1–2 days), at least two days after a rainy day. The soil measurements were carried out in three different seasons (spring, summer, and fall).

2.3. Data Analysis

The height and diameter of the dominant trees (h_{dom} and d_{dom}), defined as the mean diameter/height of the 100 largest trees per ha [6], were calculated for each species (beech and silver fir). The volume (V) of living and standing dead trees was determined by applying a double-logarithmic regression:

$$log(v) = a_0 + a_1 \times log(d) + a_2 \times log^2(d) + a_3 \times log(h) + a_4 \times log^2(h)$$

where $a_0, a_1, a_2, a_3,$ and a_4 are species-specific regression coefficients [46]. The volume of broken standing trees and lying deadwood was calculated using the volume formula of a truncated cone, considering only the portion lying within the plot boundaries [37]. The vertical structure of the two forests was described by the A index proposed in [47], which combines the Shannon indices after tree stratification into three height layers (>2/3 top height, 1/3–2/3 top height, and <1/3 top height), where the top height is the average height of the 20% largest trees per sample plot (for more details, please see [48]). All ground vegetation indices were calculated using PAST 4.12 [49]. The Renyi diversity profiles of each sample plot of both forest types were calculated using PAST 4.12 [49]. Renyi curves enable the comparison of different diversity indices (each alpha value on the y-axis corresponds to a different diversity index) [49–51]. Low evenness is reflected by an abrupt decrease along the vertical axis, while high evenness is represented by a smooth line [51]. To assess the statistical significance of differences between the two forests regarding the studied traits and ground vegetation diversity indices, we applied an analysis of variance. The differences between mean values were tested by Tukey's HSD for an unequal N' post hoc test. The normality of residuals was tested with the Kolmogorov–Smirnov test and homoscedasticity, with Levene's test. If the assumption necessary to apply ANOVA was not fulfilled, a non-parametric Mann–Whitney U-test was applied. The analyses were achieved using Statistica 12 [52].

3. Results

3.1. Structural Characteristics

Eight years after the first silvicultural intervention (i.e., 2009, group shelterwood system, first cutting to promote seeding), the mean density of living trees per hectare was greater in the VF (650 ± 46.2) than in the FVF (507 ± 53.0), but without a significant difference (Table 1). As we expected, significant differences between the two forests were found in the basal area and stand volume: the basal area was 36% smaller in the former virgin forest compared to the virgin forest, while the volume in the FVF represented 57% of that found in the VF (Table 1). The silvicultural interventions led to significantly smaller d_{dom} and h_{dom} values (Table 1) and a lower number of trees with larger dimensions. The average number of trees with dbh $\geq$ 80 cm per hectare was more than double in the virgin than in the former virgin forest (28 vs. 11, respectively), but the number of trees taller than 40 m was quite similar between the two forests (Table 1). The main species of both stands were beech and silver fir, with a very rare presence of spruce (<1% in both forest stands). The participation of the two tree species was different between the two forests when the tree number was considered (greater percentage of beech in the FVF of 71% than in the virgin forest of 57%), but more similar (about 50% for each species) when the basal area or volume was considered. The vertical diversity of the former virgin forest was similar to that of the virgin forest (Table 1).

Table 1. Basic statistical parameters for the stand characteristics of the investigated VF and FVF silver fir–beech forests. Abbreviations used: h_{dom} = top height of 100 tallest trees, d_{dom} = mean diameter of 100 thickest trees, CV = coefficient of variation. Different letters represent the significant differences between the means of stand characteristics using ANOVA–Tukey's test.

Stand Characteristics/ Statistical Parameters	Virgin Forest		Former Virgin Forest	
	Mean ± Std. Error (Min–Max)	CV%	Mean ± Std. Error (Min–Max)	CV%
No. of trees per ha	650 ± 46.2 [a] (492–932)	22	507 ± 53.0 [a] (320–776)	29
No. of trees per ha with dbh ≥ 80 cm	28.4 ± 4.6 [a] (12–48)	52	11 ± 1.8 [b] (4–16)	46
No. of trees per ha with height ≥ 40 m	14.8 ± 4.0 [a] (0–40)	85	15.5 ± 3.7 [a] (4–36)	68
Basal area ($m^2\ ha^{-1}$)	54.5 ± 3.2 [a] (44.7–77.3)	18	35.0 ± 3.6 [b] (16.6–48.8)	28
Volume ($m^3\ ha^{-1}$)	893.1 ± 55.7 [a] (655.2–1236.6)	19	510.0 ± 60.2 [b] (209.7–725.7)	33
h_{dom} (m)	38.4 ± 0.9 [a] (35.4–43.1)	7	32.4 ± 1.7 [b] (24.3–36.1)	11
d_{dom} (cm)	75.9 ± 2.9 [a] (67.6–90.4)	9	54.2 ± 3.8 [b] (37.2–66.3)	18
A index	1.53 ± 0.04 [a] (1.25–1.66)	9	1.58 ± 0.04 [a] (1.36–1.71)	8

The largest difference between tree distribution along the diameter gradient was found for the smallest diameter classes, in which the number of silver firs in the FVF was significantly lower (lower than half) than in the VF. In the FVF, the largest trees (dbh > 104 cm) were missing (Figure 2).

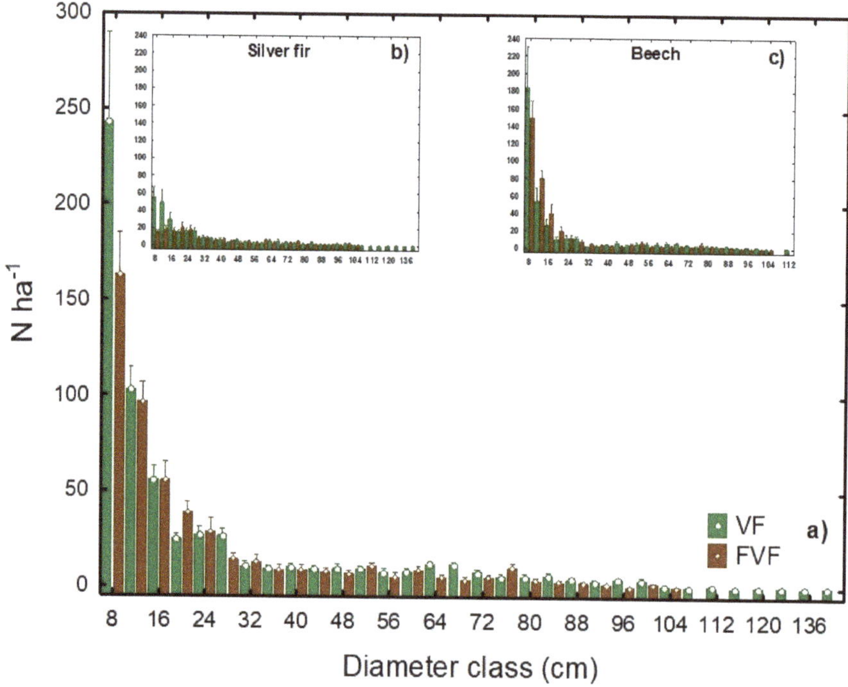

Figure 2. The distribution of the number of trees per diameter class in VF and in FVF (**a**) independent of species and separately for the two main species ((**b**)—Silver fir and (**c**)—Beech).

3.2. Deadwood

The quantity of the total deadwood (DW) in the FVF was approximately half (46%) of that in the VF (112 m^3 ha^{-1} vs. 244 m^3 ha^{-1}, respectively, Table 2). In the former virgin forest, from this total amount, a small proportion (11%) represented DW, resulting from the silvicultural interventions (e.g., woody debris or standing trees left in the forest). The proportion of lying DW was higher (71%) in the VF than in the FVF (60%), while in the FVF, the participation of standing was greater (40%) in comparison with the VF (29%). In both sites, the volume of DW consisted mostly of silver fir, which accounted for 84% of the total DW in the former VF and 63% in the FVF. The average dead-to-live wood ratio was slightly lower in the FVF (18%) than in the VF (21%).

Table 2. Mean, standard error, and range values of deadwood volume (lying and standing deadwood) per species and forest type.

Species	Beech		Silver Fir		Undefined		Total	
Forest Type	Former Virgin	Virgin	Former Virgin	Virgin	Former Virgin	Virgin	Former Virgin	Virgin
Lying deadwood (m^3 ha^{-1})	10 ± 5.9 (0–45.7)	60.2 ± 13.52 (8.6–144.3)	57 ± 28 (0.5–182.5)	114.4 ± 18.1 (12.8–214.9)	0.3 ± 0.1 (0–3.5)	0.6 ± 0.4 (0–3.5)	67.2 ± 33.7 (1–228)	174.2 ± 21.4 (71.8–277.8)
Standing deadwood (m^3 ha^{-1})	5.6 ± 4.1 (0–34.4)	29.44 ± 10.9 (0.7–117.8)	37.8 ± 8.5 (2.7–65.2)	41.8 ± 11.6 (2.9–101.7)	1.5 ± 0.6 (0–4.9)	0.0	45 ± 9.9 (8.5–80.9)	70.3 ± 14.6 (3.6–145.6)
Total deadwood volume (m^3 ha^{-1})	15.6 ± 6.2 (0.3–45.7)	89.6 ± 22.2 (10.9–262.2)	94.8 ± 30.5 (4.6–230)	154.3 ± 24.5 (15.8–248.1)	1.8 ± 0.6 (0–5)	0.6 ± 0.4 (0–3.5)	112.3 ± 34.7 (10.9–276.2)	244.2 ± 31.3 (75.5–400.4)

The deadwood structure on decay class was more or less similar in the two investigated forests, with more than half of the DW amount (70% vs. 60%) recorded in the most advanced decay classes (i.e., moldering—E class and mull wood—D class) (Figure 3). The fresh deadwood percentage was slightly higher in the FVF compared to VF (11% vs. 6%).

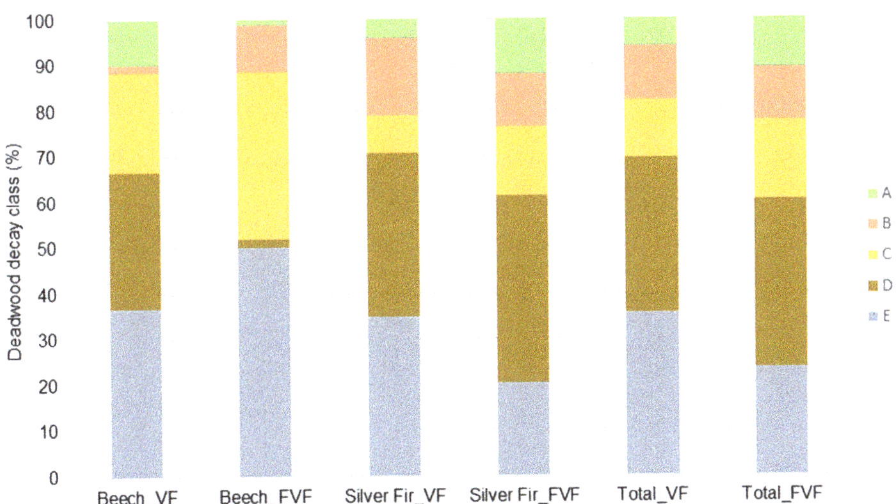

Figure 3. The deadwood structure on decay class (A—fresh, B—hard, C—rotten, D—mouldering, E—mull wood), species, and forest type (VF—virgin, FVF—former virgin forests).

3.3. Natural Regeneration and Light Conditions

We found a slightly lower regeneration density (i.e., all saplings with a height > 10 cm and dbh < 6 cm) in the VF site (6015 seedlings per ha) than in the FVF (6744 seedlings per ha), without significant differences. The main difference regarding the regeneration distribution of different height classes (Figure 4) was a higher percentage of taller saplings (h > 130 cm and dbh < 6 cm) in the FVF (32%) than in VF (29%). The species composition of the regeneration was similar between the two forests, with beech being the most common species (94% in virgin and 81% in former virgin forest). The proportion of silver fir was higher in the FVF (16%) than in the VF (5%). The percentage of the other species was slightly higher in the FVF (3% vs. 1%), aside from the spruce, sycamore maple, and pioneer species such as Eurasian aspen found in the FVF (Table S1). The proportion of damaged saplings by browsing of the leading shoot was nearly twofold in the former virgin forest (14%) as opposed to in the virgin forest (7%). The most browsed species was silver fir (double the amount of beech) in both forests. The light conditions were significantly more improved in the former virgin forest (ISF: 24.4 ± 13.2% and DSF: 26.4 ± 11.7%) than in the virgin forest (ISF: 7.2 ± 0.9% and DSF: 6.7 ± 0.8%).

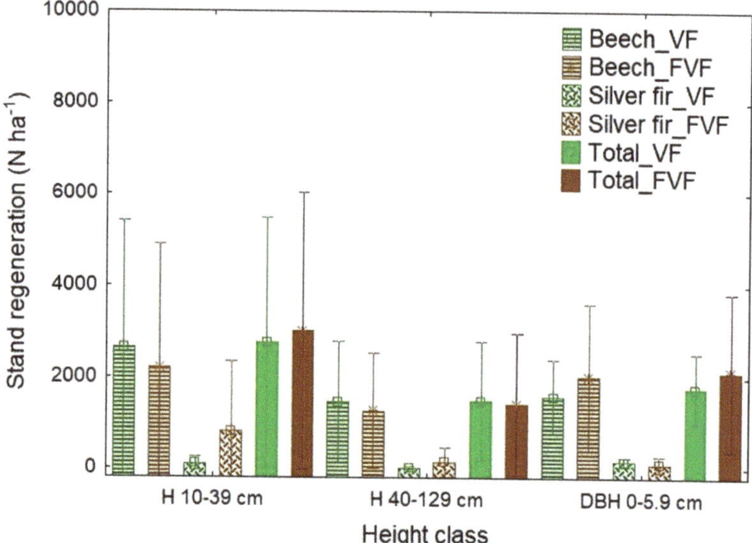

Figure 4. Density of saplings per forest type, species, and regeneration circle (RC1—5 m^2 for seedlings with height of 10–39.9 cm; RC2—10 m^2 for saplings 40–129.9 cm in height; RC3—20 m^2 for saplings with height >130 cm and dbh <6 cm). Error bars represent a 95% confidence interval. According to ANOVA–Tukey's testing, there were no significant differences found ($p > 0.05$) between the sampling category and forest type.

3.4. Soil Microclimate and Soil Respiration

Generally, the soil microclimate and soil respiration showed a seasonal pattern, especially for temperature and respiration (Figure 5). Thus, the soil temperature for all measurement periods (spring, summer, and fall, Figure 5A) was significantly higher ($p < 0.05$) in the FVF than in VF, with the largest difference in summer (5 °C) and the smallest in spring (1.7 °C). The difference between the two forests in terms of soil moisture did not show a clear pattern, being significant only for the spring measurements, with a higher average value in the virgin forest. The differences between the two forests diminished in summer, while in the fall, the soil moisture was slightly higher in the former virgin forest (Figure 5B). The soil respiration followed the seasonal pattern of the soil temperature, with the highest mean value reached in the summer and lowest in fall (Figure 5C). Significantly higher differences between the two forests were found only for summer (the largest one:

5.4 in FVF vs. 3.6 μmol CO_2 m^{-2} s^{-1} in VF) and for fall (2.82 vs. 1.99 μmol CO_2 m^{-2} s^{-1}). Generally, the soil CO_2 emissions were higher in the former virgin forest (Figure 5C).

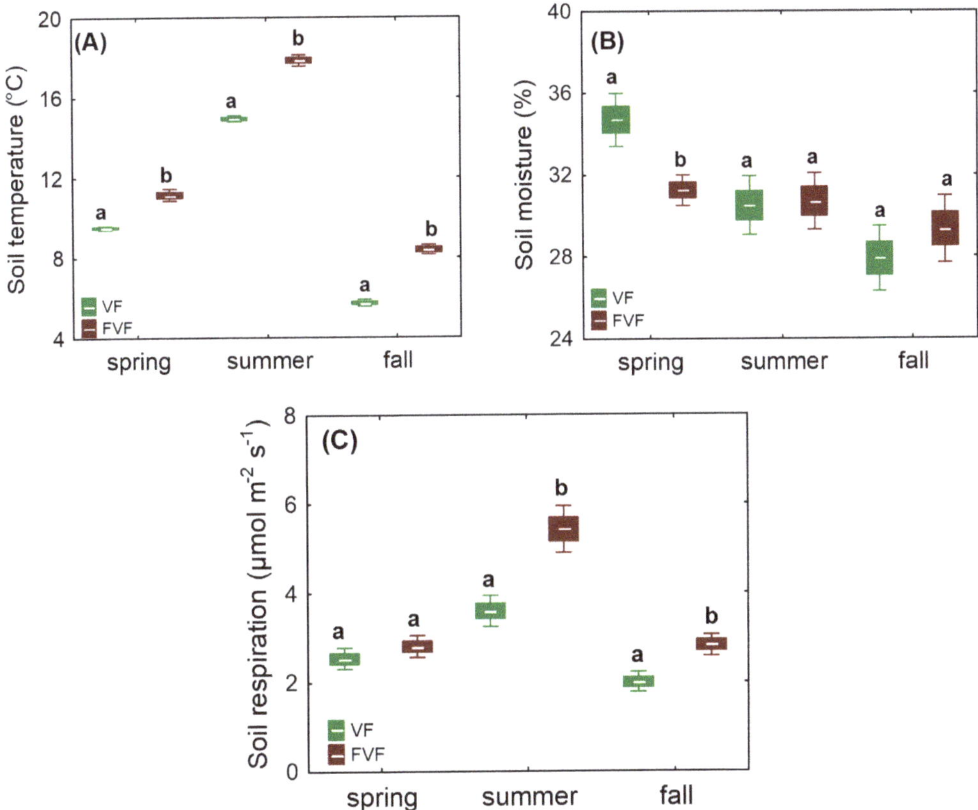

Figure 5. Boxplots of (**A**) soil temperature, (**B**) soil moisture, (**C**) and soil respiration along seasons per forest type (i.e., VF and FVF). Different letters (i.e., (**a**) and (**b**)) mean significant differences between the VF and FVF within each season. Boxes bound the standard errors of the mean. Whiskers represent the 95% confidence intervals and horizontal lines within boxes represent the mean values.

3.5. Ground Vegetation Diversity

Within each forest type, the diversity profiles of the plot curves were, in general, similar for low alpha values and diverged with an increase in the alpha up to alpha = 3, and for high alpha values (around a value of 4), they tended to converge again (Figure 6). Although as an average pattern, the decrease in lines was more abrupt for the former virgin forest (alpha varying from 0 to 2), there were also two plots among virgin plots where this decrease was pronounced.

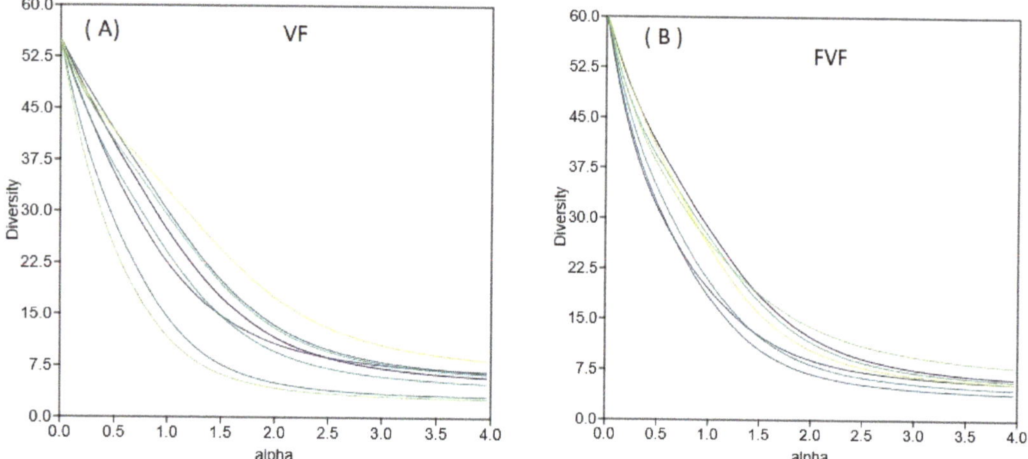

Figure 6. Diversity profiles of both forest types using the exponential of the Renyi diversity index. ((**A**)—virgin forest (VF), (**B**)—former virgin forest (FVF)). The scale alpha values: alpha = 0—provides the total number of species, alpha = 1—gives a measure directly proportional to the Shannon index, alpha = 2—similar to Simpson's index, high alpha value estimate the Berger–Parker index. Different color of lines represents the sampled plots within each forest type.

The assessment of the ground vegetation diversity in the two forests using different diversity indices showed different patterns of higher values (Margalef, Menhinick, evenness, equitability, Simpson), lower values (dominance, Brillouin, Fisher's alpha), or similar values (Shannon, Berger–Parker) in the virgin forest compared to the former virgin forest (Figure 7). Nevertheless, all tests failed to show significant diversity differences between both forest types (VF vs. FVF).

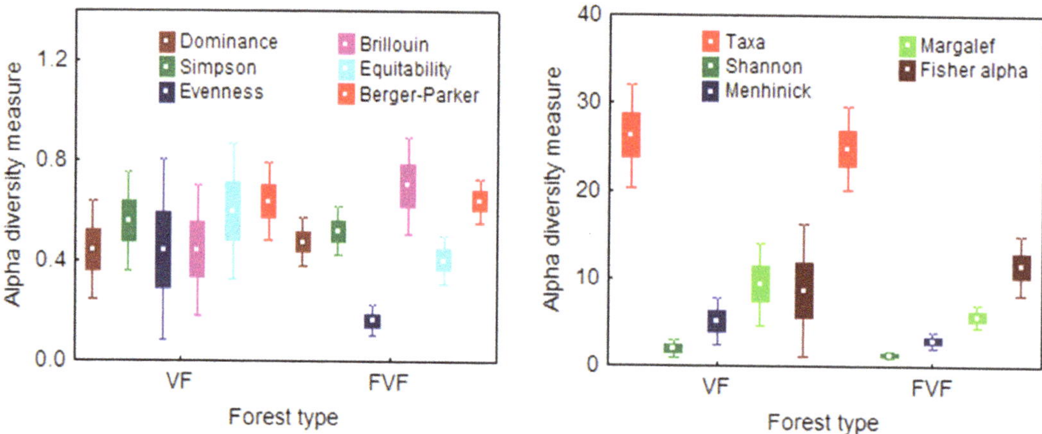

Figure 7. Ground vegetation diversity comparison between a virgin forest (VF) and former virgin forest (FVF). Alpha diversity measures: dominance, Simpson, evenness, Brillouin, equitability, Berger–Parker, taxa (richness), Shannon, Menhinick, Margalef and Fisher's alpha index. Boxes: means ± standard errors of the mean, whiskers: 95% confidence intervals.

4. Discussion

Our study found that the first silvicultural intervention in a former beech–silver fir mixed virgin forest did not significantly change the distribution of trees on diameter classes, still following an inverted J-shaped form (Figure 1) for both the silver fir and European beech species (Figure 2b,c). The inverted J shape is the typical diameter distribution for natural forests with shade-tolerant silver fir and beech species [3,12,21], suggesting a general structural stability and resilience [42,53] but also susceptibility to disturbance events of low-to-moderate severity [33,54]. This suggests that a single harvest did not change the characteristic shape of the tree distribution in our former virgin forests, as was the case in forests with long-term historical management, which showed a more bell-shaped distribution, clearly different from those of virgin forests [11]. Furthermore, the former virgin forest showed a similarly high vertical diversity. However, although the tree density was not significantly reduced in the former virgin forest, the number of trees with a dbh greater than 80 cm became significantly lower (Table 1), which is in agreement with the general removal of large-diameter trees in managed forests [35]. In the same way, some Bosnian, Croatian, and Slovenian virgin forests were transformed into managed forests approximately 120 years ago. The cutting intensity there was also around 40% for the first time, and later, it was up to 20% by applying a single-tree selection system or an uneven aged silvicultural system [21,35,55,56]. We supposed a similar evolution for our former virgin forest, which became a buffer zone for the virgin forest, after its inclusion as part of a UNESCO World Heritage Site in 2017. Related to the management goal for the forests in the buffer zone, according to the proposed (but not yet approved) guidelines (about what should be done in the buffer zones of UNESCO areas), forest managers should apply only selection systems (single-tree selection or a group selection system) nearby UNESCO reserves, which will maintain the uneven-aged structure into the future. Other structural forest characteristics such as the basal area, volume, Hdom, and Ddom were significantly reduced in the former virgin forest in comparison with the virgin forest, shown by the main differences between the virgin and managed forests [11,22]. The species composition in both forests was dominated by beech. This finding is in line with data presented in the work of Vandekerkhove et al. [57] for beech-dominated old-growth forest where silver fir and spruce were in small proportions. Previous studies [21,42,58] have reported a decline in silver fir in several old-growth forests in Central and Southeastern Europe. In contrast to these researchers, Keren et al. [59] observed, in the Dinaric Mountains of Bosnia, that fir and spruce dominated in the upper layer of the stand compared to beech. Moreover, during the last 50 years, European Beech has become the most competitive and vital tree forest species from sub-mountain to mountain regions, especially in the Central European mountains [60].

The amount of deadwood was arguably clearly different between the managed and unmanaged forests [11,61], with a significantly lower quantity in the managed forests, especially because in the past, the retention of deadwood in forests was not desired or allowed. Deadwood has been seen as a risk for the propagation of diseases and as a potential trigger of disturbance events [62], but since its crucial role in the preservation of forest biodiversity and ecosystem functioning has been scientifically proven [63], the retention of a certain amount of deadwood in managed forests from biodiversity goals has become an important task as part of numerous integrative forest management approaches [64]. The amount of deadwood in our former virgin forest reached only half of the amount in the virgin forest, but its amount (112 m^3 ha^{-1}) was still superior to that found in the managed forests. An amount of about 60–70 m^3 ha^{-1} of deadwood was found in the silver fir mixed managed forests of the Dinaric Mountains, three to five times lower than in virgin forests [11,65]. Moreover, Glatthorn et al. [15] found a much lower biomass stocked in the deadwood in managed forests in comparison with virgin Slovakian beech forests. However, the amount of deadwood in our former virgin forest was much greater than those found in the nemoral beech–oak forests of Western Romania (35–57 m^3 ha^{-1}) [63] or than the average volume in European forests (20 m^3 ha^{-1}, Forest Europe, 2020) and exceeded the quantity (30–40 m^3 ha^{-1}) proposed as the threshold for mixed montane European forests [66]. According to our study, where the proportion of DW in each forest was highest for silver fir compared with European beech (i.e., 84% of the total DW in FVF and 63% in VF, Table 2), our results confirmed that silver fir is in decline in both forests, which is consistent with numerous studies across Europe that have made observations over several decades [21,67].

A single silvicultural intervention did not significantly influence the distribution of deadwood by different decay classes (Figure 3); in both forests, the most decayed deadwood was predominant. The low percentages of "fresh" and "hard" deadwood indicate that intermediate or severe disturbances did not occur in the investigated virgin forest [42], which is in agreement with the findings of Visjnic et al. [11], while in the former virgin forests, part of the deadwood probably perceived as fresh and good was removed. Furthermore, the proportion of deadwood in relation to the total volume of trees was similar between the virgin and former virgin forest (22% vs. 27%, respectively). These small differences in the ratios of deadwood to live tree populations highlight the similar structure of FVF to VF, prior to anthropogenic disturbance. A similar proportion of DW in relation to the total wood mass was found in other virgin forests [11], while in managed forests, this ratio was significantly lower [65]. Moreover, the presence of DW for all decay classes in both sites highlights the highest level of ecosystem stabilization, especially for biodiversity conservation and soil biochemistry [42,68,69]. Despite much better light conditions, the regeneration in the former virgin forest showed only slightly higher density values than in the virgin forest (6744 vs. 6015 seedlings per ha, respectively). This may be a consequence of the short time after intervention (only seven years), but also of the increased competition of the understory vegetation [37]. However, a slightly higher density of seedlings (10–39 cm height) and of the tallest saplings (H > 1.3 m and dbh < 6 cm) in the former virgin forest suggests that the greater light availability created by tree harvesting stimulated the production of new seedlings and accelerated the growth of larger saplings, in accordance with the main objective of the group shelterwood system [40]. The better light conditions in the former virgin forest led to the improvement of regeneration diversity, even though the proportion of other species was insignificant (Acer pseudoplatanus, 0.2%; Picea abies, 0.3%, Populus tremula, 1.0%). In the future, it is expected that the larger gaps, specifically made by silvicultural practices (i.e., group shelter wood system) in FVF, will accelerate the tree recruitment process with consequences for its structure and composition [65]. Beech was the main species of the regeneration layer in both forests, but the proportion of silver fir regeneration in the former virgin forest was three times higher than in the virgin forest (Table S1). This tendency could have been a consequence of both soil microclimate conditions and the spectral composition of the light which can influence seedling emergence. A prior study [70] observed that the natural regeneration of fir trees is hindered by a low proportion of the blue component and a surplus of the green and red components of the light spectrum. Nevertheless, the higher proportion of silver fir seedlings in managed forests is considered as a result of silvicultural interventions [71], which induced more economically valuable conifer species such silver fir compared to beech, as is a common practice in many European beech–conifer mixtures [11,72]. One of the main factors controlling seed germination may be set by the soil microclimate. The dynamics of a soil microclimate along seasonal growth vegetation, and especially the soil temperature, which was significantly higher in the former virgin forest (Figure 5A), can accelerate the dynamics of natural regeneration with consequences in the structure and composition of seedlings or recruits. Forest microclimatic buffering capacity increases in periods with warmer and drier years. For instance, intact forest canopies mitigated increases in vapor pressure deficits and temperature extremes by 20% and 15%, respectively, compared to disturbed forests [3]. Another factor that can influence regeneration is browsing. As in other studies [42,73], we found beech to be less susceptible to browsing by ungulates (i.e., roe deer and red deer) than silver fir. A greater proportion of damaged saplings was found in the FVF (almost two times more) than in VF. Therefore, silver fir seedlings can lose the competitive advantage of early height growth, mainly by beech or pioneer species [74]. In addition, changes in the forest canopy, mainly due to human disturbance (e.g., silvicultural interventions), could greatly affect not only the soil microclimate but also soil respiration [75,76]. Furthermore, silvicultural interventions (e.g., group shelterwood system according to a forest management plan) can lead to soil compaction, which could thus restrict root growth and microbial activities. On the one hand, the significant differences in soil temperature between forest types, especially in summer (Figure 5C), was the main factor contributing to increased soil respiration in the FVF. On the other hand, silvicultural interventions could reduce competition for soil moisture and nutrients, which would help to stimulate the root growth of seedlings and recruits [77]. In addition, wood harvesting and organic matter mixed into the mineral soil would increase substrate availability, which could increase microbial activity and accelerate the decomposition of litter and soil organic matter [78]. Even though harvesting operations started less than ten years ago, it was still observed that adventive geoelements [43] appeared in the FVF, unlike in the VF. The category of adventive geoelements is characteristic of species (e.g., *Festuca drymea*, *Erigeron annuus*, *Erectites hieracifolia*) that are massively dispersed by humans [79]. Therefore, the presence of these geoelements is justified in forests where forest management is practiced, because only within them is the anthropogenic factor present.

5. Conclusions

The effect of the disturbance regime did not considerably change the shape of the distribution of trees per diameter class between each forest type, showing structural similarities between the two forest types before disturbance events. Furthermore, the differences in soil microclimate and the proportion of regenerated species between forests (i.e., VF and FVF) may affect the future of the natural forest type, such as its composition. On the one hand, this temporal fluctuation in natural regeneration will allow foresters to take beneficial and rapid measures to preserve natural species composition over time and space to, for instance, promote more silver fir. On the other hand, turning virgin forests into managed forests is not the right way to go. However, on the basis of the obtained results, the forest management applied here (i.e., group shelterwood system) was not always "bad". It promoted the regeneration and recruitment of silver fir, which according to the latest research, will disappear in many European old-growth forests or its share will fall to a negligible percentage in these forests in the near future. This is what makes sense in terms of our study area for future ecological research in the next 10–20 years, at least regarding silver fir dynamics.

Supplementary Materials: The following supporting information can be downloaded at: https://www.mdpi.com/article/10.3390/f14040742/s1, Table S1: Natural regeneration per forest type, species, and regeneration circle (RC1—5 m^2 for seedlings with height of 10–39.9 cm; RC2—10 m^2 for saplings 40–129.9 cm in height; RC3—20 m^2 for saplings with height >130 cm and dbh <6 cm) (N ha^{-1} mean $\pm$ standard error). Minimum and maximum values (indicated in brackets) correspond to the average of the four regeneration sub-plots per plot.

Author Contributions: Conceptualization, C.I.B. and A.M.P.; methodology, C.I.B. and A.M.P.; formal analysis, C.I.B., A.M.P. and I.C.P.; data curation, V.S. and D.V.; writing—original draft preparation, C.I.B. and A.M.P.; writing—review and editing, C.I.B., V.E.C., I.C.P., D.V., V.S. and G.L.; supervision, A.M.P., V.E.C. and I.C.P.; funding acquisition, A.M.P. All authors have read and agreed to the published version of the manuscript.

Funding: This project was financed by the Ministry of Research, Innovation and Digitization, under the project number PN 23090301 (Contract No. 12N/2023), within the FORCLIMSOC program (Sustainable Forest Management Adapted to Climate Change and Societal Challenges), and under the project number PN 16330204 (Romanian National Authority for Scientific Research and Innovation (ANCIS)).

Data Availability Statement: Data that support the findings of this study are available from the corresponding author upon reasonable request.

Acknowledgments: We thank Sorin Urdea, for providing permission to perform the fieldwork in the Sinca virgin forest. The authors would like to express their gratitude to both anonymous reviewers for their useful recommendations to improving the former draft of the manuscript.

Conflicts of Interest: The authors declare no conflict of interest.

References

1. Parviainen, J.; Little, D.; Doyle, M.; O'Sullivan, A.; Kettunen, M.; Korhonen, M. Metsäntutkimuslaitos Research in Forest Reserves and Natural Forests in European Countries: Country Reports for the COST Action E4: Forest Reserves Research Network. 1999, p. 304, ISBN 952-9844-31-X, ISSN 1237-8801. Available online: https://efi.int/sites/default/files/files/publication-bank/2018/proc16_net.pdf (accessed on 8 February 2023).
2. Kaplan, J.O.; Krumhardt, K.M.; Zimmermann, N. The Prehistoric and Preindustrial Deforestation of Europe. *Quat. Sci. Rev.* **2009**, *28*, 3016–3034. [CrossRef]
3. Stillhard, J.; Hobi, M.L.; Brang, P.; Brändli, U.-B.; Korol, M.; Pokynchereda, V.; Abegg, M. Structural Changes in a Primeval Beech Forest at the Landscape Scale. *For. Ecol. Manag.* **2022**, *504*, 119836. [CrossRef]
4. Sabatini, F.M.; Bluhm, H.; Kun, Z.; Aksenov, D.; Atauri, J.A.; Buchwald, E.; Burrascano, S.; Cateau, E.; Diku, A.; Duarte, I.M.; et al. European Primary Forest Database v2.0. *Sci. Data* **2021**, *8*, 220. [CrossRef]
5. Catalogul Padurilor Virgine. 2022. Available online: https://www.mmediu.ro/articol/catalogul-padurilor-virgine-sicvasivirgine-din-romania/5550 (accessed on 8 February 2023).
6. Commarmot, B.; Bachofen, H.; Bundziak, Y.; Bürgi, A.; Ramp, B.; Shparyk, Y.; Sukhariuk, D.; Viter, R.; Zingg, A. Structures of Virgin and Managed Beech Forests in Uholka (Ukraine) and Sihlwald (Switzerland): A Comparative Study. *For. Snow Landsc. Res.* **2005**, *79*, 45–56.
7. Frankovič, M.; Janda, P.; Mikoláš, M.; Čada, V.; Kozák, D.; Pettit, J.L.; Nagel, T.A.; Buechling, A.; Matula, R.; Trotsiuk, V.; et al. Natural Dynamics of Temperate Mountain Beech-Dominated Primary Forests in Central Europe. *For. Ecol. Manag.* **2021**, *479*, 118522. [CrossRef]
8. Veen, P.; Fanta, J.; Raev, I.; Biriş, I.-A.; de Smidt, J.; Maes, B. Virgin Forests in Romania and Bulgaria: Results of Two National Inventory Projects and Their Implications for Protection. *Biodivers. Conserv.* **2010**, *19*, 1805–1819. [CrossRef]
9. Sabatini, F.M.; Burrascano, S.; Keeton, W.S.; Levers, C.; Lindner, M.; Pötzschner, F.; Verkerk, P.J.; Bauhus, J.; Buchwald, E.; Chaskovsky, O.; et al. Where Are Europe's Last Primary Forests? *Divers. Distrib.* **2018**, *24*, 1426–1439. [CrossRef]
10. Gibson, L.; Lee, T.M.; Koh, L.P.; Brook, B.W.; Gardner, T.A.; Barlow, J.; Peres, C.A.; Bradshaw, C.J.A.; Laurance, W.F.; Lovejoy, T.E.; et al. Primary Forests Are Irreplaceable for Sustaining Tropical Biodiversity. *Nature* **2011**, *478*, 378–381. [CrossRef]
11. Višnjić, Ć.; Solaković, S.; Mekić, F.; Balić, B.; Vojniković, S.; Dautbašić, M.; Gurda, S.; Ioras, F.; Ratnasingam, J.; Abrudan, I.V. Comparison of Structure, Regeneration and Dead Wood in Virgin Forest Remnant and Managed Forest on Grmeč Mountain in Western Bosnia. *Plant Biosyst. Int. J. Deal. Asp. Plant Biol.* **2013**, *147*, 913–922. [CrossRef]
12. Keren, S.; Medarević, M.; Obradović, S.; Zlokapa, B. Five Decades of Structural and Compositional Changes in Managed and Unmanaged Montane Stands: A Case Study from South-East Europe. *Forests* **2018**, *9*, 479. [CrossRef]
13. Knohl, A.; Schulze, E.-D.; Kolle, O.; Buchmann, N. Large Carbon Uptake by an Unmanaged 250-Year-Old Deciduous Forest in Central Germany. *Agric. For. Meteorol.* **2003**, *118*, 151–167. [CrossRef]
14. Luyssaert, S.; Schulze, E.-D.; Börner, A.; Knohl, A.; Hessenmöller, D.; Law, B.E.; Ciais, P.; Grace, J. Old-Growth Forests as Global Carbon Sinks. *Nature* **2008**, *455*, 213–215. [CrossRef] [PubMed]
15. Glatthorn, J.; Feldmann, E.; Pichler, V.; Hauck, M.; Leuschner, C. Biomass Stock and Productivity of Primeval and Production Beech Forests: Greater Canopy Structural Diversity Promotes Productivity. *Ecosystems* **2018**, *21*, 704–722. [CrossRef]
16. Watson, J.E.M. The Exceptional Value of Intact Forest Ecosystems. *Nat. Ecol. Evol.* **2018**, *2*, 599–610. [CrossRef] [PubMed]
17. Franklin, J.F.; Van Pelt, R. Spatial Aspects of Structural Complexity in Old-Growth Forests. *J. For.* **2004**, *102*, 22–28.
18. Thom, D.; Seidl, R. Natural Disturbance Impacts on Ecosystem Services and Biodiversity in Temperate and Boreal Forests. *Biol. Rev.* **2016**, *91*, 760–781. [CrossRef]
19. Bauhus, J.; Puettmann, K.; Messier, C. Silviculture for Old-Growth Attributes. *For. Ecol. Manag.* **2009**, *258*, 525–537. [CrossRef]
20. Keeton, W.S.; Chernyavskyy, M.; Gratzer, G.; Main-Knorn, M.; Shpylchak, M.; Bihun, Y. Structural Characteristics and Aboveground Biomass of Old-growth Spruce–Fir Stands in the Eastern Carpathian Mountains, Ukraine. *Plant Biosyst.-Int. J. Deal. Asp. Plant Biol.* **2010**, *144*, 148–159. [CrossRef]
21. Diaci, J.; Rozenbergar, D.; Anic, I.; Mikac, S.; Saniga, M.; Kucbel, S.; Visnjic, C.; Ballian, D. Structural Dynamics and Synchronous Silver Fir Decline in Mixed Old-Growth Mountain Forests in Eastern and Southeastern Europe. *Forestry* **2011**, *84*, 479–491. [CrossRef]
22. Matović, B.; Stjepanović, S.; Kneginjić, I.; Stojanović, D.; Kisin, B.; Koprivica, M. Comparison of Stand Structure in Managed and Virgin European Beech Forests in Serbia. *Šumar. List* **2018**, *142*, 57. [CrossRef]
23. Wirth, C.; Messier, C.; Bergeron, Y.; Frank, D.; Kahl, A. Old-Growth Forest Definitions: A Pragmatic View. *Old-Growth For.* **2009**, *207*, 11–33. [CrossRef]
24. Seidl, R.; Schelhaas, M.-J.; Rammer, W.; Verkerk, P.J. Increasing Forest Disturbances in Europe and Their Impact on Carbon Storage. *Nat. Clim. Chang.* **2014**, *4*, 806–810. [CrossRef] [PubMed]

25. Schelhaas, M.-J.; Nabuurs, G.-J.; Schuck, A. Natural Disturbances in the European Forests in the 19th and 20th Centuries. *Global Change Biol.* **2003**, *9*, 1620–1633. [CrossRef]
26. Ulanova, N. The Effects of Windthrow on Forests at Different Spatial Scales: A Review. *Forest Ecol. Manag.* **2000**, *135*, 155–167. [CrossRef]
27. Janda, P.; Trotsiuk, V.; Mikoláš, M.; Bače, R.; Nagel, T.A.; Seidl, R.; Seedre, M.; Morrissey, R.C.; Kucbel, S.; Jaloviar, P.; et al. The Historical Disturbance Regime of Mountain Norway Spruce Forests in the Western Carpathians and Its Influence on Current Forest Structure and Composition. *For. Ecol. Manag.* **2017**, *388*, 67–78. [CrossRef] [PubMed]
28. Hirschmugl, M.; Gallaun, H.; Dees, M.; Datta, P.; Deutscher, J.; Koutsias, N.; Schardt, M. Methods for Mapping Forest Disturbance and Degradation from Optical Earth Observation Data: A Review. *Curr. For. Rep.* **2017**, *3*, 32–45. [CrossRef]
29. Svoboda, M.; Nagel, T. Gap Disturbance Regime in an Old-Growth *Fagus-Abies* Forest in the Dinaric Mountains, Bosnia-Herzegovina. *Can. J. For. Res.* **2008**, *38*, 2728–2737. [CrossRef]
30. Petritan, A.M.; Bouriaud, O.; Frank, D.C.; Petritan, I.C. Dendroecological Reconstruction of Disturbance History of an Old-Growth Mixed Sessile Oak–Beech Forest. *J. Veg. Sci.* **2017**, *28*, 117–127. [CrossRef]
31. Any Mary Petritan, Robert S Nuske, Ion Catalin Petritan, Nicu Constantin Tudose Gap Disturbance Patterns in an Old-Growth Sessile Oak (*Quercus petraea* L.)–European Beech (*Fagus sylvatica* L.) Forest Remnant in the Carpathian Mountains, Romania. *For. Ecol. Manag.* **2013**, *308*, 67–75. [CrossRef]
32. Nagel, T.A.; Firm, D.; Rozman, A. Intermediate Disturbances Are a Key Driver of Long-Term Tree Demography across Old-Growth Temperate Forests. *Ecol. Evol.* **2021**, *11*, 16862–16873. [CrossRef]
33. Willim, K.; Ammer, C.; Seidel, D.; Annighöfer, P.; Schmucker, J.; Schall, P.; Ehbrecht, M. Short—Term Dynamics of Structural Complexity in Differently Managed and Unmanaged European Beech Forests. *Trees For. People* **2022**, *8*, 100231. [CrossRef]
34. Boncina, A. Comparison of Structure and Biodiversity in the Rajhenav Virgin Forest Remnant and Managed Forest in the Dinaric Region of Slovenia. *Glob. Ecol. Biogeogr.* **2000**, *9*, 201–211. [CrossRef]
35. Keren, S.; Diaci, J.; Motta, R.; Govedar, Z. Stand Structural Complexity of Mixed Old-Growth and Adjacent Selection Forests in the Dinaric Mountains of Bosnia and Herzegovina. *For. Ecol. Manag.* **2017**, *400*, 531–541. [CrossRef]
36. Tauber, F. Contributii la sintaxonomia Fagetelor Carpato Dacice (Symphyto- Fagenalia Subordo Novum). *Trib. Bot.* **1987**, *27*, 179–191.
37. Vasile, D.; Petritan, A.M.; Tudose, N.C.; Toiu, F.L.; Scarlatescu, V.; Petritan, I.C. Structure and Spatial Distribution of Dead Wood in Two Temperate Old-Growth Mixed European Beech Forests. *Not. Bot. Horti Agrobot. Cluj-Napoca* **2017**, *45*, 639–645. [CrossRef]
38. Petritan, I.C.; Mihăilă, V.-V.; Bragă, C.I.; Boura, M.; Vasile, D.; Petritan, A.M. Litterfall Production and Leaf Area Index in a Virgin European Beech (*Fagus sylvatica* L.)–Silver Fir (*Abies alba* Mill.) Forest. *Dendrobiology* **2020**, *83*, 75–84. [CrossRef]
39. Nicolescu, V.-N.; Ghinescu, M.N.; Mihăilescu, G. Surprinzătoarea simplitate a tratamentelor adecvate stejăretelor pure și amestecurilor cu stejar. *Bucov. For.* **2021**, *21*, 183–197. [CrossRef]
40. Niculescu, V.N. *The Practice of Silviculture*; Aldus: Brasov, Romania, 2018.
41. Keller, M. Swiss National Forest Inventory. In *Manual of the Field Survey 2004–2007*; Swiss Federal Research Institute WSL Birmendsdorf, CH: Zürich, Switzerland, 2011.
42. Petritan, I.C.; Commarmot, B.; Hobi, M.L.; Petritan, A.M.; Bigler, C.; Abrudan, I.V.; Rigling, A. Structural Patterns of Beech and Silver Fir Suggest Stability and Resilience of the Virgin Forest Sinca in the Southern Carpathians, Romania. *For. Ecol. Manag.* **2015**, *356*, 184–195. [CrossRef]
43. Cristea, V.; Denayer, S. *De La Biodiversitate La OMG-Uri?* Eikon: Cluj-Napoca, Romania, 2004; ISBN 973-7987-77-2.
44. Van der Maarel, E.; Franklin, J. *Vegetation Ecology*; Wiley-Blackwell: Oxford, UK, 2013.
45. Curiel Yuste, J.; Flores-Rentería, D.; García-Angulo, D.; Hereş, A.-M.; Braga, C.; Petritan, A.M.; Petritan, I. Cascading Effects Associated with Climate-Change-Induced Conifer Mortality in Mountain Temperate Forests Result in Hot-Spots of Soil CO_2 Emissions. *Soil Biol. Biochem.* **2019**, *133*, 50–59. [CrossRef]
46. Giurgiu, V.; Drăghiciu, D.; Decei, I. *Metode Si Tabele de Productie*; Ceres: Austin, TX, USA, 2004.
47. Pretzsch, H. Strukturvielfalt Als Ergebnis Waldbaulichen Handelns. *Allg. Forst-Und Jagdztg.* **1996**, *167*, 213–221.
48. Petritan, A.M.; Biris, I.A.; Merce, O.; Turcu, D.O.; Petritan, I.C. Structure and Diversity of a Natural Temperate Sessile Oak (*Quercus petraea* L.)—European Beech (*Fagus sylvatica* L.) Forest. *For. Ecol. Manag.* **2012**, *280*, 140–149. [CrossRef]
49. Hammer, O.; Harper, D.A.T.; Ryan, P.D. PAST: Paleontological Statistics Software Package for Education and Data Analysis. *Palaeontol. Electron.* **2001**, *4*, 9.
50. Tóthmérész, B. Comparison of Different Methods for Diversity Ordering. *J. Veg. Sci.* **1995**, *6*, 283–290. [CrossRef]
51. Andrade, E.R.; Jardim, J.G.; Santos, B.A.; Melo, F.P.L.; Talora, D.C.; Faria, D.; Cazetta, E. Effects of Habitat Loss on Taxonomic and Phylogenetic Diversity of Understory Rubiaceae in Atlantic Forest Landscapes. *For. Ecol. Manag.* **2015**, *349*, 73–84. [CrossRef]
52. StatSoft, Inc. *STATISTICA (Data Analysis Software System)*; Version 12; 2013. Available online: http://www.statsoft.com (accessed on 8 February 2023).
53. Alessandrini, A.; Biondi, F.; Di Filippo, A.; Ziaco, E.; Piovesan, G. Tree Size Distribution at Increasing Spatial Scales Converges to the Rotated Sigmoid Curve in Two Old-Growth Beech Stands of the Italian Apennines. *For. Ecol. Manag.* **2011**, *262*, 1950–1962. [CrossRef]
54. Šamonil, P.; Antolík, L.; Svoboda, M.; Adam, D. Dynamics of Windthrow Events in a Natural Fir-Beech Forest in the Carpathian Mountains. *For. Ecol. Manag.* **2009**, *257*, 1148–1156. [CrossRef]

55. Čavlović, J.; Andabaka, M.; Božić, M.; Teslak, K.; Beljan, K. Current Status and Recent Stand Structure Dynamics in Mixed Silver Fir—European Beech Forests in Croatian Dinarides: Are There Differences between Managed and Unmanaged Forests? *Sustainability* **2021**, *13*, 9179. [CrossRef]

56. Mason, W.L.; Diaci, J.; Carvalho, J.; Valkonen, S. Continuous Cover Forestry in Europe: Usage and the Knowledge Gaps and Challenges to Wider Adoption. *For. Int. J. For. Res.* **2022**, *95*, 450. [CrossRef]

57. Vandekerkhove, K.; Vanhellemont, M.; Vrška, T.; Meyer, P.; Tabaku, V.; Thomaes, A.; Leyman, A.; De Keersmaeker, L.; Verheyen, K. Very Large Trees in a Lowland Old-Growth Beech (*Fagus sylvatica* L.) Forest: Density, Size, Growth and Spatial Patterns in Comparison to Reference Sites in Europe. *For. Ecol. Manag.* **2018**, *417*, 1–17. [CrossRef]

58. Šamonil, P.; Vrška, T. Trends and Cyclical Changes in Natural Fir-Beech Forests at the North-Western Edge of the Carpathians. *Folia Geobot.* **2007**, *42*, 337–361. [CrossRef]

59. Keren, S.; Motta, R.; Govedar, Z.; Lucic, R.; Medarevic, M.; Diaci, J. Comparative Structural Dynamics of the Janj Mixed Old-Growth Mountain Forest in Bosnia and Herzegovina: Are Conifers in a Long-Term Decline? *Forests* **2014**, *5*, 1243–1266. [CrossRef]

60. Kulla, L.; Roessiger, J.; Bošeľa, M.; Kucbel, S.; Murgaš, V.; Vencurik, J.; Pittner, J.; Jaloviar, P.; Šumichrast, L.; Saniga, M. Changing Patterns of Natural Dynamics in Old-Growth European Beech (*Fagus sylvatica* L.) Forests Can Inspire Forest Management in Central Europe. *For. Ecol. Manag.* **2023**, *529*, 120633. [CrossRef]

61. Debeljak, M. Coarse Woody Debris in Virgin and Managed Forest. *Ecol. Indic.* **2006**, *6*, 733–742. [CrossRef]

62. Radu, S. The Ecological Role of Deadwood in Natural Forests. In *Nature Conservation*; Environmental Science and Engineering; Springer: Berlin/Heidelberg, Germany, 2006; pp. 137–141. [CrossRef]

63. Öder, V.; Petritan, A.M.; Schellenberg, J.; Bergmeier, E.; Walentowski, H. Patterns and Drivers of Deadwood Quantity and Variation in Mid-Latitude Deciduous Forests. *For. Ecol. Manag.* **2021**, *487*, 118977. [CrossRef]

64. Spînu, A.P.; Asbeck, T.; Bauhus, J. Combined Retention of Large Living and Dead Trees Can Improve Provision of Tree-Related Microhabitats in Central European Montane Forests. *Eur. J. For. Res.* **2022**, *141*, 1105–1120. [CrossRef]

65. Keren, S.; Diaci, J. Comparing the Quantity and Structure of Deadwood in Selection Managed and Old-Growth Forests in South-East Europe. *Forests* **2018**, *9*, 76. [CrossRef]

66. Christensen, M.; Hahn, K.; Mountford, E.P.; Ódor, P.; Standovár, T.; Rozenbergar, D.; Diaci, J.; Wijdeven, S.; Meyer, P.; Winter, S.; et al. Dead Wood in European Beech (*Fagus sylvatica*) Forest Reserves. *For. Ecol. Manag.* **2005**, *210*, 267–282. [CrossRef]

67. Diaci, J.; Adamic, T.; Fidej, G.; Roženbergar, D. Toward a Beech-Dominated Alternative Stable State in Dinaric Mixed Montane Forests: A Long-Term Study of the Pecka Old-Growth Forest. *Front. For. Glob. Change* **2022**, *5*, 937404. [CrossRef]

68. Chivulescu, Ș.; Pitar, D.; Apostol, B.; Leca, Ș.; Badea, O. Importance of Dead Wood in Virgin Forest Ecosystem Functioning in Southern Carpathians. *Forests* **2022**, *13*, 409. [CrossRef]

69. Dynamics of Dead Wood Decay in Swiss Forests | Forest Ecosystems. Available online: https://forestecosyst.springeropen.com/articles/10.1186/s40663-020-00248-x (accessed on 6 February 2023).

70. Vrška, T.; Adam, D.; Hort, L.; Kolář, T.; Janík, D. European Beech (*Fagus sylvatica* L.) and Silver Fir (*Abies alba* Mill.) Rotation in the Carpathians—A Developmental Cycle or a Linear Trend Induced by Man? *For. Ecol. Manag.* **2009**, *258*, 347–356. [CrossRef]

71. Horvat, V.; García De Vicuña, J.; Biurrun, I.; García-Mijangos, I. Managed and Unmanaged Silver Fir-Beech Forests Show Similar Structural Features in the Western Pyrenees. *iFor.—Biogeosci. For.* **2018**, *11*, 698. [CrossRef]

72. Abrudan, I.V. Natural and Semi-Natural Mixed Stands in the Romanian Carpathians. In *Forests in Sustainable Mountain Development. A State of Knowledge Report for 2000*; Price, M.F., Butt, N., Eds.; CABI Publishing: Wallingford, UK, 2000; pp. 208–209.

73. Senn, J.; Suter, W. Ungulate Browsing on Silver Fir (*Abies alba*) in the Swiss Alps: Beliefs in Search of Supporting Data. *For. Ecol. Manag.* **2003**, *181*, 151–164. [CrossRef]

74. Schulze, E.D.; Bouriaud, O.B.; Wäldchen, J.; Eisenhauer, N.; Walentowski, H.; Seele, C.; Heinze, E.; Pruschitzki, U.P.; Dănilă, G.; Marin, G.; et al. Ungulate Browsing Causes Species Loss in Deciduous Forests Independent of Community Dynamics and Silvicultural Management in Central and Southeastern Europe. *Ann. For. Res.* **2014**, *57*, 267–288. [CrossRef]

75. Schlesinger, W.H.; Andrews, J.A. Soil Respiration and the Global Carbon Cycle. *Biogeochemistry* **2000**, *48*, 7–20. [CrossRef]

76. Zhang, Y.; Zou, J.; Dang, S.; Osborne, B.; Ren, Y.; Ju, X. Topography Modifies the Effect of Land-Use Change on Soil Respiration: A Meta-Analysis. *Ecosphere* **2021**, *12*, e03845. [CrossRef]

77. Peng, Y.; Thomas, S.; Tian, D. Forest Management and Soil Respiration: Implications for Carbon Sequestration. *Environ. Rev.* **2008**, *16*, 93–111. [CrossRef]

78. Jandl, R.; Lindner, M.; Vesterdal, L.; Bauwens, B.; Baritz, R.; Hagedorn, F.; Johnson, D.W.; Minkkinen, K.; Byrne, K.A. How Strongly Can Forest Management Influence Soil Carbon Sequestration? *Geoderma* **2007**, *137*, 253–268. [CrossRef]

79. Vasile, D.; Lazăr, G.; Cojocariu, D.; Enescu, R.; Crișan, V.; Scărlătescu, V.; Ienășoiu, G.; Petrițan, A.M. Comparative phytosociology study between a virgin mixed forest and a managed forest from the Șinca area. *Rev. Silvic. Cineg.* **2017**, *22*, 78–85.

forests

Article

Relation between Topography and Gap Characteristics in a Mixed Sessile Oak–Beech Old-Growth Forest

Nicu Constantin Tudose [1,*], Ion Catalin Petritan [2], Florin Lucian Toiu [1], Any-Mary Petritan [1] and Mirabela Marin [1]

[1] National Institute of Research and Development in Forestry 'Marin Dracea', Eroilor Bulevard, No. 128, 077190 Voluntari, Romania

[2] Department of Forest Engineering, Forest Management Planning and Terrestrial Measurements, Faculty of Silviculture and Forest Engineering, Transilvania University of Brasov, 500123 Brasov, Romania

[*] Correspondence: cntudose@yahoo.com

Abstract: The interest to assess the relationship between forest gap characteristics and topography features has been growing in the last decades. However, such an approach has not been studied in undisturbed mixed sessile oak–beech old-growth forests. Therefore, the present study carried out in one of the best-preserved sessile oak–beech old-growth forests in Europe, aims to assess the influence of topographic features (slope, altitude and aspect) on *(i) some characteristics* of canopies and expanded gaps (surface, diameter and perimeter) and *(ii) the proportion of beech and sessile oak as bordering trees, gap fillers and gap makers*. Through a complete gap survey on an area of 32 ha, 321 gaps were identified and mapped. The largest gaps and also the highest gap frequency (140) was found in the slope class (15.1–20°), while the gap frequency increased with altitude, with 99 gaps being recorded at 601–650 m a.s.l. The size and perimeter of the canopy and expanded gaps, as well as the number of gap makers, were negatively related to the slope and altitude. The expanded gap to canopy gap size ratio decreased with the slope and was positively related to the altitude, while a significant negative decrease in gap filler density with altitude was encountered. The sessile oak participation ratio as bordering trees forming the gap increased not only with the altitude but also with the slope. The topography plays an important role in the formation of gaps as well as in the characteristics of the future stand. This study provides valuable insights into the relationship between canopy gap characteristics and topography, which is useful information for forest owners that pursue the design of forest management toward nature-based solutions.

Keywords: natural reserve; mixed sessile oak–beech forest; old-growth forest; topography features; gap characteristics

Citation: Tudose, N.C.; Petritan, I.C.; Toiu, F.L.; Petritan, A.-M.; Marin, M. Relation between Topography and Gap Characteristics in a Mixed Sessile Oak–Beech Old-Growth Forest. *Forests* **2023**, *14*, 188. https://doi.org/10.3390/f14020188

Academic Editor: Dirk Landgraf

Received: 19 December 2022
Revised: 4 January 2023
Accepted: 13 January 2023
Published: 18 January 2023

1. Introduction

Old-growth forests represent ecosystems where human interventions are missing or slight [1]. Even if old-growth forests cover only 3% of total European forests, these are rich in biodiversity, thereby contributing to climate change mitigation and delivering fundamental ecosystem services [1]. However, forest characteristics change over time, and these changes are foreseeable only a few years after they occur or when old-growth forests are replaced by secondary forests [2,3].

Scarce old-growth forests composed of mixed beech (*Fagus sylvatica* L.) and sessile oak (*Quercus petraea* (MATT.) Liebl.)—two very important European species from an ecological and economic point of view—are of special interest to explore the main processes occurring naturally in such forest ecosystems. Investigating the changes in forest characteristics provides valuable information on forest behavior under different stress factors [4–6]. Topographic elements such as altitude, aspect and slope influence the availability of soil moisture, nutrients and daily insolation, which are determining factors for species suitability and composition [7]. Moreover, under a changing climate, those effects can be magnified [8,9] and can trigger modifications in the forest structure [10,11]. The topographic features that represent key drivers in defining forest structure and regeneration are still poorly understood [4,10,12]. Knowing the manner in which topography influences forest structure, composition and regeneration is essential to design sustainable management strategies focused on maximizing forest resilience [9].

These patterns can be identified within the development stages of old-growth forests characterized by a natural dynamic [13,14]. A key role in enhancing forest sustainability is played by silvicultural treatments [15], particularly those that imply the creation of canopy gaps [16–18]. With a major objective concerning preserving the naturalness in uneven-aged structures, the "close-to-nature" silvicultural practice represents a widely used approach that has lately gained increased interest due to its strong nature conservation concept [17,19–22]. The management of forests in a natural way underpins an improved understanding beyond other factors, patterns and processes, which happen naturally in forests [23], such as the effect of topography on gap characteristics. Investigating this aspect is important for a better understanding of the gap dynamics [24].

Some topographic features, such as altitude, wind exposure, slope, aspect and soil conditions, characterize the site of the gap [25,26]. Numerous soil properties (e.g., moisture conditions, nutrient amounts, soil temperature, etc.) are controlled by topographic features such as slope, aspect and altitude [27–30] and influence not only stand richness and diversity but also the gap characteristics [28,30]. In areas with steep slopes, the long axes of gaps are mostly oriented downslope, thus, influencing the size of the gaps [25,31]. Additionally, gap densities are related to slope and altitude, being more numerous at high altitudes and on steep slopes [32].

Tree species composition and productivity are controlled by the spatial and temporal distribution of climatic factors such as radiation, precipitation and temperature [26,33–35]. One of the major factors influencing species composition is light, with different regimes in accordance with gap size, canopy height and aspect [36–38]. Topographic factors exhibit a straightforward influence on forest dynamics, particularly in mountainous regions [39–41].

Forest canopy gaps are known as small openings created by natural and artificial disturbances, which damage tree species, thereby causing their death or injury [42,43]. Nevertheless, tree death does not always lead to gap formation, with their occurrence also being controlled by the sub-canopy structure [24]. However, canopy gaps are extremely important for the regeneration and growth process, especially for shade-intolerant species [12,19,21,37,44]. Canopy gaps also control light conditions, nutrient availability, soil moisture and biological properties, thus, influencing forest ecology, diversity and dynamics by creating suitable conditions for tree species development [30,37,45,46]. Tree species composition is influenced by the environment inside the gap and by the gap size [47,48]. While large gaps favor the installation of shade-intolerant or intermediate species, small gaps are more favorable for the establishment of shade-tolerant species [48]. Gap size distribution should be considered, particularly in natural regeneration or in mixed-species forests where stands dominated by a single tree species over large areas may hamper biodiversity [47]. Gap dynamics are controlled especially by small gaps, which close faster than large ones, and this contributes to forest management development when the stand canopy turnover rates are evaluated [49,50]. In this sense, forest managers should pay attention to species associations when designing a silvicultural system that emulates natural patterns [48].

For this reason, lately, interest in canopy gaps has increased in sustainable forest management practices [51]. Hence, it is important to have a better understanding of natural disturbance regimes and gap dynamics [16,37,48]. There is a wide range of causes that create gaps, such as wind damage, snow break, trees snapped by wind, tree mortality due to old age, large branches breaking, wildfires, landslides, tree fall and uprooting. Among these, the most common are uprooting, standing dead and snapping [19,32,45,52].

Europe's forests were substantially exploited in the past, and only a few old-growth forests still exist [53]. Most of these forests (90%) are located in countries from Northern to South-eastern Europe [1,54,55]. These ecosystems are dominated by European beech trees (*Fagus sylvatica* L.) and a mixture of other tree species, such as *Abies alba* Miller and *Picea abies* (L.) Karst [45]. Unfortunately, in broadleaved forests, particularly in the central part of Europe, the European beech and the sessile oak prevail (Quercus petraea (Matt.) Liebl.) [56] but were very exploited in the past [57]. Many studies carried out in Europe have focused on the structural characteristics of the stands, such as the size of the canopy gap distribution [48,58–63], and only a few have focused on the relationship between gap characteristics and topography [9,26,64]. However, numerous studies regarding canopy gaps have focused on gaps formed near ground-level [25,26,44,60,65,66]. These kinds of gaps are important for forest regeneration and the sapling growth of light-demanding species [44,49]. It is difficult to measure the gaps formed higher up in the canopy, but these gaps are important for the higher growth of larger trees [26,44,67].

Romania is the country with the largest extent of primary and virgin forests in the European Union [54,55,68]. These ecosystems have minimal human influence and provide favorable circumstances to explore natural disturbance regimes [19]. Having the highest naturalness levels is important to understand the influence of topographic features on stand development, composition, growth rates, age structure and regeneration within these types of ecosystems. Therefore, in this study, we pursue the investigation of the influence of topographic features on gap characteristics in one of the best-preserved old-growth mixed beech–sessile oak forests of Europe. Our assessment focuses on testing the influence of topographic features (slope, altitude and aspect) on (i) some characteristics of the canopy and expanded gaps (e.g., surface, diameter and perimeter) and (ii) on the proportion of beech and sessile oak as bordering trees, gap fillers and gap makers.

2. Materials and Methods

2.1. Study Site

The study site was located in Arad County, the western part of Romania. The Runcu-Grosi Natural Reserve situated in the Zarand Mountains is framed by 46°11′ northern latitude and 22°07′ eastern longitude [19]. The total area of the reserve is 262.6 ha. The altitude of the study site ranges between 334 and 686 m above sea level, while the slopes vary from 1 to 34 degrees (Figure 1).

The climate is temperate continental with a slight Mediterranean influence. Mean annual temperature varies between 7.6 °C and 9.4 °C. Mean annual precipitation is estimated at 750 and 925 mm/year, 60% of which falls during the growing season according to records from the Monorostia hydrometric station (150 m above sea level), which is the nearest to the study site location.

Soils are well drained and with good nutrient supply and predominantly belong to Cambisols and Luvisol. The forest canopy is dominated by *Fagus sylvatica* and *Quercus petraea*, but there are also some other species present, such as *Carpinus betulus*, *Quercus frainetto*, *Quercus cerris*, *Acer pseudoplatanus*, *Prunus avium*, *Ulmus glabra*, *Tillia cordata* and *Sorbus torminalis* [69].

The Runcu-Grosi Natural Reserve exhibits some important features. First of all, it represents one of the best-preserved natural beech–sessile oak forests at the European level [69]. Second, it is a remnant of ancient natural forests on the territory of our country [70]. Third, it is included entirely in the ROSCI0070 Drocea site and hosts three types of forest habitats of community interest: Asperulo–Fagetum beech forests, Galio–Carpinetum oak–hornbeam forests and Dacian oak and hornbeam forests [19,71]. Fourth, it has meaningful practical importance due to having the highest taxonomic and genetic heterogeneity as a result of its association with plants with bio-historical value [19]. Finally, it represents a unique old-growth oak mixed forest in Europe that hosts trees of impressive size and quality [19,69].

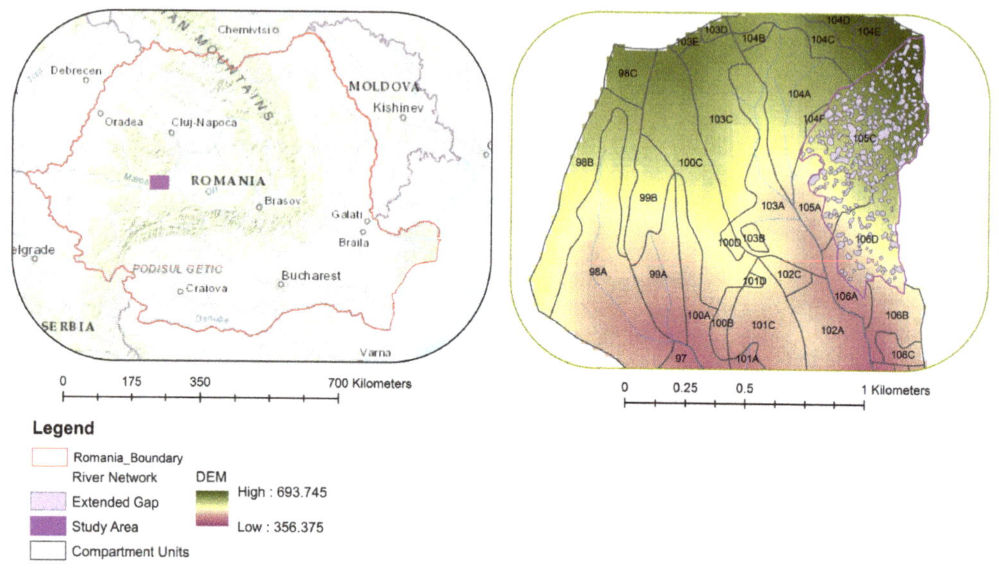

(a) **(b)**

Figure 1. Location of Runcu-Grosi Natural Reserve. (**a**) Studied area with the canopy gaps mapped terrestrially. (**b**) (Sources of left panels: Esri, DigitalGlobe, GeoEye, i-cubed, USDA FSA, USGS, AEX, Getmapping, Aerogrid, IGN, IGP, swisstopo and the GIS User Community).

2.2. Field Measurements

All canopy and expanded gaps were assessed in the best-preserved part of the Nature Reserve on 32.3 ha of forest, using the Field Map Data Collector [72]. We considered the canopy gap an opening in the stand with an area exceeding 10 m^2 characterized by the death of one or a group of trees with remnants of the gap maker still detectable [19]. The canopy gaps were defined as the area confined by the vertical projection of the crowns of the bordering trees, and the expanded gaps were defined as the area delimited by the position of their trunks [25]. The mapping of each gap was measured as the radii from the approximate gap center to the edge of the tree crowns and to the trunks of the trees bordering the gap [19], which were identified as species. A total number of 321 gaps were identified and mapped over the entire studied area of 32.3 ha, and the bordering trees, which formed the expanded gap and the gap makers, were identified as species. In 70 gaps (all expanded gaps greater than 800 m^2 and a random sample of gaps with an area lower than this threshold), all the gap fillers (all trees >7 cm diameter at breast height (DBH) and <20 m height) and gap saplings (taller than 1.3 m and smaller than 7 cm in DBH) were recorded as species (for more details, see [19]).

Canopy gaps and expanded gaps were collected in Field Map [73] in the vectorial format and were exported in ArcGIS for further processing using ArcGIS tools. The Digital Terrain Model with 1 m^2 pixel resolution was also integrated into the ArcGIS spatial database. Querying the database allowed us to determine the canopy gap characteristics (e.g., surface, diameter and perimeter) and their topographic features (e.g., slope, altitude, aspect, etc.). The database was explored to also calculate the frequency of the gaps, species proportion, gap fillers, gap saplings and expanded gaps at different classes of slope and altitude. The slope classes that we considered through dividing the slope in 5-degree steps were 0–5°, 5.1–10°, 10.1–15°, 15.1–20° and >21°. The altitude classes that resulted by framing the altitude in 50 m steps were 451–500 m, 501–550 m, 551–600 m, 601–650 m and 651–700 m. The aspect categories used in the analysis were N, E, NW, NE, S, SE, SW and W.

2.3. Data Analysis

All analyses were performed using the data obtained from all sampled canopy gaps (n = 321). The data regarding slope, altitude and aspect were extracted from the DTM based on the tools provided by ArcGIS. A discriminant analysis (DA) was carried out in PAST 4.11 (Natural History Museum, University of Oslo, Norway) in order to evaluate the grouping accuracy of the gaps regarding the altitude and slope. The testing of differences among the aspect categories for gap characteristics was performed by applying Mann–Whitney U-test (assumptions of ANOVA were not validated). To test the relation between topographic features (slope and altitude) and gap characteristics (size, perimeter, tree species proportion of bordering gap trees, gap fillers, gap saplings and the number of gap makers), multiple regression was performed using STATISTICA 13.5.0.17.

3. Results

The first tested topographic feature was the aspect of the terrain. No significant differences in all investigated gap characteristics among the aspect categories were detected. In Figure 2, several examples are shown (canopy gap area, expanded gap area, canopy gap perimeter and expanded gap perimeter). We focused, in particular, on the difference in testing between southern and northern exposures, and on southern exposures, better regeneration and the growing of sessile oak trees were expected compared to the northern slopes, where beech gap saplings and gap fillers could be at an advantage.

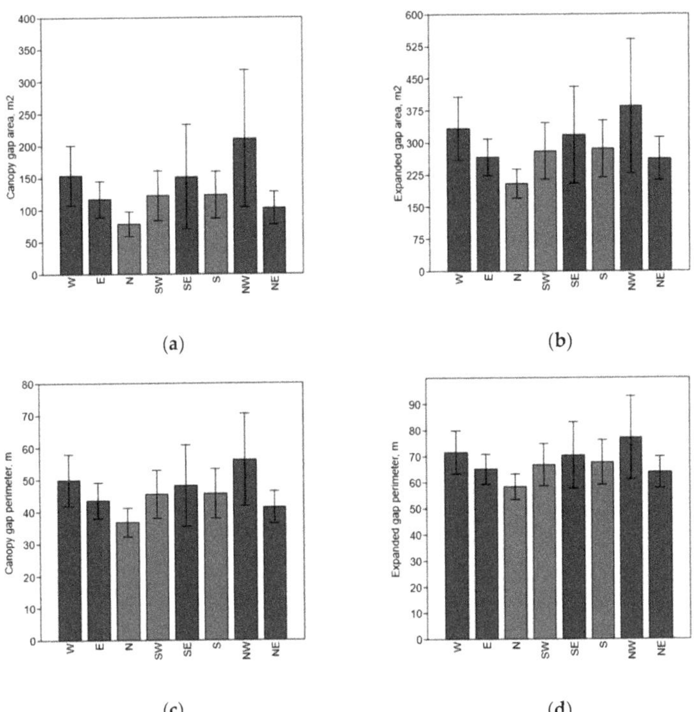

Figure 2. Canopy gap area (**a**), expanded gap area (**b**), canopy gap perimeter (**c**) and expanded gap perimeter (**d**) on aspect categories.

The highest frequency of gaps (140) was detected at slopes between 15.1 and 20°, where gaps with the largest surface were found. More than 50% of gaps were identified on slopes greater than 15.1° and only 7% on slopes between 0 and 5° (Figure 3b). More than 90% of canopy gaps were encountered on slopes that exceeded 10.1°.

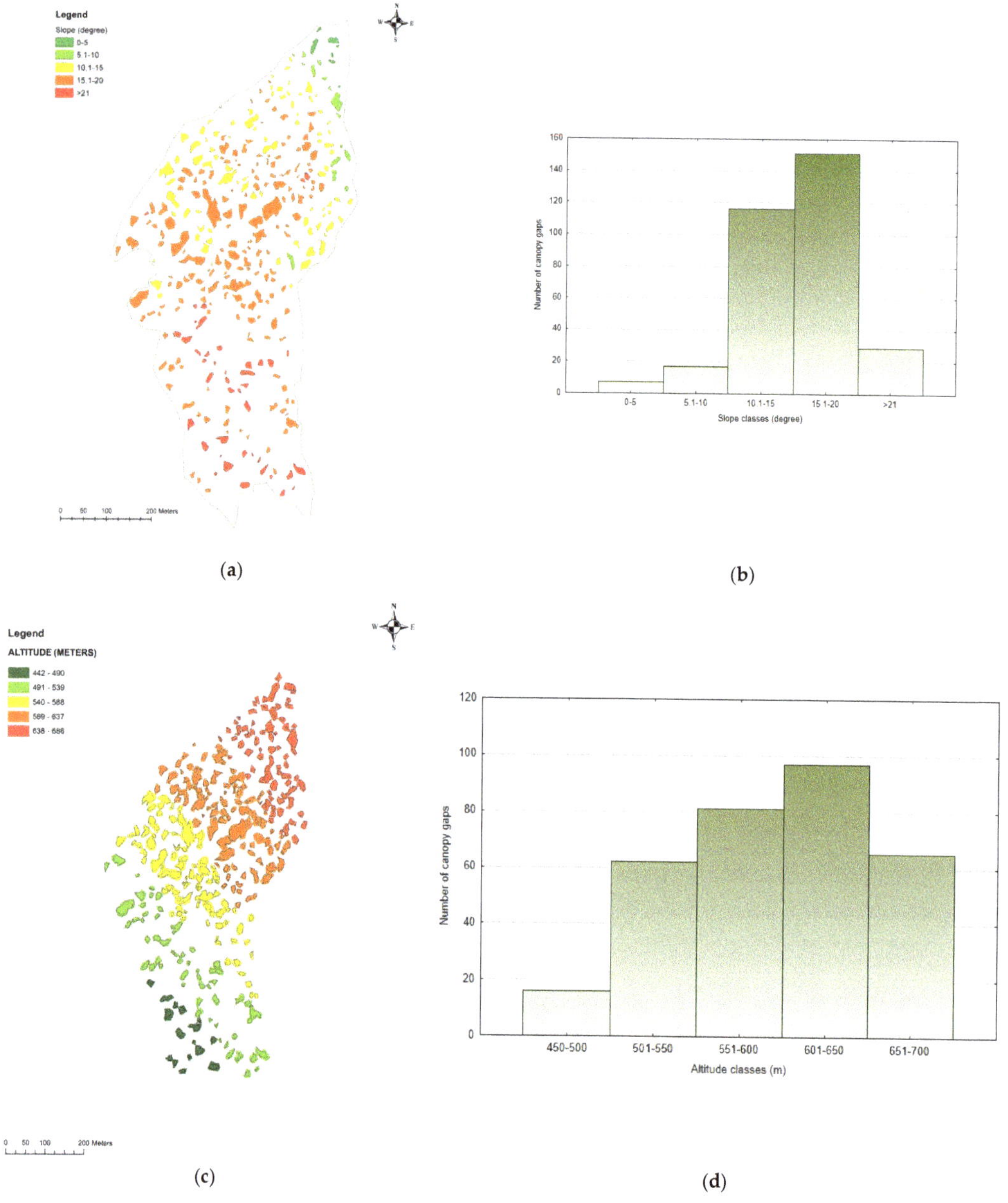

Figure 3. Canopy gap distribution on slope and altitude is shown as (**a**) the spatial distribution of gaps in slope classes, (**b**) the distribution of canopy gap numbers in slope classes, (**c**) the spatial distribution of gaps in altitude classes and (**d**) the distribution of canopy gap numbers in altitude classes.

We noticed that gap frequency increased with altitude (Figure 3d). Therefore, the highest gap frequencies of 25% and 30% were identified on slopes between 10.1 and 20°, respectively. At altitudes between 601 and 650 m, 99 gaps were encountered, while at altitudes between 450 and 500 m, only 18 gaps were found. The largest gap detected was 1387 m^2 in size (at 578 m), while the smallest had a minimum size of 11 m^2 (at 471 m). The high gap fraction value of 25% was recorded at 551–600 m altitude, followed by 601–650 m with a value of 20%.

The first discriminant axis explained 94.96% of the total variance in the centroids, while the second axis explained 4.13% (Figure 4). On the first axis, the highest contributor is the altitude factor, but the expanded gap size to canopy gap size ratio contributes at a similar level to the DA1 axis. However, the expanded gap size and perimeter and the canopy gap size and perimeter are stronger contributors to the DA2, on which the slope variable is the most associated important topographic factor. As expected, the number of saplings is positively correlated with the DA2 (the axis of gap dimensions), and the number of gap fillers is negatively correlated with the altitude and slope interaction.

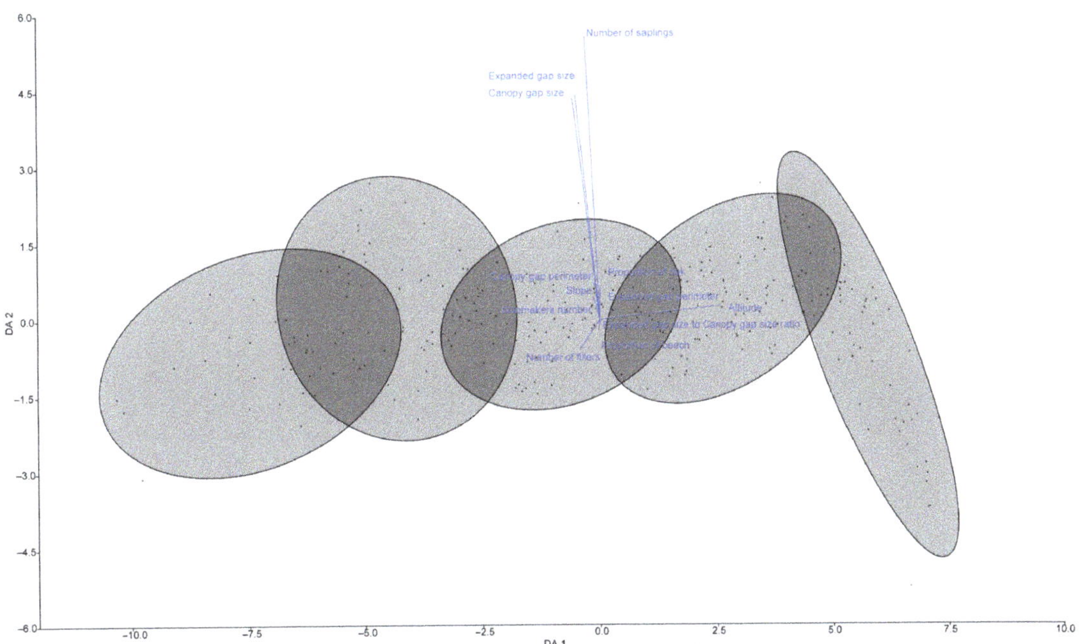

Figure 4. Discriminant analysis (DA) contributions biplot of gaps.

A total of 94% gaps were correctly classified initially using the five altitude classes (Table 1).

Table 1. Confusion matrix associated with the performed discriminant analysis.

Predicted/Given Groups	3	4	2	1	5	Total
3	75	4	6	0	0	85
4	2	92	0	0	2	96
2	0	0	57	3	0	60
1	0	0	0	29	0	29
5	0	2	0	0	49	51
Total	77	98	63	32	51	321

The gap characteristics, such as size, perimeter, the proportion of species and the number of gap makers were appraised in previous research [19]. Herein, we tested the influence of topographic features (slope and altitude) on some characteristics of sampled gaps (surface and perimeter) and on the proportion of sessile oaks as bordering trees, gap fillers, gap saplings and gap makers. As shown in Table 2, we found a significant influence of altitude on almost all analyzed gap characteristics (surface, parameter, the expanded gap to canopy gap ratio and the proportion of sessile oaks as bordering trees ($P_b < 0.05$)). Additionally, we found that gap characteristics (e.g., expanded gap perimeter, canopy gap perimeter, the proportion of sessile oaks as bordering gap trees and the number of gap fillers) were linked significantly to the slope ($P_c < 0.05$). However, a nonsignificant influence of slope on expanded gap size and canopy gap surface was detected ($P_c > 0.05$). Additionally, the expanded gap maker variables were not significantly linked to topographic features. In addition, from the data acquired, we can also see that there was a significant influence, regarding the interaction between slope and altitude, on the number of saplings regenerated within the gaps.

Table 2. Influence of altitude and slope on gap characteristics. a, b and c represent the model coefficients (gap characteristic $= a + b \times Altitude + c \times Slope$), and p_a, p_b and p_c represent the probability (*p*-value) for the two-tailed *t*-test (prob $> |t|$) used to test the significance of the model coefficients.

Gap Characteristics	a	$P_a > |t|$	b	$P_b > |t|$	C	$P_c > |t|$
Canopy gap size (m^2)	457.844	0.004	−0.448	0.033	−4.054	0.163
Expanded gap size (m^2)	818.295	0.001	−0.666	0.038	−8.612	0.053
Expanded gap size to canopy gap size ratio	1.241	0.346	0.003	0.037	−0.027	0.024
Canopy gap perimeter (m)	118.015	0.001	−0.097	0.005	−0.965	0.044
Expanded gap perimeter (m)	139.291	0.001	−0.089	0.016	−1.230	0.017
Proportion of sessile oaks as bordering trees (%)	−97.833	0.078	0.154	0.028	3.294	0.006
Number of gap makers	9.136	0.003	−0.007	0.085	−0.068	0.221
Number of gap fillers	332.414	0.215	−0.505	0.141	11.690	0.041
Number of gap saplings	4542.484	0.128	−3.319	0.382	−93.648	0.139

Within the research area, the number of gap fillers ranged from 0 to 584 gap fillers/ha, while the average density varied between 233 ha^{-1} and 1070 ha^{-1}. The main tree species that fill the formed gaps was beech (91% of all gap fillers). The remaining 9% comprised *Carpinus betulus*, while other species, such as *Quercus petraea*, *Tillia cordata*, *Prunus avium* and *Sorbus aucuparia*, occupied less than 1% [19]. The expanded gaps were formed by 2475 trees that were subsequently used for estimating the canopy composition. The relationships between the ratio of expanded gap to the canopy gap size and topographic features (slope and altitude) are shown in Figure 5. The expanded gap to canopy gap size ratio decreased with an increase in slope ($r = -0.1989$, $p < 0.05$ (Figure 5a) but was positively related to altitude ($r = 0.2202$, $p < 0.05$ (Figure 5b).

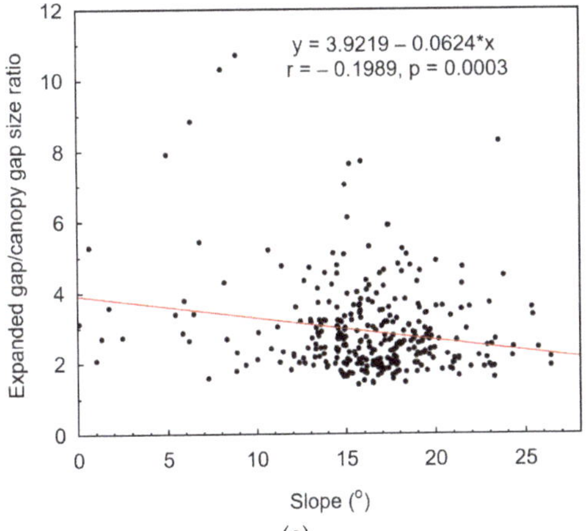
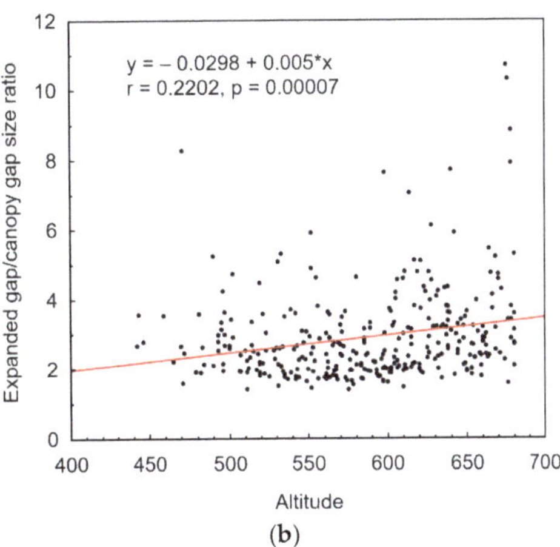

Figure 5. Variation in the expanded gap/canopy gap size ratio with slope (**a**) and altitude (**b**).

Higher values in the expanded gap to canopy gap size ratio were found on slopes of 10.1–20° and started to decrease on slopes greater than 21°. On one hand, this may be due to the proportion of a certain tree species surrounding the expanded gap [19]. On the other hand, the expanded gap to canopy gap size ratio increased with altitude, starting with altitudes higher than 500 m. A higher value in the expanded gap to canopy gap size ratio means that the trees forming the gap have larger crowns.

4. Discussion

Our study showed the existence of a relationship between terrain morphology and gap frequency. This was demonstrated through a significant field effort of financial and human resources and not by using recent technologies, such as remote sensing, which enable quick, automatic mapping of canopy gaps [74]. Our work suggests that more than 50% of the gaps are encountered on slopes of 15.1–20°, and only 7% of gaps are encountered at slopes between 0 and 5° (Figure 3a,b). This result confirms the findings of other studies [32,75] regarding the higher frequency of gaps on high-slope terrains. On these sloping terrains, the gaps provide the optimum light requirements for species development and, thus, exhibit an important role in the regeneration process of tree species [75]. The frequency distribution of gap sizes, from small gaps to large gaps, is also related to the slope [75]. Hence, we noticed that as the slope increases, the size of the gaps increases (Table 2), unlike the results of [31], which state that gentle slopes host the largest gaps. More than 90% of the canopy gaps were identified on slopes that exceeded 10.1°. These results are in line with the findings of [75,76], thus, confirming that gap size increases with slope. This is due to the fact that with the occurrence of increasing slope and altitude, the soils become shallower, and this aspect increases the instability of trees and their exposure to disturbance factors, such as wind. Wind is recognized as the main factor that triggers disturbance within central European forests [48,77]. Wind exposure favors the death of trees through uprooting and snapping, and therefore, the process of gap formation is triggered [19]. In altitudinal classes (Figure 3c,d), we found that the most frequent gaps are aggregated at 601–650 m altitude, which is consistent with other research [31,78] that considered wind exposure as a trigger factor. This disturbance factor generates an increasing number of canopy gaps and influences their openings, thus, creating larger gaps. The changes in gap disturbance patterns are related to features that act both at coarser (e.g., forest age) and finer scales (e.g., slope) [76]. However, some studies suggest that dissimilarities between the gap disturbance and gap size frequency distribution within a forest may be encountered [79–83]. Concerning the distribution of gap fillers with altitude, we obtained a weak correlation, like the findings of [84], which state that altitude did not influence gap fillers. However, gap closure depends on different drivers, such as the density of seedlings, sapling diameter and the height of gap-filling trees in each layer [45]. Similar to our findings, species of the genus Fagus have been reported as the main species for gap closure in other research [45,46,48,85,86]. This situation is mainly due to the fact that beech is a shade-tolerant tree species, with highly developed and interconnected root systems and advanced growth, which overwhelm other species [51]. Regarding the ratio of the expanded gap to canopy gap size and topography (Figure 5a), we obtained a negative correlation with the slope, mainly because the studied species have a very good root system, are well anchored in the soil and, thus, are highly resilient to wind damage. Our result is in disagreement with the findings of [32], which reported a positive correlation of expanded gap ratio with slope. However, unlike our study, the authors focused on other tree species with shallow root systems that are more exposed to uprooting, which can trigger canopy damage.

Additionally, another important factor is hydrology, which is influenced by terrain aspects [87]. Plateaus and ridges have drier soils, while steeper slopes retain more moisture during dry seasons [88]. Hydrological conditions also influence the gap size distribution [30], with the larger gaps being aggregated in areas exposed to flooding [31]. Additionally, hydrology has been linked to mortality rates, which are lower on moister soils during droughts [89]. Additionally, most of the analyzed characteristics of sampled gaps are directly influenced by the topographic features (e.g., altitude, slope and so on). Topographic differences, which influence and affect the gap size distribution, are expected to trigger changes in the forest structure or composition [26,76]. The configuration of forest structure and dynamics is driven by canopy gaps, which favor abundant natural regeneration and environmental heterogeneity [37,89]. Canopy gaps are important when an orientation toward sustainable and ecological management is desired, as well as for preserving forest biodiversity, enhancing functional diversity [37,46,90] and improving forest adaptability [91]. These initiatives need appropriate management and exploitation practices to ensure maximized species richness [92]. Investigating the relation between canopy gaps and topography will enhance knowledge about forest spatial and structural variations and successional processes at the ecosystem level [93]. Moreover, it provides a basis for forest managers that pursue reproducing models of natural disturbances [93].

The creation of natural gap openings at high altitudes directly exposes seedlings or seeds to disturbance factors. However, natural competition between seedlings and the formation of strong roots are favored [94–96]. The relevance of gaps formed near ground level has been shown as important not only for species regeneration but also for the growth of light-demanding species saplings [44,49]. In larger gaps, species competition ability increases because they develop strong root systems that favor better uptake of nutrients from soil and a higher resistance to biotic and abiotic factors [16,97]. In the case of sessile oak, the regeneration process has meaningful importance for forest managers, mainly due to the fact that, very often, this tree species is overwhelmed by beech [19,98]. Species dispersal is not only a key factor for forest succession [46] but also species shade tolerance. Although sessile oak regeneration is installed in small gaps, to grow and compete with more shade-tolerant species such as beech, they need brighter conditions, with gaps greater than at least 300 m^2 [99].

The altitude also plays an important role in gap formation. Our study reveals that the frequency of gaps increases with altitude. The highest frequency of gaps (30% and 25%) was identified at altitudes of 601–650 m followed by altitudes between 551 and 600 m, respectively. Those findings are in agreement with the results reported by [32], which obtained higher gap distribution at higher altitudes. We found that on slopes that exceed 21° and 450–500 m altitudes, the gap frequency decreases, while on slopes between 10.1 and 15° and 551–650 m altitudes, the gap frequency increases. This result is in disagreement with the findings of [32], which reported higher gap distribution at steep slopes. However, a converse relationship between the proportions of high canopy gaps (5–10 m height) and topographic features was also found in a previous study [26], which confirms that the proportion of low canopy gaps (≤ 5 m height) increases as altitude decreases in valleys and on plateaus but decreases on ridges and upper slopes. Additionally, this study also confirms that medium slopes and ridges have a small ratio of low canopy gaps.

Moreover, the topography is an important factor that also controls carbon densities within forests [100], with old-growth forests being the ecosystems that store the greatest amounts of carbon [101,102]. Therefore, further research is necessary to better understand how topographic aspects influence species regeneration within the gaps, mainly because these openings are crucial not only for forest succession but also for carbon storage.

Concerning the limitations of this study, in almost all of the gap characteristics (canopy gap surface, expanded gap area, canopy gap perimeter, expanded gap perimeter, expanded gap size to canopy gap size ratio, gap makers and gap saplings), the multiple regression model with slope and altitude as predictor variables explained around 5% of the dependent variable variation ($R^2 < 0.05$). Only for the proportion of sessile oaks as gap bordering trees and gap fillers, the tested multiple regression model explained more than 10% ($R^2 = 0.11$ for the proportion of sessile oak and 0.21 for gap fillers). Although the reduced proportion of variance explained by the multiple regression model implies the involvement of other variables in the analysis, in the current study, we focused only on the topographic traits used as independent variables (altitude and slope), which can be very easily obtained from the digital terrain model. To prove that topographic features are important in the natural disturbance regime, they have to be analyzed in the broader picture of natural mortality. Considering more factors in the model, we would analyze and prove the importance of topography on a broader scale. Why did the gap trees die? Was it related to storms, beetles, diseases or climate change? We do not have another more accurate variable available at this stage to be included in the analysis. For example, the lack of information related to the dominant wind direction and intensity in this region or how frequently insects attacked in the past and at what intensity makes it very difficult to take them into consideration. From the observations of another study (Petritan et al., 2013) [19], sessile oak—the main gap maker species—died mostly by uprooting, but sometimes while standing (it is difficult to know accurately why these gap trees died—whether by storms, beetles or diseases—and recognize precisely if the uprooting occurred directly, mainly due to a storm event, or if the tree died a long time before uprooting, mainly due to biotic factors, and remained standing dead for several decades).

5. Conclusions

Canopy gaps emulate patterns of natural variability in species and are the driving force of forest continuity, being widely applied within sustainable forest management practices. They contribute to the regeneration process and are fundamental, particularly for light-demanding species or those with a low dispersion of seedlings, such as sessile oak. Due to the meaningful importance of canopy gaps in forest ecology, herein we proposed the assessment of the influence of topographic features on gap characteristics within one of the best-preserved natural sessile oak–beech mixed forests in Europe. Hence, our study reveals that the distribution of gaps is influenced by slope steepness and altitude levels. Additionally, we found that gap characteristics are correlated with topographic features. Regarding gap closure, we noticed that gap filler distribution is independent of the altitude but controlled by species characteristics. Hence, we can conclude that topographic features exhibit meaningful importance for the aspect and proportion of the gaps. Additionally, these openings ensure forest succession and continuity. Concerning gaps' ability to simulate forest dynamics, this study provides valuable insights about the relationship between canopy gap characteristics and topography to be considered when designing sustainable management practices.

Author Contributions: Conceptualization, N.C.T. and I.C.P.; methodology, N.C.T. and I.C.P.; software, N.C.T.; validation, I.C.P. and A.-M.P.; formal analysis, N.C.T.; investigation, N.C.T., F.L.T. and I.C.P.; resources, N.C.T. and I.C.P.; data curation, N.C.T. and I.C.P.; writing—original draft preparation, M.M. and N.C.T.; writing—review and editing, I.C.P. and A.-M.P.; visualization, N.C.T., I.C.P. and

A.-M.P.; supervision, I.C.P. and A.-M.P.; project administration, I.C.P.; funding acquisition, A.-M.P. All authors have read and agreed to the published version of the manuscript.

Funding: This project was financed by the Romanian National Authority for Scientific, project number PN-II-RU-TE-2011-3-0075; by FORCLIMSOC – *Management forestier sustenabil adaptat schimbărilor climatice și provocărilor societale* (PN 23090203 and PN23090301) and by CresPerfInst – *Increasing the institutional capacity and performance of INCDS 'Marin Drăcea' in the activity of RDI*, contract no. 34PFE./30.12.2021, financed by the Ministry of Research, Innovation and Digitization through Program 1—Development of the national research development system and Subprogramme 1.2—Institutional performance projects to finance excellence in RDI.

Data Availability Statement: Datasets generated and/or analyzed during the current study are available from the corresponding authors on request.

Acknowledgments: The authors would like to express their gratitude to the editor and anonymous reviewers for their useful advice that helped to substantially improve the manuscript.

Conflicts of Interest: The authors declare no conflict of interest.

References

1. Barredo, J.I.; Brailescu, C.; Teller, A.; Sabatini, F.M.; Mauri, A. *Mapping and Assessment of Primary and Old-Growth Forests in Europe*; Publications Office of the European Union: Luxembourg, 2021; ISBN 9789276342304.
2. Lin, S.Y.; Shaner, P.J.L.; Lin, T.C. Characteristics of Old-Growth and Secondary Forests in Relation to Age and Typhoon Disturbance. *Ecosystems* **2018**, *21*, 1521–1532. [CrossRef]
3. Albrich, K.; Rammer, W.; Seidl, R. Climate change causes critical transitions and irreversible alterations of mountain forests. *Glob. Chang Biol.* **2020**, *26*, 4013–4027. [CrossRef] [PubMed]
4. Muñoz Mazón, M.; Klanderud, K.; Finegan, B.; Veintimilla, D.; Bermeo, D.; Murrieta, E.; Delgado, D.; Sheil, D. How forest structure varies with elevation in old growth and secondary forest in Costa Rica. *For. Ecol. Manag.* **2020**, *469*, 118191. [CrossRef]
5. McDowell, N.G.; Allen, C.D.; Anderson-Teixeira, K.; Aukema, B.H.; Bond-Lamberty, B.; Chini, L.; Clark, J.S.; Dietze, M.; Grossiord, C.; Hanbury-Brown, A.; et al. Pervasive shifts in forest dynamics in a changing world. *Science (80-)* **2020**, *368*, eaaz9463. [CrossRef] [PubMed]
6. Chivulescu, S.; García-Duro, J.; Pitar, D.; Ștefan, L.; Badea, O. Past and Future of Temperate Forests State under Climate Change Effects in the Romanian Southern Carpathians. *Forests* **2021**, *12*, 885. [CrossRef]
7. Wang, Z.; Lyu, L.; Liu, W.; Liang, H.; Huang, J.; Zhang, Q. Bin Topographic patterns of forest decline as detected from tree rings and NDVI. *Catena* **2021**, *198*, 105011. [CrossRef]
8. Clark, D.B.; Hurtado, J.; Saatchi, S.S. Tropical Rain Forest Structure, Tree Growth and Dynamics along a 2700-m Elevational Transect in Costa Rica. *PLoS ONE* **2015**, *10*, e0122905. [CrossRef]
9. Jucker, T.; Bongalov, B.; Burslem, D.F.R.P.; Nilus, R.; Dalponte, M.; Lewis, S.L.; Phillips, O.L.; Qie, L.; Coomes, D.A. Topography shapes the structure, composition and function of tropical forest landscapes. *Ecol. Lett.* **2018**, *21*, 989–1000. [CrossRef]
10. Seidl, R.; Thom, D.; Kautz, M.; Martin-Benito, D.; Peltoniemi, M.; Vacchiano, G.; Wild, J.; Ascoli, D.; Petr, M.; Honkaniemi, J.; et al. Forest disturbances under climate change. *Nat. Clim. Chang.* **2017**, *7*, 395–402. [CrossRef]
11. Tudose, N.; Ungurean, C.; Davidescu, Ș.; Cheval, S.; Marin, M. Information Tailored to the Needs of Stakeholders in the Romanian Case Study. Deliverable 4.3. CLISWELN Project. Available online: https://www.hzg.de/ms/clisweln/075105/index.php.en (accessed on 7 December 2021).
12. González, M.E.; Donoso, P.J.; Szejner, P. Tree-fall gaps and patterns of tree recruitment and growth in Andean old-growth forests in south-central Chile. *Bosque* **2015**, *36*, 383–394. [CrossRef]
13. Král, K.; Daněk, P.; Janík, D.; Krůček, M.; Vrška, T. How cyclical and predictable are Central European temperate forest dynamics in terms of development phases? *J. Veg. Sci.* **2018**, *29*, 84–97. [CrossRef]
14. Chivulescu, S.; Ciceu, A.; Leca, S.; Apostol, B.; Popescu, O.; Badea, O. Development phases and structural characteristics of the Penteleu-Viforâta virgin forest in the Curvature Carpathians. *IForest* **2020**, *13*, 389–395. [CrossRef]
15. Brzeziecki, B.; Bielak, K.; Bolibok, L.; Drozdowski, S.; Zajączkowski, J.; Żybura, H. Structural and compositional dynamics of strictly protected woodland communities with silvicultural implications, using Białowieża Forest as an example. *Ann. For. Sci.* **2018**, *75*, 1–15. [CrossRef]
16. Kern, C.C.; Burton, J.I.; Raymond, P.; D'Amato, A.W.; Keeton, W.S.; Royo, A.A.; Walters, M.B.; Webster, C.R.; Willis, J.L. Challenges facing gap-based silviculture and possible solutions for mesic northern forests in North America. *For. Int. J. For. Res.* **2017**, *90*, 4–17. [CrossRef]
17. Lu, D.; Wang, G.G.; Yu, L.; Zhang, T.; Zhu, J. Seedling survival within forest gaps: The effects of gap size, within-gap position and forest type on species of contrasting shade-tolerance in Northeast China. *For. Int. J. For. Res.* **2018**, *91*, 470–479. [CrossRef]
18. Mazdi, R.A.; Mataji, A.; Fallah, A. Canopy Gap Dynamics, Disturbances, and Natural Regeneration Patterns in a Beech-Dominated Hyrcanian Old-Growth Forest. *Balt. For.* **2021**, *27*, 535. [CrossRef]

19. Petritan, A.M.; Nuske, R.S.; Petritan, I.C.; Tudose, N.C. Gap disturbance patterns in an old-growth sessile oak (*Quercus petraea* L.)–European beech (*Fagus sylvatica* L.) forest remnant in the Carpathian Mountains, Romania. *For. Ecol. Manag.* **2013**, *308*, 67–75. [CrossRef]
20. Schütz, J.-P.; Saniga, M.; Diaci, J.; Vrška, T. Comparing close-to-naturesilviculture with processes in pristine forests: Lessons from Central Europe. *Ann. For. Sci.* **2016**, *73*, 911–921. [CrossRef]
21. Mölder, A.; Meyer, P.; Nagel, R.V. Integrative management to sustain biodiversity and ecological continuity in Central European temperate oak (Quercus robur, Q. petraea) forests: An overview. *For. Ecol. Manag.* **2019**, *437*, 324–339. [CrossRef]
22. Kuuluvainen, T.; Angelstam, P.; Frelich, L.; Jõgiste, K.; Koivula, M.; Kubota, Y.; Lafleur, B.; Macdonald, E. Natural Disturbance-Based Forest Management: Moving Beyond Retention and Continuous-Cover Forestry. *Front. For. Glob. Chang.* **2021**, *4*, 629020. [CrossRef]
23. Manning, D.B. *Stand Structure, Gap Dynamics and Regeneration of a Semi-Natural Mixed Beech Forest on Limestone in Central Europe—A Case Study*; Sustainable Land Management View Project; Waldbau-Institut: Freiburg im Breisgau, Germany, 2007.
24. Rabins, G. Canopy Gap Characteristics, Their Size-Distribution and Spatial Pattern in a Mountainous Cool Temperate Forest of Japan. Master's Thesis, University of Helsinki, Helsinki, Finland, 2019; p. 60.
25. Runkle, J.R. *Guidelines and Salmpe Protocol for Sampling Forest Gaps*; Gen. Tech. Rep. PNW-GTR-283; U.S. Department of Agriculture, Forest Service, Pacific Northwest Research Station: Portland, OR, USA, 1992; p. 44.
26. Gale, N. The Relationship between Canopy Gaps and Topography in a Western Ecuadorian Rain Forest on JSTOR. *Biotropica* **2000**, *32*, 653–661. [CrossRef]
27. Pickett, S.T.A. *The Ecology of Natural Disturbance and Patch Dynamics*; White, P.S., Ed.; Academic Press: London, UK, 1985; ISBN 0125545215.
28. de Lima, R.A.F.; de Moura, L.C. Gap disturbance regime and composition in the Atlantic Montane Rain Forest: The influence of topography. *Plant Ecol.* **2008**, *197*, 239–253. [CrossRef]
29. Florinsky, I.V. Influence of Topography on Soil Properties. In *Digital Terrain Analysis in Soil Science and Geology*; Elsevier: Amsterdam, The Netherlands, 2012; pp. 145–149.
30. Tang, F.; Quan, W.; Li, C.; Huang, X.; Wu, X.; Yang, Q.; Pan, Y.; Xu, T.; Qian, C.; Gu, Y. Effects of Small Gaps on the Relationship Among Soil Properties, Topography, and Plant Species in Subtropical Rhododendron Secondary Forest, Southwest China. *Int. J. Environ. Res. Public Health* **2019**, *16*, 1919. [CrossRef] [PubMed]
31. Goulamoussène, Y.; Bedeau, C.; Descroix, L.; Linguet, L.; Hérault, B. Environmental control of natural gap size distribution in tropical forests. *Biogeosciences* **2017**, *14*, 353–364. [CrossRef]
32. Zhu, C.; Zhu, J.; Zheng, X.; Lu, D.; Li, X. Comparison of gap formation and distribution pattern induced by wind/snowstorm and flood in a temperate secondary forest ecosystem, Northeast China. *Silva Fenn.* **2017**, *51*, 15. [CrossRef]
33. Stage, A.R.; Salas, C. Interactions of Elevation, Aspect, and Slope in Models of Forest Species Composition and Productivity. *For. Sci.* **2007**, *53*, 486–492.
34. Chen, S.; Wen, Z.; Ma, M.; Wu, S. Disentangling Climatic Factors and Human Activities in Governing the Old and New Forest Productivity. *Remote Sens.* **2021**, *13*, 3746. [CrossRef]
35. Ehbrecht, M.; Seidel, D.; Annighöfer, P.; Kreft, H.; Köhler, M.; Zemp, D.C.; Puettmann, K.; Nilus, R.; Babweteera, F.; Willim, K.; et al. Global patterns and climatic controls of forest structural complexity. *Nat. Commun.* **2021**, *12*, 1–12. [CrossRef]
36. Gendreau-Berthiaume, B.; Kneeshaw, D. Influence of Gap Size and Position within Gaps on Light Levels. *Int. J. For. Res.* **2009**, *2009*, 1–8. [CrossRef]
37. Muscolo, A.; Bagnato, S.; Sidari, M.; Mercurio, R. A review of the roles of forest canopy gaps. *J. For. Res.* **2014**, *25*, 725–736. [CrossRef]
38. Farhadur Rahman, M.; Onoda, Y.; Kitajima, K. Forest canopy height variation in relation to topography and forest types in central Japan with LiDAR. *For. Ecol. Manag.* **2022**, *503*, 119792. [CrossRef]
39. Ohkubo, T.; Tani, M.; Noguchi, H.; Yamakura, T.; Itoh, A.; Kanzaki, M.; Seng, H.L.; Tan, S.; Ashton, P.S.; Ogino, K.; et al. Spatial and topographic patterns of canopy gap formation in a mixed dipterocarp forest in Sarawak, Malaysia. *Tropics* **2007**, *16*, 151–163. [CrossRef]
40. Måren, I.E.; Karki, S.; Prajapati, C.; Yadav, R.K.; Shrestha, B.B. Facing north or south: Does slope aspect impact forest stand characteristics and soil properties in a semiarid trans-Himalayan valley? *J. Arid Environ.* **2015**, *121*, 112–123. [CrossRef]
41. Bałazy, R.; Kamińska, A.; Ciesielski, M.; Socha, J.; Pierzchalski, M. Modeling the Effect of Environmental and Topographic Variables Affecting the Height Increment of Norway Spruce Stands in Mountainous Conditions with the Use of LiDAR Data. *Remote Sens.* **2019**, *11*, 2407. [CrossRef]
42. Schliemann, S.A.; Bockheim, J.G. Methods for studying treefall gaps: A review. *For. Ecol. Manag.* **2011**, *261*, 1143–1151. [CrossRef]
43. Zhu, J.; Lu, D.; Zhang, W. Effects of gaps on regeneration of woody plants: A meta-analysis. *J. For. Res.* **2014**, *25*, 501–510. [CrossRef]
44. Demeter, L.; Bede-Fazekas, Á.; Molnár, Z.; Csicsek, G.; Ortmann-Ajkai, A.; Varga, A.; Molnár, Á.; Horváth, F. The legacy of management approaches and abandonment on old-growth attributes in hardwood floodplain forests in the Pannonian Ecoregion. *Eur. J. For. Res.* **2020**, *139*, 595–610. [CrossRef]
45. Bottero, A.; Garbarino, M.; Dukić, V.; Govedar, Z.; Lingua, E.; Nagel, T.A.; Motta, R. Gap-phase dynamics in the old-growth forest of Lom, Bosnia and Herzegovina. *Silva Fenn.* **2011**, *45*, 875–887. [CrossRef]

46. Valerio, M.; Ib, R. The Role of Canopy Cover Dynamics over a Decade of Changes in the Understory of an Atlantic Beech-Oak Forest. *Forests* **2021**, *12*, 938. [CrossRef]

47. Schall, P.; Gossner, M.M.; Heinrichs, S.; Fischer, M.; Boch, S.; Prati, D.; Jung, K.; Baumgartner, V.; Blaser, S.; Böhm, S.; et al. The impact of even-aged and uneven-aged forest management on regional biodiversity of multiple taxa in European beech forests. *J. Appl. Ecol.* **2018**, *55*, 267–278. [CrossRef]

48. Jaloviar, P.; Sedmáková, D.; Pittner, J.; Danková, L.J.; Kucbel, S.; Sedmák, R.; Saniga, M. Gap Structure and Regeneration in the Mixed Old-Growth Forests of National Nature Reserve. *Forests* **2020**, *11*, 81. [CrossRef]

49. Zhu, C.; Zhu, J.; Wang, G.G.; Zheng, X.; Lu, D.; Gao, T. Dynamics of gaps and large openings in a secondary forest of Northeast China over 50 years. *Ann. For. Sci.* **2019**, *76*, 1–10. [CrossRef]

50. Hunter, M.O.; Keller, M.; Morton, D.; Cook, B.; Lefsky, M.; Ducey, M.; Saleska, S.; De Oliveira, R.C.; Schietti, J.; Zang, R. Structural dynamics of tropical moist forest gaps. *PLoS ONE* **2015**, *10*, e0132144. [CrossRef] [PubMed]

51. Feldmann, E.; Glatthorn, J.; Ammer, C. Regeneration Dynamics Following the Formation of Understory Gaps in a Slovakian Beech Virgin Forest. *Forests* **2020**, *11*, 585. [CrossRef]

52. Zeibig, A.; Diaci, J.; Wagner, S. Gap disturbance patterns of a Fagus sylvatica virgin forest remnant in the mountain vegetation belt of Slovenia Gap disturbance patterns of a Fagus sylvatica virgin forest remnant in the mountain vegetation belt of Slovenia. *For. Snow Landsc. Res.* **2005**, *79*, 69–80.

53. Freer-Smith, P.; Muys, B.; Bozzano, M.; Drössler, L.; Farrelly, N.; Jactel, H.; Korhonen, J.; Minotta, G.; Nijnik, M.; Orazio, C. *Plantation Forests in Europe: Challenges and Opportunities. From Science to Policy 9*; European Forest Institute: Joensuu, Finland, 2019. [CrossRef]

54. Veen, P.; Fanta, J.; Raev, I.; Biris, I.-A.; Biris, B.; De Smidt, J.; Maes, B. Virgin forests in Romania and Bulgaria: Results of two national inventory projects and their implications for protection. *Biodivers. Conserv.* **2010**, *19*, 1805–1819. [CrossRef]

55. Spînu, A.P.; Petrițan, I.C.; Mikoláš, M.; Janda, P.; Vostarek, O.; Čada, V.; Svoboda, M. Moderate-to High-Severity Disturbances Shaped the Structure of Primary Picea Abies (L.) Karst. Forest in the Southern Carpathians. *Forests* **2020**, *11*, 1315. [CrossRef]

56. Bolte, A.; Czajkowski, T.; Kompa, T. The north-eastern distribution range of European beech—A review. *For. Int. J. For. Res.* **2007**, *80*, 413–429. [CrossRef]

57. Parviainen, J.; Little, D.; Doyle, M.; O'sullivan, A.; Kettunen, M.; Korhonen, M. *Research in Forest Reserves and Natural Forests in European Countries—Country Reports for the COST Action E4: Forest Reserves Research Network*; European Forest Institute: Joensuu, Finland, 1999.

58. Gálhidy, L.; Mihók, B.; Hagyó, A.; Rajkai, K.; Standovár, T. Effects of gap size and associated changes in light and soil moisture on the understorey vegetation of a Hungarian beech forest. *Plant Ecol.* **2006**, *183*, 133–145. [CrossRef]

59. Nagel, T.A.; Svoboda, M. Gap disturbance regime in an old-growth Fagus-Abies forest in the Dinaric Mountains, Bosnia-Herzegovina. *Can. J. For. Res.* **2008**, *38*, 2728–2737. [CrossRef]

60. Garbarino, M.; Mondino, B.; Nagel, T.; Borgogno Mondino, E.; Lingua, E.; Nagel, T.A.; Dukić, V.; Govedar, Z.; Motta, R.; Garbarino, M.; et al. Gap disturbances and regeneration patterns in a Bosnian old-growth forest: A multispectral remote sensing and ground-based approach Gap disturbances and regeneration patterns in a Bosnian old-growth forest: A multispectral remote sensing and ground-based. *Herzegovina Ann. For. Sci.* **2011**, *69*, 617–625. [CrossRef]

61. Zellweger, F.; Braunisch, V.; Morsdorf, F.; Baltensweiler, A.; Abegg, M.; Roth, T.; Bugmann, H.; Bollmann, K. Disentangling the effects of climate, topography, soil and vegetation on stand-scale species richness in temperate forests. *For. Ecol. Manag.* **2015**, *349*, 36–44. [CrossRef]

62. Drössler, L.; Feldmann, E.; Glatthorn, J.; Annighöfer, P.; Kucbel, S.; Tabaku, V.; Drössler, L.; Feldmann, E.; Glatthorn, J.; Annighöfer, P.; et al. What Happens after the Gap?— Size Distributions of Patches with Homogeneously Sized Trees in Natural and Managed Beech Forests in Europe. *Open J. For.* **2016**, *6*, 177–190. [CrossRef]

63. Orman, O.; Dobrowolska, D. Gap dynamics in the Western Carpathian mixed beech old-growth forests affected by spruce bark beetle outbreak. *Eur. J. For. Res.* **2017**, *136*, 571–581. [CrossRef]

64. Kenderes, K.; Aszalós, R.; Ruff, J.; Barton, Z.; Standovár, T. Effects of topography and tree stand characteristics on susceptibility of forests to natural disturbances (ice and wind) in the Börzsöny Mountains (Hungary). *Community Ecol.* **2007**, *8*, 209–220. [CrossRef]

65. Van der Meer, P.; Bongers, F.; Chatrou, L.; Riera, B. Defining canopy gaps in a tropical rain forest: Effects on gap size and turnover time. *Acta Oecologica-Int. J. Ecol.* **1994**, *15*, 701–714.

66. Gutiérrez, Á.G.; Chávez, R.O.; Díaz-Hormazábal, I. Canopy gap structure as an indicator of intact, old-growth temperate rainforests in the valdivian ecoregion. *Forests* **2021**, *12*, 1183. [CrossRef]

67. Fransson, P.; Brännström, Å.; Franklin, O. A tree's quest for light—Optimal height and diameter growth under a shading canopy. *Tree Physiol.* **2021**, *41*, 1–11. [CrossRef]

68. Sabatini, F.M.; Burrascano, S.; Keeton, W.S.; Levers, C.; Lindner, M.; Pötzschner, F.; Verkerk, P.J.; Bauhus, J.; Buchwald, E.; Chaskovsky, O.; et al. Where are Europe's last primary forests? *Divers. Distrib.* **2018**, *24*, 1426–1439. [CrossRef]

69. Petritan, A.M.; Biris, I.A.; Merce, O.; Turcu, D.O.; Petritan, I.C. Structure and diversity of a natural temperate sessile oak (*Quercus petraea* L.)—European Beech (*Fagus sylvatica* L.) forest. *For. Ecol. Manag.* **2012**, *280*, 140–149. [CrossRef]

70. Merce, O.; Turcu, D.; Cantar, I. The structure of a natural mixed beech—Sessile oak forest in Runcu Grosi Natural Reserve. *J. Hortic. For. Biotechnol.* **2012**, *16*, 131–138.

71. Merce, O.; Turcu, D.O. Natura 2000 forest habitats of the "Runcu-Groși" Nature Reserve. *For. Biotechnol.* **2016**, *20*, 140–144.

72. IFER. Field-Map—Tool Designed for Computer Aided Field Data Collection. Available online: https://www.fieldmap.cz/ (accessed on 11 August 2022).

73. IFER. Available online: https://fieldmap.cz/?page=fmsoftware (accessed on 29 November 2022).

74. Jucker, T. Deciphering the fingerprint of disturbance on the three-dimensional structure of the world's forests. *New Phytol.* **2022**, *233*, 612–617. [CrossRef] [PubMed]

75. Bianchini, E.; Garcia, C.C.; Pimenta, J.A.; Torezan, J.M.D. Slope variation and population structure of tree species from different ecological groups in South Brazil. *An. Acad. Bras. Cienc.* **2010**, *82*, 643–652. [CrossRef] [PubMed]

76. Lobo, E.; Dalling, J.W. Effects of topography, soil type and forest age on the frequency and size distribution of canopy gap disturbances in a tropical forest. *Biogeosciences* **2013**, *10*, 6769–6781. [CrossRef]

77. Splechtna, B.E.; Gratzer, G.; Black, B.A. Disturbance History of a European Old-Growth Mixed-Species Forest: A Spatial Dendro. *Source J. Veg. Sci.* **2005**, *16*, 511–522.

78. Negrón-Juárez, R.I.; Chambers, J.Q.; Hurtt, G.C.; Annane, B.; Cocke, S.; Powell, M.; Stott, M.; Goosem, S.; Metcalfe, D.J.; Saatchi, S.S. Remote Sensing Assessment of Forest Disturbance across Complex Mountainous Terrain: The Pattern and Severity of Impacts of Tropical Cyclone Yasi on Australian Rainforests. *Remote Sens.* **2014**, *6*, 5633–5649. [CrossRef]

79. Lertzman, K.P.; Sutherland, G.D.; Inselberg, A.; Saunders, S.C. Canopy gaps and the landscape mosaic in a coastal temperate rain forest. *Ecology* **1996**, *77*, 1254–1270. [CrossRef]

80. Baker, P.J.; Bunyavejchewin, S.; Oliver, C.D.; Ashton, P.S. Disturbance history and historical stand dynamics of a seasonal tropical forest in western Thailand. *Ecol. Monogr.* **2005**, *75*, 317–343. [CrossRef]

81. Fisher, J.I.; Hurtt, G.C.; Thomas, R.Q.; Chambers, J.Q. Clustered disturbances lead to bias in large-scale estimates based on forest sample plots. *Ecol. Lett.* **2008**, *11*, 554–563. [CrossRef]

82. Foster, J.R.; Townsend, P.A.; Zganjar, C.E. Spatial and temporal patterns of gap dominance by low-canopy lianas detected using EO-1 Hyperion and Landsat Thematic Mapper. *Remote Sens. Environ.* **2008**, *112*, 2104–2117. [CrossRef]

83. Chambers, J.Q.; Robertson, A.L.; Carneiro, V.M.C.; Lima, A.J.N.; Smith, M.L.; Plourde, L.C.; Higuchi, N. Hyperspectral remote detection of niche partitioning among canopy trees driven by blowdown gap disturbances in the Central Amazon. *Oecologia* **2009**, *160*, 107–117. [CrossRef]

84. King, S.L.; Antrobus, T.J. Relationships between Gap Makers and Gap Fillers in an Arkansas. *J. Veg. Sci.* **2005**, *16*, 471–478. [CrossRef]

85. Rentch, J.S.; Schuler, T.M.; Nowacki, G.J.; Beane, N.R.; Ford, W.M. Canopy gap dynamics of second-growth red spruce-northern hardwood stands in West Virginia. *For. Ecol. Manag.* **2010**, *260*, 1921–1929. [CrossRef]

86. Amolikondori, A.; Abrari Vajari, K.; Feizian, M. Assessing the effects of forest gaps on beech (*Fagus orientalis* L.) trees traits in the logged temperate broad-leaf forest. *Ecol. Indic.* **2021**, *127*, 107689. [CrossRef]

87. Cielo-Filho, R.; Gneri, M.A.; Martins, F.R. Position on slope, disturbance, and tree species coexistence in a Seasonal Semideciduous Forest in SE Brazil. *Plant Ecol.* **2007**, *190*, 189–203. [CrossRef]

88. Becker, P.; Rabenold, P.E.; Idol, J.R.; Smith, A.P. Water potential gradients for gaps and slopes in a Panamanian tropical moist forest's dry season. *J. Trop. Ecol.* **1988**, *4*, 173–184. [CrossRef]

89. Condit, R.; Hubbell, S.P.; Foster, R.B. Mortality rates of 205 neotropical tree and shrub species and the impact of a severe drought. *Ecol. Monogr.* **1995**, *65*, 419–439. [CrossRef]

90. Reis, C.R.; Jackson, T.D.; Gorgens, E.B.; Dalagnol, R.; Jucker, T.; Nunes, M.H.; Ometto, J.P.; Aragão, L.E.O.C.; Rodriguez, L.C.E.; Coomes, D.A. Forest disturbance and growth processes are reflected in the geographical distribution of large canopy gaps across the Brazilian Amazon. *J. Ecol.* **2022**, *110*, 1–13. [CrossRef]

91. Larsen, J.B.; Angelstam, P.; Bauhus, J.; Carvalho, J.F.; Diaci, J.; Dobrowolska, D.; Gazda, A.; Gustafsson, L.; Krumm, F.; Knoke, T.; et al. Closer-to-Nature Forest Management. In *From Science to Policy 12*; European Forest Institute: Joensuu, Finland, 2022. [CrossRef]

92. Poudyal, B.H.; Maraseni, T.; Cockfield, G. Impacts of forest management on tree species richness and composition: Assessment of forest management regimes in Tarai landscape Nepal. *Appl. Geogr.* **2019**, *111*, 102078. [CrossRef]

93. Goodbody, T.R.H.; Tompalski, P.; Coops, N.C.; White, J.C.; Wulder, M.A.; Sanelli, M. Uncovering spatial and ecological variability in gap size frequency distributions in the Canadian boreal forest. *Sci. Rep.* **2020**, *10*, 1–12. [CrossRef]

94. Dusan, R.; Stjepan, M.; Igor, A.; Jurij, D. Gap regeneration patterns in relationship to light heterogeneity in two old-growth beech–fir forest reserves in South East Europe. *For. Int. J. For. Res.* **2007**, *80*, 431–443. [CrossRef]

95. Zhu, J.; Jiang, L.; Zhu, D.-H.; Xing, C.; Jin, M.-R.; Liu, J.-F.; He, Z.-S. Forest gaps regulate seed germination rate and radicle growth of an endangered plant species in a subtropical natural forest. *Plant Divers.* **2021**, *44*, 445–454. [CrossRef]

96. Margreiter, V.; Walde, J.; Erschbamer, B. Competition-free gaps are essential for the germination and recruitment of alpine species along an elevation gradient in the European Alps. *Alp. Bot.* **2021**, *131*, 135–150. [CrossRef]

97. Harmer, R.; Robertson, M. Seedling root growth of six broadleaved tree species grown in competition with grass under irrigated nursery conditions. *Ann. For. Sci.* **2003**, *60*, 601–608. [CrossRef]

98. Tinya, F.; Kovács, B.; Aszalós, R.; Tóth, B.; Csépányi, P.; Németh, C.; Ódor, P. Initial regeneration success of tree species after different forestry treatments in a sessile oak-hornbeam forest. *For. Ecol. Manag.* **2020**, *459*, 117810. [CrossRef]

99. Lüpke, V.B. Überschirmungstoleranz von Stiel- und Traubeneichen als Voraussetzung für Verjüngungsverfahren unter Schirm. *Mitt. ad Forstl. Vers. Anst. Rheinland-Pfalz. Hrsg. von Forstliche Versuchsanstalt Rheinland-Pfalz* **1995**, *34/95*, 141–160.

100. Zhang, C.; Deng, Q.; Liu, A.; Liu, C.; Xie, G. Effects of Stand Structure and Topography on Forest Vegetation Carbon Density in Jiangxi Province. *Forests* **2021**, *12*, 1483. [CrossRef]
101. Pugh, T.A.M.; Lindeskog, M.; Smith, B.; Poulter, B.; Arneth, A.; Haverd, V.; Calle, L. Role of forest regrowth in global carbon sink dynamics. *Proc. Natl. Acad. Sci. USA* **2019**, *116*, 4382–4387. [CrossRef]
102. Martin-Benito, D.; Pederson, N.; Férriz, M.; Gea-Izquierdo, G. Old forests and old carbon: A case study on the stand dynamics and longevity of aboveground carbon. *Sci. Total Environ.* **2021**, *765*, 142737. [CrossRef] [PubMed]

MDPI
St. Alban-Anlage 66
4052 Basel
Switzerland
www.mdpi.com

Forests Editorial Office
E-mail: forests@mdpi.com
www.mdpi.com/journal/forests